GALE DIRECTORY OF PUBLICATIONS AND BROADCAST MEDIA

Update

ISSN 1048-7972

143rd Edition

GALE DIRECTORY OF PUBLICATIONS AND BROADCAST MEDIA

Update

An Interedition Service Providing New Listings
and Updates to Listings in the Main Volumes
See Introduction for Details

Louise Gagne, Project Editor

GALE
CENGAGE Learning™

Detroit • New York • San Francisco • New Haven, Conn • Waterville, Maine • London

Gale Directory of Publications and Broadcast Media, 143rd Edition Supplement

Project Editor: Louise Gagne

Editorial: Verne Thompson

Editorial Support Services: Manny Barrido

Composition and Electronic Prepress: Gary Leach

Manufacturing: Rita Wimberley

Product Manager: Michele P. LaMeau

ISBN-13: 978-0-7876-9675-7
ISBN-10: 0-7876-9675-9675-7

ISSN 1048-7972

Gale
27500 Drake Rd.
Farmington Hills, MI, 48331-3535

Printed in the United States of America
2 3 4 5 6 7 12 11 10 09 08

Contents

Introduction

This *Supplement* to the *Gale Directory of Publications and Broadcast Media (GDPBM),* published between main editions of the *Directory,* provides new listings of newspapers, magazines, journals, radio stations, television stations, and cable systems.

Preparation, Content, and Arrangement

Sent free to all *Directory* subscribers, the *Supplement* contains full-text listings on newly established or newly identified media outlets. Information provided in *GDPBM* is obtained primarily through research of publication and publisher websites and questionnaire responses from organizations listed. Some clarification and verification of data is obtained through telephone calls. Other published sources are used to verify some information, such as the audited circulation data in publication listings. More detail on the scope and coverage of *Supplement* listings is contained in the User's Guide following this Introduction.

Available in Electronic Formats

Licensing. *Gale Directory of Publications and Broadcast Media* is available for licensing. The complete database is provided in a fielded format and is deliverable on such media as disk or CD-ROM. For more information, contact Gale's Business Development Group at 1-800-877-GALE, or visit our web site at www.gale.cengage.com/bizdev.

Online. *Gale Directory of Publications and Broadcast Media* (along with *Directories in Print* and *Newsletters in Print*) is accessible as File 469: *Gale Database of Publications and Broadcast Media* through the Dialog Corporation's DIALOG service. *GDPBM* is also accessible as PUBBRD through LexisNexis. For more information, contact The Dialog Corporation, 11000 Regency Parkway, Ste. 10, Cary, NC 27511; phone: (919) 462-8600; toll-free: 800-3-DIALOG; or LexisNexis, P.O. Box 933, Dayton, OH 45401-0933; phone (937) 865-6800; toll-free: 800-227-9597.

The *Directory* is also available online as part of the *Gale Directory Library.* For more information, call 1-800-877-GALE.

Comments and Suggestions Welcome

If you have questions, concerns, or comments about *Gale Directory of Publications and Broadcast Media,* and/or suggestions for new listings, please contact:

Project Editor

Gale Directory of Publications and Broadcast Media

Gale

27500 Drake Rd.

Farmington Hills, MI 48331-3535

Phone: (248) 699-4253

Toll-free: 800-347-GALE

Fax: (248) 699-8075

Email: Louise.Gagne@cengage.com

URL: http://www.gale.cengage.com

The samples and notes below offer details on specific *GDPBM* content and how to use the *Supplement*'s listings and indexes. Please note that entry information appearing in this section has been fabricated.

Sample Entries

In the samples that follow, each numbered section designates information that might appear in a *GDPBM* listing. The numbered items are explained in the descriptive paragraphs following each sample.

Sample Publication Listing

▌1▌ 222 ▪ ▌2▌ **American Computer Review**
▌3▌ Jane Doe Publishing Company, Inc.
▌4▌ 199 E 49th St. PO Box 724866 Salem, NY 10528-5555
▌5▌ Phone: (518)555-9277
▌6▌ Fax: (518)555-9288
▌7▌ Free: 800-555-5432
▌8▌ Publication E-mail: acr@jdpci.com
▌9▌ Publisher E-mail: jdpci@jdpci.com
▌10▌ Magazine for users of Super Software Plus products. ▌11▌ **Subtitle:** The Programmer's Friend. ▌12▌ **Founded:** June 1979. ▌13▌ **Freq:** Monthly (combined issue July/Aug.). ▌14▌ **Print Method:** Offset. ▌15▌ **Trim Size:** 8/12 x 11. ▌16▌ **Cols./Page:** 3. ▌17▌ **Col. Width:** 24 nonpareils. ▌18▌ **Col. Depth:** 294 agate lines. ▌19▌ **Key Personnel:** Ian Smith, Editor, phone (518)555-1201, fax (518)555-1202, ismith@jdpci.com; James Newman, Publisher; Steve Jones Jr., Advertising Mgr. ▌20▌ **ISSN:** 5555-6226. ▌21▌ **Subscription Rates:** $25; $30 Canada; $2.50 single issue. ▌22▌ **Remarks:** Color advertising not accepted. ▌23▌ **Online:** Lexis-Nexis **URL:** http://www.acrmagazine.com. ▌24▌ **Alternate Format(s):** Braille; CD-ROM; Microform. ▌25▌ **Formerly:** Computer Software Review (Dec. 13, 1986). ▌26▌ **Feature Editors:** Ann Walker, *Consumer Affairs, Editorials,* phone (518)555-2306, fax (518)555-2307, aw@jdpci.com. ▌27▌ **Additional Contact Information: Advertising:** 123 Main St., New York, NY 10016, (201)555-1900, fax: (201)555-1908. ▌28▌ **Ad Rates:** BW: $850, PCI: $.75 ▌29▌ **Circulation:** 25,000

Description of Numbered Elements

▌1▌ **Symbol/Entry Number.** Entries are numbered sequentially. Entry numbers, rather than page numbers, are used to refer to listings.

▌2▌ **Publication Title.** Publication names are listed *as they appear on the masthead or title page,* as provided by respondents.

▌3▌ **Publishing Company.** The name of the commercial publishing organization, association, or academic institution, as provided by respondents.

▌4▌ **Address.** Full mailing address information is provided wherever possible. This may include: street address; post office box; city; state or province; and ZIP or postal code. ZIP plus-four numbers are provided when known.

▌5▌ **Phone.** Phone numbers listed in this section are usually the respondent's switchboard number.

▌6▌ **Fax.** Facsimile numbers are listed when provided.

▌7▌ **Free.** Toll-free numbers are listed when provided.

▌8▌ **Publication E-mail.** Electronic mail addresses for the publication are included as provided by the listee.

▌9▌ **Publisher E-mail:** Electronic mail addresses for the publishing company are included as provided by the listee.

▌10▌ **Description.** Includes the type of publication (i.e., newspaper, magazine) as well as a brief statement of purpose, intended audience, or other relevant remarks.

▌11▌ **Subtitle.** Included as provided by the listee.

▌12▌ **Founded.** Date the periodical was first published.

▌13▌ **Frequency.** Indicates how often the publication is issued—daily, weekly, monthly, quarterly, etc. Explanatory remarks sometimes accompany this information (e.g., for weekly titles, the day of issuance; for collegiate titles, whether publication is limited to the academic year; whether certain issues are combined.)

▌14▌ **Print Method.** Though offset is most common, other methods are listed as provided.

▌15▌ **Trim Size.** Presented in inches unless otherwise noted.

▌16▌ **Number of Columns Per Page.** Usually one figure, but some publications list two or more, indicating a variation in style.

▌17▌ **Column Width.** Column sizes are given exactly as supplied, whether measured in inches, picas (6 picas to an inch), nonpareils (each 6 points, 72 points to an inch), or agate lines (14 to an inch).

▌18▌ **Column Depth.** Column sizes are given exactly as supplied, whether measured in inches, picas (6 picas to an inch), nonpareils (each 6 points, 72 points to an inch), or agate lines (14 to an inch).

▌19▌ **Key Personnel.** Presents the names and titles of

contacts at each publication. May include phone, fax, and e-mail addresses if different than those for the publication and company.

▮20▮ International Standard Serial Number (ISSN). Included when provided. Occasionally, United States Publications Serial (USPS) numbers are reported rather than ISSNs.

▮21▮ Subscription Rates. Unless otherwise stated, prices shown in this section are the individual annual subscription rate. Other rates are listed when known, including multiple-year rates, prices outside the United States, discount rates, library/institution rates, and single copy prices.

▮22▮ Remarks. Information listed in this section further explains the Ad Rates.

▮23▮ Online. If a publication is accessible online via computer, that information is listed here. If the publication is available online, but the details of the URL or vendor are not known, the notation "Available Online" will be listed.

▮24▮ Alternate Format(s). Lists additional mediums in which a publication may be available (other than online), including CD-ROM and microform.

▮25▮ Variant Name(s). Lists former or variant names of the publication, including the year the change took place, when known.

▮26▮ Feature Editors. Lists the names and beats of any feature editors employed by the publication.

▮27▮ Additional Contact Information. Includes mailing, advertising, news, and subscription addresses and phone numbers when different from the editorial/publisher address and phone numbers.

▮28▮ Ad Rates. Respondents may provide non-contract (open) rates in any of six categories:

GLR = general line rate

BW = one-time black & white page rate

4C = one-time four-color page rate

SAU = standard advertising unit rate

CNU = Canadian newspaper advertising unit rate

PCI = per column inch rate

Occasionally, explanatory information about other types of advertising appears in the Remarks section of the entry.

▮29▮ Circulation. Figures represent various circulation numbers; the figures are accompanied by a symbol (except for sworn and estimated figures). Following are explanations of the eight circulation classifications used by *GDPBM*; the corresponding symbols, if any, are listed at the bottom of each right-hand page. All circulation figures *except* publisher's reports and estimated figures appear in boldface type.

These audit bureaus are independent, nonprofit organizations (with the exception of VAC, which is for-profit) that verify circulation rates. Interested users may contact the associations for more information.

- **ABC:** Audit Bureau of Circulations, 900 N Meacham Rd., Schaumburg, IL 60173; (847)605-0909
- **CAC:** Certified Audit of Circulations, Inc., 155 Willowbrook Blvd., 4th Fl., Wayne, NJ 07470-7036; (973)785-3000
- **CCAB:** Canadian Circulations Audit Board, 90 Eglinton Ave. E, Ste. 980, Toronto, ON Canada M4P 2Y3; (416)487-2418
- **VAC:** Verified Audit Circulation, 900 Larkspur Landing Cir., Ste. 295, Lakespur, CA 94939; (800)775-3332
- **Post Office Statement:** These figures were verified from a U.S. Post Office form.
- **Publisher's Statement:** These figures were accompanied by the signature of the editor, publisher, or other officer.
- **Sworn Statement:** These figures, which appear in **boldface** without a symbol, were accompanied by the notarized signature of the editor, publisher, or other officer of the publication.
- **Estimated Figures:** These figures, which are shown in lightface without a symbol, are the unverified report of the listee.

The footer on every odd-numbered page contains a key to circulation and entry type symbols, as well as advertising abbreviations.

Sample Broadcast Listing

▮1▮ 111 ▪ **▮2▮ WCAF-AM—1530**
▮3▮ 199 E 49th St.
PO Box 724866
Salem, NY 10528-5555
▮4▮ Phone: (518)555-9277
▮5▮ Fax: (518)555-9288
▮6▮ Free: 800-555-5432
▮7▮ E-mail: wcafwcaf.com
▮8▮ Format: Classical. **▮9▮ Simulcasts:** WCAF-FM. **▮10▮ Network(s):** Westwood One Radio; ABC. **▮11▮ Owner: Affici Communications, Inc., at above address. ▮12▮ Founded:** 1996. **▮13▮ Formerly:** WCAH-AM (1992). **▮14▮ Operating Hours:** Continuous; 90% local, 10% network. **▮15▮ ADI:** Elmira, NY. **▮16▮ Key Personnel:** James Smith, General Mgr., phone (518)555-1002, fax (518)555-1010, jsmithwcaf.com; Don White, Program Dir. **▮17▮ Cities Served:** Salem, NY. **▮18▮ Postal Areas Served:** 10528; 10529. **▮19▮ Local Programs:** Who's Beethoven? Clement Goebel, Contact, (518)555-1301, fax (518)555-1320. **▮20▮ Wattage:** 5000. **▮21▮ Ad Rates:** Underwriting available. $10-15 for 30 seconds; $30-35 for 60 seconds. Combined advertising rates available with WCAF-FM. **▮22▮ Additional Contact Information:** Mailing address: PO Box 555, Elmira, NY 10529. **▮23▮ URL:** http://www.wcaf.com.

Description of Numbered Elements

▮1▮ Entry Number. Entries are numbered sequentially. Entry numbers (rather than page numbers) are used to refer to listings.

▮2▮ Call Letters and Frequency/Channel or **Cable Company Name.**

▮3▮ Address. Location and studio addresses appear as supplied by the respondent. If provided, alternate addresses are listed in the Additional Contact Information section of the entries (see item 22 below).

▮4▮ Phone. Telephone numbers are listed as provided.

▌5▌ Fax. Facsimile numbers are listed when provided.

▌6▌ Free. Toll-free numbers are listed when provided.

▌7▌ E-mail. Electronic mail addresses are included as provided by the listee.

▌8▌ Format. For television station entries, this subheading indicates whether the station is commercial or public. Radio station entries contain industry-defined (and, in some cases, station-defined) formats as indicated by the listee.

▌9▌ Simulcasts. Lists stations that provide simulcasting.

▌10▌ Network(s). Notes national and regional networks with which a station is affiliated. The term "independent" is used if indicated by the listee.

▌11▌ Owner. Lists the name of an individual or company, supplemented by the address and telephone number, when provided by the listee. If the address is the same as that of the station or company, the notation "at above address" is used, referring to the station or cable company address.

▌12▌ Founded. In most cases, the year the station/company began operating, regardless of changes in call letters/names and ownership.

▌13▌ Variant Name(s). For radio and television stations, former call letters and the years in which there were changes are presented as provided by the listee. Former cable company names and the years in which they were changed are also noted when available.

▌14▌ Operating Hours. Lists on-air hours and often includes percentages of network and local programming.

▌15▌ ADI (Area of Dominant Influence). The Area of Dominant Influence is a standard market region defined by the Arbitron Ratings Company for U.S. television stations. Some respondents also list radio stations as having ADIs.

▌16▌ Key Personnel. Presents the names and titles of contacts at each station or cable company.

▌17▌ Cities Served. This heading is primarily found in cable system entries and provides information on channels and the number of subscribers.

▌18▌ Postal Areas Served. This heading is primarily found in cable system entries and provides information on the postal (zip) codes served by the system.

▌19▌ Local Programs. Lists names, air times, and contact personnel of locally-produced television and radio shows.

▌20▌ Wattage. Applicable to radio stations, the wattage may differ for day and night in the case of AM stations. Occasionally a station's ERP (effective radiated power) is given in addition to, or instead of, actual licensed wattage.

▌21▌ Ad Rates. Includes rates for 10, 15, 30, and 60 seconds as provided by respondents. Some stations price advertisement spots "per unit" regardless of length; these units vary.

▌22▌ Additional Contact Information. Includes mailing, advertising, news, and studio addresses and phone numbers when different from the station, owner, or company address and phone numbers.

▌23▌ Online. If a radio station or cable company is accessible online via computer, that information is listed here. If the station or company is available online but the details of the URL or vendor are not known, the notation "Available Online" will be listed.

Index Notes

The Master Name and Keyword Index provides access to all entries in the *Supplement*. Citations in the index are listed alphabetically regardless of media type.

Publication citations include the following:

- titles
- keywords within titles
- former titles
- foreign-language titles
- alternate titles

Broadcast media citations include the following:

- station call letters
- cable company names (U.S. and Canada)
- former call letters
- former cable company names (U.S. and Canada)

Indexing is word-by-word rather than letter-by-letter. Thus, "New York" is listed before "News." Current listings in the Index include geographic information and entry number.

Geographic Abbreviations

U.S. State and Territory Postal Codes

AK	Alaska
AL	Alabama
AR	Arkansas
AZ	Arizona
CA	California
CO	Colorado
CT	Connecticut
DC	District of Columbia
DE	Delaware
FL	Florida
GA	Georgia
HI	Hawaii
IA	Iowa
ID	Idaho
IL	Illinois
IN	Indiana
KS	Kansas
KY	Kentucky
LA	Louisiana
MA	Massachusetts
MD	Maryland
ME	Maine
MI	Michigan
MN	Minnesota
MO	Missouri
MS	Mississippi
MT	Montana
NC	North Carolina
ND	North Dakota
NE	Nebraska
NH	New Hampshire
NJ	New Jersey
NM	New Mexico
NV	Nevada
NY	New York
OH	Ohio
OK	Oklahoma
OR	Oregon
PA	Pennsylvania
PR	Puerto Rico
RI	Rhode Island
SC	South Carolina
SD	South Dakota
TN	Tennessee
TX	Texas
UT	Utah
VA	Virginia
VT	Vermont
WA	Washington
WI	Wisconsin
WV	West Virginia
WY	Wyoming

Canadian Province and Territory Postal Codes

AB	Alberta
BC	British Columbia
MB	Manitoba
NB	New Brunswick
NL	Newfoundland and Labrador
NS	Nova Scotia
NT	Northwest Territories
ON	Ontario
PE	Prince Edward Island
QC	Quebec
SK	Saskatchewan
YT	Yukon Territory

Australian State and Territory Codes

ACT	Australian Capitol Territory
NSW	New South Wales
NT	Northern Territory
QLD	Queensland
SA	South Australia
TAS	Tasmania
VIC	Victoria
WA	Western Australia

Chinese Province and Region Codes

AN	Anhui
FJ	Fujian
GS	Gansu
GD	Guangdong
GZ	Guangxi Zhuangzu
GH	Guizhou
HB	Hebei
HL	Heilongjiang
HN	Henan
HU	Hubei
HA	Hunan
JS	Jiangsu
JX	Jiangxi
JI	Jilin
LI	Liaoning
NM	Nei Monggol Zizhiqu
NH	Ningxia Huizu
QI	Qinghai
SH	Shaanxi
SD	Shandong
SX	Shanxi
SI	Sichuan
XU	Xinjiang Uygur Zizhigu
XZ	Xizang
YU	Yunnan
ZH	Zhejiang

Indian State and Territory Codes

AN	Andaman and Nicobar
AP	Andhra Pradesh
AR	Arunachal Pradesh
AS	Assam
BH	Bihar
CH	Chandigarh
DN	Dadra and Nagar Haveli
DH	Delhi
GD	Goa Daman and Diu
GJ	Gujarat
HY	Haryana
HP	Himachal Pradesh
JK	Jammu and Kashmir
KA	Karnataka
KE	Kerala
LC	Laccadive Minicoy and Amindivi
MP	Madhya Pradesh
MH	Maharashtra
MN	Manipur
MG	Meghalaya

MZ	Mizoram
MY	Mysore
NG	Nagaland
OR	Orissa
PN	Pondicherry
PJ	Punjab
RJ	Rajasthan
SK	Sikkim
TN	Tamil Nadu
TR	Tripura
UP	Uttar Pradesh
WB	West Bengal (W. Bengal)

Irish County Codes

CV	Cavan
CA	Carlow
CL	Clare
CK	Cork
DO	Donegal
DU	Dublin
GL	Galway
KR	Kerry
KL	Kildare
KK	Kilkenny
LA	Laoighis
LE	Leitrim
LI	Limerick
LO	Longford
LU	Louth
MA	Mayo
ME	Meath
MO	Monaghan
OF	Offaly
RO	Roscommon
SL	Sligo
TP	Tipperary
WA	Waterford
WE	Westmeath
WX	Wexford
WI	Wicklow

Mexican State Codes

AG	Aguascalientes
BN	Baja California Norte
BS	Baja California Sur
CM	Campeche
CP	Chiapas
CH	Chihuahua
CO	Coahuila
CL	Colima
DF	Distritto Federal
DU	Durango
GJ	Guanajuato
GU	Guerrero
HD	Hidalgo
JA	Jalisco
ME	Mexico
MI	Michoacan
MO	Morelos
NY	Nayarit
NL	Nuevo Leon
OX	Oaxaca
PU	Puebla
QT	Queretaro
QR	Quintana Roo
SP	San Luis Potosi
SN	Sinaloa
SR	Sonora
TB	Tabasco
TM	Tamaulipas
TL	Tlaxcala
VC	Veracruz
YU	Yucatan
ZA	Zacatecas

Nigerian States

AN	Anambra
BA	Bauchi
BE	Bendel
BN	Benue
BR	Borno
CR	Cross River
GO	Gongola
IM	Imo
KD	Kaduna
KN	Kano
KW	Kwara
LG	Lagos
NG	Niger
OG	Ogun
ON	Ondo
OY	Oyo
PL	Plateau
RV	Rivers
SK	Sokoto

Country Abbreviations

For England, Northern Ireland, Scotland, and Wales, please see United Kingdom (GBR).

ALB	Albania
ALG	Algeria
ANG	Angola
AIA	Anguilla
ATG	Antigua-Barbuda
ARG	Argentina
AMA	Armenia
AUS	Australia
AUT	Austria
AJN	Azerbaijan
BHS	Bahamas
BHR	Bahrain
BGD	Bangladesh
BRB	Barbados
BLR	Belarus
BEL	Belgium
BLZ	Belize
BEN	Benin
BMU	Bermuda
BTN	Bhutan
BOL	Bolivia
BWA	Botswana
BRZ	Brazil
BUL	Bulgaria
BFA	Burkina Faso
BDI	Burundi
CMB	Cambodia
CMR	Cameroon
CYM	Cayman Islands
CHL	Chile
CHN	People's Republic of China
COL	Colombia
CRI	Costa Rica
COT	Cote d'Ivoire
CTA	Croatia
CUB	Cuba
CYP	Cyprus
CZE	Czech Republic
DEN	Denmark
DMA	Dominica
DOM	Dominican Republic
ECU	Ecuador
EGY	Egypt
ELS	El Salvador
EST	Estonia
ETH	Ethiopia
FAR	Faroe Islands
FIJ	Fiji
FIN	Finland
FRA	France
GAB	Gabon
GBR	United Kingdom
GMB	Gambia
GRG	Georgia
GER	Germany
GHA	Ghana
GRC	Greece
GUM	Guam
GTM	Guatemala
GIN	Guinea
GUY	Guyana
HBO	Bosnia-Hercegovina
HND	Honduras
HKG	Hong Kong
HUN	Hungary
ICE	Iceland
IND	India
IDN	Indonesia
IRN	Iran
IRQ	Iraq
IRL	Ireland
ISR	Israel
ITA	Italy
JAM	Jamaica
JPN	Japan
JOR	Jordan
KAZ	Kazakhstan
KEN	Kenya
KGA	Kirgizstan
KOR	Republic of Korea

KWT	Kuwait
LAT	Latvia
LBN	Lebanon
LIT	Lithuania
LUX	Luxembourg
MEC	Macedonia
MWI	Malawi
MYS	Malaysia
MDV	Maldives
MLI	Mali
MAL	Malta
MUS	Mauritius
MEX	Mexico
MDI	Moldova
MCO	Monaco
MNG	Mongolia
MON	Montenegro
MOR	Morocco
MOZ	Mozambique
NAM	Namibia
NPL	Nepal
NLD	Netherlands
NAT	Netherlands Antilles
NCL	New Caledonia
NZL	New Zealand
NCG	Nicaragua
NER	Niger
NGA	Nigeria
NOR	Norway
OMN	Oman
PAK	Pakistan
PAN	Panama
PNG	Papua New Guinea
PER	Peru
PHL	Philippines
POL	Poland
PRT	Portugal
ROM	Romania
RUS	Russia
RWA	Rwanda
SLC	St. Lucia
SAU	Saudi Arabia
SEN	Senegal
SER	Serbia
SYC	Seychelles
SLE	Sierra Leone
SGP	Singapore
SLK	Slovakia
SVA	Slovenia
SLM	Solomon Islands
SAF	Republic of South Africa
SPA	Spain
SRI	Sri Lanka
SDN	Sudan
SWZ	Swaziland
SWE	Sweden
SWI	Switzerland
SYR	Syrian Arab Republic
TWN	Taiwan
TDN	Tajikistan
TZA	United Republic of Tanzania
THA	Thailand
TGO	Togo
TGA	Tonga
TTO	Trinidad and Tobago
TUN	Tunisia
TUR	Turkey
UGA	Uganda
URE	Ukraine
UAE	United Arab Emirates
URY	Uruguay
UZN	Uzbekistan
VAT	Vatican City
VEN	Venezuela
VNM	Vietnam
BVI	British Virgin Islands
VIR	Virgin Islands of the United States
YEM	Yemen
ZMB	Zambia
ZWE	Zimbabwe

Miscellaneous Abbreviations

&	And
4C	One-Time Four Color Page Rate
ABC	Audit Bureau of Circulations
Acad.	Academy
Act.	Acting
Adm.	Administrative, Administration
Admin.	Administrator
AFB	Air Force Base
AM	Amplitude Modulation
Amer.	American
APO	Army Post Office
Apt.	Apartment
Assn.	Association
Assoc.	Associate
Asst.	Assistant
Ave.	Avenue
Bldg.	Building
Blvd.	Boulevard
boul.	boulevard
BPA	Business Publications Audit of Circulations
BTA	Best Time Available
BW	One-time Black & White Page Rate
C	Central
CAC	Certified Audit of Circulations
CCAB	Canadian Circulations Audit Board
CEO	Chief Executive Officer
Chm.	Chairman
Chwm.	Chairwoman
CNU	Canadian Newspaper Advertising Unit Rate
c/o	Care of
Col.	Column
Coll.	College
Comm.	Committee
Co.	Company
COO	Chief Operating Officer
Coord.	Coordinator
Corp.	Corporation
Coun.	Council
CP	Case Postale
Ct.	Court
Dept.	Department
Dir.	Director
Div.	Division
Dr.	Doctor, Drive
E	East
EC	East Central
ENE	East Northeast
ERP	Effective Radiated Power
ESE	East Southeast
Eve.	Evening
Exec.	Executive
Expy.	Expressway
Fed.	Federation
Fl.	Floor
FM	Frequency Modulation
FPO	Fleet Post Office
Fri.	Friday
Fwy.	Freeway
Gen.	General
GLR	General Line Rate
Hd.	Head
Hwy.	Highway
Inc.	Incorporated
Info.	Information
Inst.	Institute
Intl.	International
ISSN	International Standard Serial Number
Jr.	Junior
Libn.	Librarian
Ln.	Lane
Ltd.	Limited
Mgr.	Manager
mi.	miles
Mktg.	Marketing
Mng.	Managing
Mon.	Monday
Morn.	Morning
N	North
NAS	Naval Air Station
Natl.	National
NC	North Central
NE	Northeast
NNE	North Northeast
NNW	North Northwest
No.	Number
NW	Northwest
Orgn.	Organization
PCI	Per Column Inch Rate

Pkwy.	Parkway
Pl.	Place
PO	Post Office
Pres.	President
Prof.	Professor
Rd.	Road
RFD	Rural Free Delivery
Rm.	Room
ROS	Run of Schedule
RR	Rural Route
Rte.	Route
S	South
Sat.	Saturday
SAU	Standard Advertising Unit Rate
SC	South Central
SE	Southeast
Sec.	Secretary
Soc.	Society
Sq.	Square
Sr.	Senior
SSE	South Southeast
SSW	South Southwest
St.	Saint, Street
Sta.	Station
Ste.	Sainte, Suite
Sun.	Sunday
Supt.	Superintendent
SW	Southwest
Terr.	Terrace
Thurs.	Thursday
Tpke.	Turnpike
Treas.	Treasurer
Tues.	Tuesday
Univ.	University
USPS	United States Publications Serial
VAC	Verified Audit Circulation
VP	Vice President
W	West
WC	West Central
Wed.	Wednesday
WNW	West Northwest
WSW	West Southwest
x/month	Times per Month
x/week	Times per Week
x/year	Times per Year

ALABAMA

ANNISTON

1 ■ WHOG-AM - 1120
Radio Bldg., Sta. 411
1330 Noble St., Ste. 25
Anniston, AL 36201
Phone: (256)236-6484
Fax: (256)236-5338
E-mail: whog@whog1120.com

Format: Hip Hop; Blues. **URL:** http://whog1120.com/index.html.

ASHLAND

2 ■ WCKF-FM - 100.7
518 Mountain View Rd.
Ashland, AL 36251
Phone: (256)354-1444
Fax: (256)354-1445
Free: (866)634-1007

Format: Information; News. **URL:** http://www.alabama1007fm.com/1.html.

AUBURN

3 ■ WGZZ-FM - 100.3
197 E University Dr.
Auburn, AL 36830
Phone: (334)826-2929
Fax: (334)826-9151

Format: Classical. **Owner:** Auburn Network, PO Box 950, Auburn, AL 36831-0950. **Key Personnel:** Mike Hubbard, Pres./CEO, hubbard@aunetwork.com; Chris Hines, Sen. VP, hines@aunetwork.com; Drew McCracken, Traffic Mgr., dmccracken@aunetwork.com. **URL:** http://www.wingsfm.com/.

BIRMINGHAM

4 ■ Pro Bull Rider
Grand View Media Group Inc.
200 Croft St., Ste. 1
Birmingham, AL 35242
Phone: (205)408-3797
Fax: (205)408-3797
Free: (888)431-2877
Publisher E-mail: webmaster@grandviewmedia.com

Magazine featuring professional bull riding. **Subtitle:** Official Magazine of the Professional Bull Riders. **Freq:** Bimonthly. **Trim Size:** 7.75 x 10.625. **Key Personnel:** Jeff Johnstone, Editor, jjohnstone@pbrnow.com; Hilary Mizelle, Managing Editor, hilary@grandviewmedia.com; Derrick Nawrocki, Publisher, derrickn@grandviewmedia.com. **Remarks:** Accepts advertising. **URL:** http://www.gvmg.com/PBR/index.htm. . **Ad Rates:** 4C: $4,450. **Circ:** (Not Reported)

5 ■ WXJC-AM - 850
120 Summit Pky., Ste. 200
Birmingham, AL 35209
Phone: (205)879-3324
Fax: (205)941-1095

Format: Gospel. **Owner:** Crawford Broadcasting Company, 2150 W 29th Ave., Ste. 300, Denver, CO 80211, (303)433-5500, Fax: (303)433-1555. **Key Personnel:** Mark Berthiaume, Operations Dir., mark_berthiaume@850wxjc.com; Bob Ratchford, Operations Mgr., bob@850wxjc.com. **URL:** http://www.850wxjc.com/.

BRIDGEPORT

6 ■ WYMR-AM - 1480
PO Box 1018
Laguna Beach, CA 92652
Fax: (469)241-6795
Free: (800)639-5433

Format: Religious. **Owner:** New Life Ministries, at above address. **Key Personnel:** Stephen Arterburn, Founder/Chm. **URL:** http://www.newlife.com/Radio/findradio.asp?id=AL&t=Alabama.

GADSDEN

7 ■ WMGJ-AM - 1240
815 Tuscaloosa Ave.
Gadsden, AL 35901
Phone: (256)546-4434
Fax: (256)546-9645

Format: News; Gospel; Blues; Urban Contemporary. **Operating Hours:** Continuous. **URL:** http://www.wmgj.com/.

HEFLIN

8 ■ WPIL-FM - 91.7
256 Brockford Rd.
Heflin, AL 36264
Phone: (256)463-4226
Fax: (256)463-4232
E-mail: wpil@wpilfm.com

Format: Gospel; Country. **URL:** http://www.wpilfm.com/.

HUNTSVILLE

9 ■ WAYH-FM - 88.1
9582 Madison Blvd., No. 8
Madison, AL 35758
Phone: (256)837-9293
Fax: (256)772-6731
Free: (888)239-2936

Format: Contemporary Christian. **Owner:** WAY-FM Media Group, PO Box 64500, Colorado Springs, CO 80962, (719)533-0300, Fax: (719)278-4339. **Key Personnel:** Thom Ewing, General Mgr.; Lisa Shelton, Devel. Dir.; Linda Cashin, Office Mgr. **URL:** http://wayh.wayfm.com/.

10 ■ WWFF-FM - 93.3
1717 US Hwy. 72 E
Huntsville, AL 35811
Phone: (256)830-8300
Fax: (256)232-6842

Format: Contemporary Hit Radio (CHR); Country. **Ad Rates:** Advertising accepted; rates available upon request. **URL:** http://www.wolf933.com/.

MADISON

WAYH-FM - See Huntsville

NORTHPORT

11 ■ WMFT-FM - 88.9
5710 Watermelon Rd., Ste. 316
Northport, AL 35473
Phone: (205)758-7900
Fax: (334)992-2637
Free: (888)624-7234
E-mail: wmft@moody.edu

Format: Religious; Gospel. **Owner:** Moody Bible Institute, 820 N LaSalle Blvd., Chicago, IL 60610. **Key Personnel:** Rob Moore, Manager. **URL:** http://www.mbn.org/GenMoody/default.asp?SectionID=426477D62803447DBE15F0D402564FEF.

PRATTVILLE

12 ■ WIQR-AM - 1410
921 E Main St.
Prattville, AL 36066
Phone: (334)358-0410

Format: Sports. **Ad Rates:** Advertising accepted; rates available upon request. **URL:** http://www.wiqr.net/.

TROY

13 ■ WAXU-FM - 91.1
PO Drawer 2440
Tupelo, MS 38803
Free: (800)326-4543

Format: Religious. **Owner:** American Family Association, at above address. **URL:** http://waxu.afr.net/index.php.

TRUSSVILLE

14 ■ WRLM-AM - 1480
PO Box 1313
Trussville, AL 35173
Phone: (205)419-5753
Free: (866)359-6303

Format: Ethnic. **Wattage:** 5000. **URL:** http://www.latinomix1480.com/.

Circulation: ★ = ABC; △ = BPA; ♦ = CAC; • = CCAB; ❑ = VAC; ⊕ = PO Statement; ‡ = Publisher's Report; Boldface figures = sworn; Light figures = estimated.

TUSCALOOSA

15 ■ International Journal for Manufacturing Science and Production
Freund Publishing House Ltd.
Center for Green Manufacturing
Department of Metallurgical & Materials Engineering
University of Alabama
Tuscaloosa, AL 35487
Publisher E-mail: h_freund@netvision.net.il

Journal covering developments and original applications in all fields related to manufacturing and production. **Freq:** 4/yr. **Key Personnel:** Prof. Ramana G. Reddy, Editor-in-Chief, rreddy@eng.ua.edu. **ISSN:** 0793-6648. **Subscription Rates:** US$250 individuals. **URL:** http://www.freundpublishing.com/JOURNALS/materials_science_and_engineering.htm.

16 ■ International Journal of Security & Networks
Inderscience Publishers
c/o Prof. Yang Xiao, Ed.-in-Ch.
University of Alabama, Department of Computer Science
101 Houser Hall
Box 870290
Tuscaloosa, AL 35487-0290
Publisher E-mail: editor@inderscience.com

Journal covering on dissemination of network security related issues. **Founded:** 2006. **Freq:** 4/yr. **Key Personnel:** Prof. Yang Xiao, Editor-in-Chief, yangxiao@cs.ua.edu. **ISSN:** 1747-8405. **Subscription Rates:** EUR470 individuals includes surface mail, print only; EUR640 individuals print & online. **URL:** http://www.inderscience.com/browse/index.php?journalCODE=ijsn.

ALASKA

ANCHORAGE

17 ■ KJLP-FM - 88.9
PO Box 1018
Laguna Beach, CA 92652
Fax: (469)241-6795
Free: (800)639-5433

Format: Religious. **Owner:** New Life Ministries, at above address. **Key Personnel:** Stephen Arterburn, Founder/Chm. **URL:** http://www.newlife.com.

18 ■ KZND-FM - 94.7
4700 Business Park Blvd., Bldg. E, Ste. 44-a
Anchorage, AK 99503
Phone: (907)522-1018

Format: Alternative/New Music/Progressive. **URL:** http://www.947kznd.com/.

FAIRBANKS

19 ■ KTDZ-FM - 103.9
819 1st Ave., Ste. A
Fairbanks, AK 99701
Phone: (907)451-5910
Fax: (907)451-5999

Format: Oldies. **Key Personnel:** Perry Walley, General Mgr., perry.walley@nnbradio.com. **Ad Rates:** Advertising accepted; rates available upon request. **URL:** http://www.mytedfm.com/.

GIRDWOOD

KWMD-FM - See Kasilof

KASILOF

20 ■ KWJG-FM - 91.5
PO Box 1121
Kasilof, AK 99610
Free: (877)455-7702

Format: Country. **Owner:** Kasilof Public Broadcasting, at above address. **Operating Hours:** Continuous. **Key Personnel:** Bill Glynn, President. **URL:** http://www.kwjg.org/.

21 ■ KWMD-FM - 90.7
PO Box 75
Girdwood, AK 99587

Format: Public Radio. **Owner:** Alaska Educational Radio System, Inc., at above address. **URL:** http://www.oneskyradio.com/.

KENAI

22 ■ KKIS-FM - 96.5
40960 Kalifornsky Beach Rd.
Kenai, AK 99611
Phone: (907)283-5821

Format: Contemporary Hit Radio (CHR). **Owner:** KSRM, Inc., at above address. **URL:** http://www.radiokenai.net/kkis/contact.asp.

ARIZONA

PHOENIX

23 ■ Billing & OSS World
Virgo Publishing Inc.
PO Box 40079
Phoenix, AZ 85067-0079
Phone: (480)990-1101
Fax: (480)990-0819

Magazine featuring coverage and analysis of the telecommunications billing and operations support services. **Freq:** 6/yr. **Key Personnel:** Tim McElligott, Editor-in-Chief, phone (480)990-1101, tmcelligott@vpico.com. **Remarks:** Accepts advertising. **URL:** http://www.billingworld.com/. . **Ad Rates:** BW: $5,725. **Circ:** ‡6,952

24 ■ Church Solutions
Virgo Publishing Inc.
PO Box 40079
Phoenix, AZ 85067-0079
Phone: (480)990-1101
Fax: (480)990-0819

Magazine featuring issues about church business administrations. **Freq:** 12/yr. **Key Personnel:** Katherine Kennedy, Publisher, phone (480)281-6786, kkennedy@vpico.com. **Subscription Rates:** US$72 individuals; US$75 Canada; US$147 other countries. **Remarks:** Accepts advertising. **URL:** http://www.churchsolutionsmag.com/. . **Ad Rates:** BW: $5,145. **Circ:** 23,000

25 ■ Culinology
Virgo Publishing Inc.
PO Box 40079
Phoenix, AZ 85067-0079
Phone: (480)990-1101
Fax: (480)990-0819

Magazine featuring new food products developed from the combination of food science practices and traditional culinary arts. **Freq:** Quarterly. **Key Personnel:** Bob Weeks, Publisher, phone (480)990-1101, rweeks@vpico.com. **Remarks:** Accepts advertising. **URL:** http://www.culinologyonline.com/. **Circ:** (Not Reported)

26 ■ Food Product Design
Virgo Publishing Inc.
PO Box 40079
Phoenix, AZ 85067-0079
Phone: (480)990-1101
Fax: (480)990-0819

Magazine featuring food industry current news and information. **Key Personnel:** Lynn A. Kuntz, Editor, phone (480)990-1101, lkuntz@vpico.com. **Remarks:** Accepts advertising. **URL:** http://www.foodproductdesign.com/. . **Ad Rates:** BW: $5,215. **Circ:** 32,000

27 ■ Immediate Care Business
Virgo Publishing Inc.
PO Box 40079
Phoenix, AZ 85067-0079
Phone: (480)990-1101
Fax: (480)990-0819

Magazine featuring urgent care accreditation. **Subtitle:** Business Solutions for the Urgent Care Market. **Freq:** Bimonthly. **Key Personnel:** Bill Eikost, Publisher, phone (480)990-1101, weikost@vpico.com. **Remarks:** Accepts advertising. **URL:** http://www.immediatecarebusiness.com/. **Circ:** 6,000

28 ■ Infection Control Today
Virgo Publishing Inc.
PO Box 40079
Phoenix, AZ 85067-0079
Phone: (480)990-1101
Fax: (480)990-0819

Magazine featuring coverage about infection control. **Subtitle:** Influential Compelling Thorough. **Key Personnel:** Bill Eikost, Publisher, phone (480)990-1101, weikost@vpico.com. **Remarks:** Accepts advertising. **URL:** http://www.infectioncontroltoday.com/. . **Ad Rates:** BW: $6,600. **Circ:** 30,001

29 ■ Inside Cosmeceuticals
Virgo Publishing Inc.
PO Box 40079
Phoenix, AZ 85067-0079
Phone: (480)990-1101
Fax: (480)990-0819

Magazine featuring dietary supplement, healthy food, and personal care products. **Key Personnel:** Somlynn Rorie, Managing Editor, phone (480)990-1101, srorie@vpico.com. **Remarks:** Accepts advertising. **URL:** http://www.insidecosmeceuticals.com/. . **Ad Rates:** BW: $3,380. **Circ:** △**18,506**

30 ■ Journal of Endovascular Therapy
Allen Press Inc.
1928 E Highland St., PMB 605, No. F104
Phoenix, AZ 85016
Phone: (602)240-6121
Fax: (602)266-6018
Publication E-mail: jevt@allenpress.com
Publisher E-mail: info@allenpress.com

Journal covering endovascular therapy. **Founded:** 1994. **Freq:** Bimonthly. **Key Personnel:** Edward B. Diethrich, Editor-in-Chief; Thomas J. Fogarty, Editor-in-Chief. **ISSN:** 1526-6028. **Subscription Rates:** US$207 individuals print and online; US$232 other countries print and online; US$293 institutions print and online; US$318 institutions, other countries print and online. **URL:** http://www.acgpublishing.com/dir_Journals/JournalofEndovascularTherapy2.asp; http://enth.allenpress.com/enthonline/?request=index-html; http://www.jevt.org.

31 ■ LOOKING FIT
Virgo Publishing Inc.
PO Box 40079
Phoenix, AZ 85067-0079
Phone: (480)990-1101
Fax: (480)990-0819

Magazine featuring news and information in the indoor tanning industry. **Key Personnel:** Mike Saxby, Gp. Publisher, phone (480)990-1101, mikes@vpico.com. **Remarks:** Accepts advertising. **URL:** http://www.lookingfit.com. . **Ad Rates:** BW: $4,048. **Circ:** ‡6,952

32 ■ Modern Car Care
Virgo Publishing Inc.
PO Box 40079
Phoenix, AZ 85067-0079
Phone: (480)990-1101
Fax: (480)990-0819

Magazine featuring latest products in car care industry. **Freq:** 12/yr. **Trim Size:** 8.125 x 10.875. **Key Personnel:** Tony Jones, Editor, phone (480)990-1101, tjones@vpico.com. **Subscription Rates:** US$73 individuals; US$93 Canada; US$150 other countries. **Remarks:** Accepts advertising. **URL:** http://www.moderncarcare.com/. . **Ad Rates:** BW: $3,670. **Circ:** (Not Reported)

33 ■ Natural Products INSIDER
Virgo Publishing Inc.
PO Box 40079
Phoenix, AZ 85067-0079
Phone: (480)990-1101
Fax: (480)990-0819

Magazine featuring knowledge and information for

global nutrition decision-makers. **Key Personnel:** Steve Myers, Managing Editor, phone (480)990-1101, smyers@vpico.com. **Remarks:** Accepts advertising. **URL:** http://www.naturalproductsinsider.com/. . **Ad Rates:** BW: $5,020. **Circ:** △**12,022**

34 ■ Natural Products Marketplace
Virgo Publishing Inc.
PO Box 40079
Phoenix, AZ 85067-0079
Phone: (480)990-1101
Fax: (480)990-0819

Magazine featuring natural product industry. **Key Personnel:** Jodi Rich, Publisher, phone (480)990-1101, jrich@vpico.com. **Remarks:** Accepts advertising. **URL:** http://www.naturalproductsmarketplace.com/. . **Ad Rates:** BW: $4,765. **Circ:** △**18,506**

35 ■ Renal Business Today
Virgo Publishing Inc.
PO Box 40079
Phoenix, AZ 85067-0079
Phone: (480)990-1101
Fax: (480)990-0819

Magazine featuring renal business. **Key Personnel:** Keith Chartier, Editor, phone (480)990-1101, kchartier@vpico.com. **Remarks:** Accepts advertising. **URL:** http://www.renalbusiness.com/. . **Ad Rates:** BW: $3,540. **Circ:** △**20,026**

36 ■ Today's SurgiCenter
Virgo Publishing Inc.
PO Box 40079
Phoenix, AZ 85067-0079
Phone: (480)990-1101
Fax: (480)990-0819

Magazine featuring Ambulatory Surgery Centers industry. **Subtitle:** Business Solutions for the ASC. **Key Personnel:** Jennifer Schraag, Managing Editor, phone (480)990-1101, jschraag@vpico.com. **Remarks:** Accepts advertising. **URL:** http://www.surgicenteronline.com/. . **Ad Rates:** BW: $2,060. **Circ:** △**22,005**

SCOTTSDALE

37 ■ KVIB-FM - 95.1
4343 N Scottsdale Rd., No. 200
Scottsdale, AZ 85251
Phone: (480)222-3300
Fax: (480)970-1759

Format: Contemporary Hit Radio (CHR). **Key Personnel:** Jose Rodiles, General Mgr. **URL:** http://www.951latinovibefm.com/.

TUCSON

38 ■ KCEE-AM - 1030
6700 N Oracle Rd., Ste. 240
Tucson, AZ 85704
Phone: (520)498-1030
E-mail: contact@1030kcee.com

Format: Classic Rock. **Owner:** Slone Broadcasting, LLC, at above address. **URL:** http://www.1030kcee.com/.

ARKANSAS

CLARKSVILLE

39 ■ KWXT-AM - 1490
PO Box 215
Clarksville, AR 72830
Phone: (479)754-3399
E-mail: kwxt1490am@yahoo.com

Format: Public Radio. **Key Personnel:** George Domerese, Contact. **URL:** http://www.kwxt1490am.com/.

CROSSETT

40 ■ KHMB-FM - 99.5
203 Fairview Rd.
Crossett, AR 71635
Phone: (870)364-4700
Fax: (870)364-4770
E-mail: qlite@arkansas.net

Format: Adult Contemporary. **URL:** http://www.qliteradio.com/.

GRAVETTE

41 ■ KBVA-FM - 106.5
Hwy. 72 SE
Gravette, AR 72736
Phone: (479)787-6411

Format: Music of Your Life. **Key Personnel:** Gayla McKenzie, Gen. Mgr. **URL:** http://www.variety1065.com/.

HELENA

42 ■ KJIW-FM - 94.5
204 Moore St.
Helena, AR 72342-3438
Phone: (870)338-2700
Fax: (870)338-3166

Format: Gospel; Religious. **Key Personnel:** Elijah Mondy, Owner, em@lordradio.com; April Monday, Oper. Dir., april@lordradio.com; Zipporah Monday, Sales & Music Dir., zm@lordradio.com. **URL:** http://www.kjiwfm.com/.

JACKSONVILLE

43 ■ KVDW-AM - 1530
204 Bucky Beaver St.
Jacksonville, AR 72076
Phone: (501)773-1530
E-mail: victory1530@yahoo.com

Format: Gospel; Contemporary Christian. **URL:** http://www.victory1530.com/.

LITTLE ROCK

44 ■ KTUV-AM - 1440
21700 Northwestern Hwy., Tower 14, Ste. 1190
Southfield, MI 48075
Phone: (248)557-3500
Fax: (248)557-2950

Format: Hispanic. **Owner:** Birach Broadcasting Corporation, at above address. **URL:** http://www.birach.com/ktuv.htm.

MARVELL

45 ■ KLMK-FM - 90.7
2351 Sunset Blvd., Ste. 170-218
Rocklin, CA 95765
Phone: (916)251-1600
Fax: (916)251-1650

Format: Contemporary Christian. **Owner:** Educational Media Foundation, at above address. **URL:** http://www.klove.com/common/lowbandwidth.aspx.

SEARCY

46 ■ KVHU-FM - 95.3
915 E Market
PO Box 10765
Searcy, AR 72149
Phone: (501)279-4411

Format: Eclectic. **URL:** http://www.harding.edu/KHCA/.

SPRINGDALE

47 ■ KSEC-FM - 95.7
2323-D S Old Missouri Rd.
Springdale, AR 72764

Format: News. **Operating Hours:** Continuous. **URL:** http://www.ezspanishmedia.com/radio.html.

CALIFORNIA

BARSTOW

48 ■ KWTH-FM - 91.3
PO Box 637
Bishop, CA 93515
Fax: (760)872-4155

Format: Religious. **URL:** http://www.kwtw.org/.

BERKELEY

49 ■ Algebra & Number Theory
Mathematical Sciences Publishers
Department of Mathematics
University of California
Berkeley, CA 94720-3840
Phone: (510)643-8638
Fax: (510)295-2608
Publisher E-mail: contact@mathscipub.org

Journal covering algebra and number theory. **Key Personnel:** Bjorn Poonen, Managing Editor, poonen@math.berkeley.edu; Paulo Ney de Souza, Production Mgr., pacific@math.berkeley.edu. **Subscription Rates:** US$120 individuals electronic only; US$180 individuals print and electronic. **URL:** http://pjm.math.berkeley.edu/ant/about/cover/cover.html; http://www.jant.org.

50 ■ Algebraic & Geometric Topology
Mathematical Sciences Publishers
Department of Mathematics
University of California
Berkeley, CA 94720-3840
Phone: (510)643-8638
Fax: (510)295-2608
Publication E-mail: agt@msp.warwick.ac.uk
Publisher E-mail: contact@mathscipub.org

Journal covering topology and advancement of mathematics. **Freq:** Annual. **Key Personnel:** Nicholas Jackson, Managing Editor; Colin Rourke, Managing Editor. **Subscription Rates:** US$260 individuals. **URL:** http://msp.warwick.ac.uk/agt/.

51 ■ Communications in Applied Mathematics and Computational Science
Mathematical Sciences Publishers
c/o John B. Bell, Mng. Ed.
Lawrence Berkeley National Laboratory
1 Cyclotron Rd.
MS 50A-1148
Berkeley, CA 94720-8142
Phone: (510)486-5391
Publisher E-mail: contact@mathscipub.org

Journal covering applied mathematics and computational science. **Founded:** May 2005. **Key Personnel:** John B. Bell, Managing Editor, jbbell@lbl.gov; Paulo Ney de Souza, Production Mgr., pacific@math.berkeley.edu. **Subscription Rates:** US$15 libraries for print; US$10 libraries for electronic; Free for individual. **URL:** http://pjm.math.berkeley.edu/camcos/about/cover/cover.html; http://www.camcos.org/.

52 ■ Journal of Mechanics of Materials and Structures
Mathematical Sciences Publishers
Department of Mathematics
University of California
Berkeley, CA 94720-3840
Phone: (510)643-8638
Fax: (510)295-2608
Publisher E-mail: contact@mathscipub.org

Journal covering areas of engineering, materials, and biology, the mechanics of solids, materials, and structures. **Key Personnel:** Charles R. Steele, Editor-in-Chief, chasst@stanford.edu; Paulo Ney de Souza, Production Mgr., pacific@math.berkeley.edu. **Subscription Rates:** US$500 individuals. **URL:** http://pjm.math.berkeley.edu/jomms/about/cover/cover.html; http://www.jomms.org.

Circulation: ★ = ABC; △ = BPA; ♦ = CAC; • = CCAB; ❑ = VAC; ⊕ = PO Statement; ‡ = Publisher's Report; Boldface figures = sworn; Light figures = estimated.

BISHOP

KWTD-FM - See Ridgecrest

KWTH-FM - See Barstow

53 ■ KWTM-FM - 90.9
127 S Main St.
Bishop, CA 93514
Fax: (760)872-4155
Free: (866)466-KWTW

Format: Religious. **Owner:** Living Proof Broadcasting, at above address. **URL:** http://www.kwtw.org/.

ESCONDIDO

54 ■ Interweave Crochet
Interweave Press L.L.C.
PO Box 469076
Escondido, CA 92046-9076
Phone: (760)291-1531
Fax: (760)291-1567
Free: (888)403-5986
Publication E-mail: interweavecrochet@pcspublink.com
Publisher E-mail: customerservice@interweave.com

Magazine featuring ideas and articles about crochet. **Freq:** 4/yr. **Key Personnel:** Kim Werker, Founder. **Subscription Rates:** US$21.95 individuals; US$38.95 two years. **Remarks:** Accepts advertising. **URL:** http://www.interweavecrochet.com/. **Circ:** (Not Reported)

55 ■ Interweave Knits
Interweave Press L.L.C.
PO Box 469117
Escondido, CA 92046-9117
Phone: (760)291-1531
Fax: (760)291-1567
Free: (800)835-6187
Publication E-mail: interweaveknits@pcspublink.com
Publisher E-mail: customerservice@interweave.com

Magazine featuring step-by-step instructions and illustrations for knitting. **Freq:** 4/yr. **Key Personnel:** Laura Rintala, Managing Editor. **Subscription Rates:** US$24 individuals; US$39 two years. **Remarks:** Accepts advertising. **URL:** http://www.interweaveknits.com/. **Circ:** (Not Reported)

56 ■ Stringing
Interweave Press L.L.C.
PO Box 469126
Escondido, CA 92046-9126
Phone: (760)291-1531
Fax: (760)291-1567
Free: (800)782-1054
Publisher E-mail: customerservice@interweave.com

Magazine featuring jewelry design. **Freq:** 4/yr. **Key Personnel:** Leigh Trotter, Contact. **Subscription Rates:** US$21.95 individuals; US$39.95 two years. **Remarks:** Accepts advertising. **URL:** http://www.stringingmagazine.com/. **Circ:** (Not Reported)

FAIR OAKS

57 ■ KJPG-AM - 1050
7956 California Ave.
Fair Oaks, CA 95628
Free: (866)77H-EART

Format: Religious. **Owner:** Immaculate Heart Radio, at above address. **Key Personnel:** Rev. John T. Steinbock, Contact. **URL:** http://www.ihradio.org/stations/detail/3.

FERNDALE

58 ■ KWPT-FM - 100.3
1400 Main St., Ste. 104
Ferndale, CA 95536
Phone: (707)786-5104
Fax: (707)786-5100

Format: Classic Rock. **Owner:** Lost Coast Communications, Inc., at above address. **URL:** http://www.kwpt.com/.

FRESNO

59 ■ KLBN-FM - 105.0
1110 E Olive Ave.
Fresno, CA 93728
Phone: (559)497-1100

Format: Hispanic. **Owner:** Lotus Communications Corp., at above address. **Ad Rates:** Advertising accepted; rates available upon request. **URL:** http://www.1051labuena.com/.

60 ■ KRDA-FM - 107.5
1981 N Gateway, Ste. 101
Fresno, CA 93727
Phone: (559)456-4000
Fax: (559)251-9555

Format: Hispanic; Adult Contemporary. **Owner:** Univision Communications Inc., at above address. **URL:** http://corporate.univision.com/corp/en/radio_station_fresno.jsp.

HIDDEN VALLEY LAKE

61 ■ KPDO-FM - 89.3
18815 Glencove Ct., No. 1
Hidden Valley Lake, CA 95467
Phone: (707)987-9761

Format: Information. **Key Personnel:** Celeste Worden, President. **URL:** http://www.kpdo.org/.

IRVINE

62 ■ African Journal of Environmental Science & Technology
Academic Journals
c/o Prof. Oladele A. Ogunseitan, PhD, Ed.
Industrial Ecology Research Group
University of California
Irvine, CA 92697-7070
Publication E-mail: ajest@academicjournals.org
Publisher E-mail: service@academicjournals.org

Journal covering all areas environmental science and technology. **Freq:** Monthly. **Key Personnel:** Prof. Oladele A. Ogunseitan, PhD, Editor; Suping Zhou, Assoc. Ed. **ISSN:** 1996-0786. **Remarks:** Accepts advertising. **URL:** http://www.academicjournals.org/AJEST/index.htm. **Circ:** (Not Reported)

JOSHUA TREE

63 ■ KXCM-FM - 96.3
PO Box 1437
Joshua Tree, CA 92252
Phone: (760)362-4264

Format: Country. **Owner:** Copper Mountain Broadcasting, at above address. **URL:** http://www.joshuatreevillage.com/209/209.htm.

LAGUNA BEACH

KFYL-FM - See La Grande, Oregon
KJHV-FM - See Killeen, Texas
KJLP-FM - See Anchorage, Alaska
KKTT-FM - See Winnemucca, Nevada
KMAB-FM - See Madras, Oregon
KPAR-FM - See Dickinson, North Dakota
KRGR-FM - See Paradise
KSHC-FM - See Saint Helena
KTPJ-FM - See Pueblo, Colorado
KXIV-FM - See Daingerfield, Texas
KYHO-FM - See Poplar Bluff, Missouri
WASD-FM - See Aiken, South Carolina
WCVX-AM - See Cincinnati, Ohio
WDJD-FM - See Elizabethtown, North Carolina
WGGR-FM - See Carrollton, Georgia
WGLY-AM - See Plattsburgh, New York
WJHE-FM - See Newark, Ohio
WJNU-FM - See Cookeville, Tennessee
WKPJ-FM - See Athens, Tennessee
WLOJ-FM - See Calhoun, Georgia
WMLY-FM - See Marshall, Michigan
WSVV-FM - See Mastic Beach, New York
WTRL-FM - See Vonore, Tennessee
WVBL-FM - See Salem, West Virginia
WVXR-FM - See Richmond, Indiana
WVXW-FM - See West Union, Ohio
WYMR-AM - See Bridgeport, Alabama
WYSG-FM - See Hinckley, Minnesota

LOMPOC

64 ■ KLWG-FM - 88.1
1551-B E Laurel
Lompoc, CA 93436
Phone: (805)735-1511

Format: Religious; Gospel. **Owner:** Calvary Chapel Lompoc, at above address. **Key Personnel:** Mark Galvan, Contact. **URL:** http://www.cclompoc.com/.

LONG BEACH

65 ■ Cosmetic/Personal Care Packaging
Canon Communications L.L.C.
PO Box 91115
Long Beach, CA 90809-1115
Publisher E-mail: feedback@cancom.com

Magazine featuring package design for cosmetic, fragrance, and personal care products. **Founded:** 1996. **Key Personnel:** Daphne Allen, Editor, daphne.allen@cancom.com. **Subscription Rates:** Free; US$150 other countries. **Remarks:** Accepts advertising. **URL:** http://www.canonmediakit.com/publications/index.php?catId=3; http://www.cpcpkg.com/home/. **Circ:** △**12,500**

LOS ANGELES

66 ■ Advanced Design & Manufacturing
Canon Communications L.L.C.
11444 W Olympic Blvd., Ste. 900
Los Angeles, CA 90064
Phone: (310)445-4200
Fax: (310)445-4299
Publisher E-mail: feedback@cancom.com

Magazine featuring medical device industry. **Key Personnel:** Ricke Carter, Editor-in-Chief, rick.carter@cancom.com. **URL:** http://www.canonmediakit.com/publications/detail.php?publd=47.

67 ■ Medical Device Technology
Canon Communications L.L.C.
11444 W Olympic Blvd., Ste. 900
Los Angeles, CA 90064
Phone: (310)445-4200
Fax: (310)445-4299
Publisher E-mail: feedback@cancom.com

Magazine featuring practical information and in-depth technical detail on how to design and manufacture medical devices. **Key Personnel:** Annie Ellerton, Editor-in-Chief, annie.ellerton@cancom.com. **Remarks:** Accepts advertising. **URL:** http://www.canonmediakit.com/publications/detail.php?publd=39; http://www.devicelink.com/mdt/. **Circ:** △**18,040**

MERCED

68 ■ KBKY-FM - 94.1
450 Grogan Ave., Ste. A
Merced, CA 95340

Format: Sports. **Key Personnel:** Mike Meroney, Gen. Mgr., phone (209)385-9994, mmeroney941@vtlnet.com; Matt Stone, Program Dir., phone (209)385-9994, mstone941@vtlnet.com; Dave Putonen, Sales Mgr., phone (209)385-9994, david_putonen@yahoo.com. **URL:** http://www.foxsportsmerced.com/.

OAKLAND

KBFR-FM - See Bismarck, North Dakota
KEGR-FM - See Fort Dodge, Iowa
KFRD-FM - See Butte, Montana
KFRW-FM - See Black Eagle, Montana
KQFE-FM - See Springfield, Oregon

WEFR-FM - See Erie, Pennsylvania

WFRH-FM - See Kingston, New York

WFRP-FM - See Ameribus, Georgia

OCEANSIDE

69 ■ TransWorld Business
Bonnier Corporation
353 Airport Rd.
Oceanside, CA 92054

Magazine featuring board-sports news and information. **Key Personnel:** Rob Campbell, Publisher. **Remarks:** Accepts advertising. **URL:** http://www.bonniercorp.com/brands/Trans-World-Business.html. **Circ:** (Not Reported)

ORANGE

70 ■ Cottages and Bungalows
Action Pursuit Group
265 S Anita Dr., Ste. 120
Orange, CA 92868
Phone: (714)939-9991
Fax: (714)939-9909

Magazine featuring cottages and bungalow. **Freq:** 6/yr. **Key Personnel:** Craig Nickerson, Pres./CEO. **Subscription Rates:** US$24.95 individuals; US$32.95 two years. **Remarks:** Accepts advertising. **URL:** http://www.apg-media.com/content/view/26/36/. **Circ:** (Not Reported)

71 ■ Elite Fighter
Action Pursuit Group
265 S Anita Dr., Ste. 120
Orange, CA 92868
Phone: (714)939-9991
Fax: (714)939-9909

Magazine featuring elite fighters. **Freq:** 12/yr. **Key Personnel:** Craig Nickerson, Pres./CEO. **Subscription Rates:** US$27 individuals; US$37 two years. **Remarks:** Accepts advertising. **URL:** http://www.elitefightermag.com/; http://www.apg-media.com/content/view/26/36/. **Circ:** (Not Reported)

PARADISE

72 ■ KRGR-FM - 101.3
PO Box 1018
Laguna Beach, CA 92652
Fax: (469)241-6795
Free: (800)639-5433

Format: Religious. **Owner:** New Life Ministries, at above address. **Key Personnel:** Stephen Arterburn, Founder/Chm. **URL:** http://www.newlife.com.

QUARTZ HILL

73 ■ KGBB-FM - 103.9
42010 50th St. W
Quartz Hill, CA 93536
Phone: (661)718-1552
Fax: (661)718-1553

Format: Adult Contemporary. **Owner:** Adelman Broadcasting, 731 N Balsam St., Ridgecrest, CA 93555-3510, (760)371-1700. **URL:** http://www.bobfm1039.com/.

RIDGECREST

74 ■ KWTD-FM - 91.9
PO Box 637
Bishop, CA 93515
Fax: (760)872-4155

Format: Religious. **Owner:** Living Proof Broadcasting, at above address. **URL:** http://www.kwtw.org/.

ROCKLIN

KAIK-FM - See Tillamook, Oregon

KAIP-FM - See Davenport, Iowa

75 ■ KLKA-FM - 88.5
2351 Sunset Blvd., Ste. 170-218
Rocklin, CA 95765
Phone: (916)251-1600
Fax: (916)251-1650

Format: Contemporary Christian. **Owner:** Educational Media Foundation, at above address. **URL:** http://www.klove.com/Music/StationList.aspx.

KLMK-FM - See Marvell, Arkansas

KLOF-FM - See Gillette, Wyoming

KLRH-FM - See Reno, Nevada

KLWC-FM - See Casper, Wyoming

KNAR-FM - See San Angelo, Texas

KQRI-FM - See Albuquerque, New Mexico

KVRA-FM - See Bend, Oregon

KWKL-FM - See Lawton, Oklahoma

WGKV-FM - See Pulaski, New York

WKVP-FM - See Cherry Hill, New Jersey

WKYJ-FM - See Rouses Point, New York

WLAI-FM - See Danville, Kentucky

WLRK-FM - See Greenville, Mississippi

WLVZ-FM - See Hattiesburg, Mississippi

WNKV-FM - See New Orleans, Louisiana

WOAR-FM - See South Vienna, Ohio

WPRZ-FM - See Newton Grove, North Carolina

WVDA-FM - See Valdosta, Georgia

SACRAMENTO

76 ■ KLIB-AM - 1110
3463 Ramona Ave.
Sacramento, CA 95826

Format: Ethnic. **Owner:** Multicultural Radio Broadcasting, Inc., 449 Broadway, New York, NY 10013, (212)966-1059, Fax: (212)966-9580. **URL:** http://www.mrbi.net/radiogroup.htm.

SAINT HELENA

77 ■ KSHC-FM - 106.5
PO Box 1018
Laguna Beach, CA 92652
Fax: (469)241-6795
Free: (800)639-5433

Format: Religious. **Owner:** New Life Ministries, at above address. **Key Personnel:** Stephen Arterburn, Founder/Chm. **URL:** http://www.newlife.com/Radio/findradio.asp?id=CA&t=California.

SAN DIEGO

78 ■ KSCF-FM - 103.7
1515 Broadway
New York, NY 10036
Phone: (212)846-3939

Format: Adult Contemporary. **Owner:** CBS Radio Stations Inc., at above address. **URL:** http://www.cbsradio.com/index.html.

SANTA ANA

WJCZ-FM - See Milford, Illinois

WTMK-FM - See Lowell, Indiana

SANTA ROSA

79 ■ KVRV-FM - 97.7
1410 Neotomas Ave., Ste. 200
Santa Rosa, CA 95405-7533
Phone: (707)636-0977
Fax: (707)571-1097

Format: Classic Rock. **URL:** http://www.977theriver.com/.

THOUSAND OAKS

80 ■ KHJL-FM - 92.7
99 Long Ct.
Thousand Oaks, CA 91360
Phone: (805)497-8511
Fax: (805)497-8514
Free: (800)497-8511

Format: Oldies; Hot Country; Contemporary Hit Radio (CHR). **Key Personnel:** Bob Christy, Gen. Mgr., bob@927jillfm.com. **URL:** http://www.927jillfm.com/.

TRUCKEE

81 ■ KTKE-FM - 101.5
12030 Donner Pass Rd.
Truckee, CA 96161
Phone: (530)587-9999
E-mail: ktkeradio@yahoo.com

Format: Alternative/New Music/Progressive. **Key Personnel:** Lindsay Romack, Manager. **Ad Rates:** Advertising accepted; rates available upon request. **URL:** http://www.ktke1015.com/.

TUSTIN

82 ■ TransWorld Ride BMX
Bonnier Corporation
1421 Edinger Ave., Ste. D
Tustin, CA 92780
Phone: (714)247-0077
Fax: (714)247-0078

Magazine featuring information for BMX bike owners. **Freq:** Quarterly. **Key Personnel:** Keith Mulligan, Editor. **Subscription Rates:** US$4.99 single issue. **Remarks:** Accepts advertising. **URL:** http://www.bonniercorp.com/brands/Trans-World-Ride-BMX.html; http://www.ridebmx.com/. **Circ:** (Not Reported)

VENTURA

83 ■ KUNX-AM - 1590
2284 Victoria Ave., Ste. 2-G
Ventura, CA 93003
Phone: (805)289-1400

Format: News. **Owner:** Gold Coast Broadcasting, at above address. **URL:** http://www.goldcoastbroadcasting.com/.

VISALIA

84 ■ KSLK-FM - 96.1
2025 E Noble, Ste. B
Visalia, CA 93292
Phone: (559)635-0961
Fax: (559)635-0964

Format: Sports. **Operating Hours:** Continuous. **Ad Rates:** Advertising accepted; rates available upon request. **URL:** http://www.1550snr.com/.

COLORADO

COLORADO SPRINGS

85 ■ KKPK-FM - 92.9
6805 Corporate Dr., Ste. 130
Colorado Springs, CO 80919
Phone: (719)593-2700
Fax: (719)593-2727

Format: Adult Contemporary. **Key Personnel:** Bobby Irwin, Operations Mgr./Prog. Dir., phone (719)593-2700, bobby.irwin@atscitcomm.com. **URL:** http://www.929peakfm.com/.

DENVER

86 ■ KNFO-FM - 106.1
1201 18th St., Ste. 200
Denver, CO 80202
Phone: (303)296-7025
Fax: (303)296-7030

Format: News; Talk; Sports. **Owner:** NRC Broadcasting Inc., at above address. **Key Personnel:** Colleen Barill, Gen. Mgr., phone (970)925-5776, cbarill@

Circulation: ★ = ABC; △ = BPA; ♦ = CAC; • = CCAB; ❑ = VAC; ⊕ = PO Statement; ‡ = Publisher's Report; Boldface figures = sworn; Light figures = estimated.

nrcbroadcasting.com. **URL:** http://www.nrcbroadcasting.com/mountain_stations.asp.

87 ■ KNRV-AM - 1150
2821 S Parker Rd., Ste. 1205
Denver, CO 80014

Format: News; Information. **Owner:** New Radio Venture, at above address. **Key Personnel:** Annette Lavina, Gen. Sales Mgr., phone (303)969-5981, annette.lavina@onda1150am.com. **URL:** http://www.onda1150am.com/Mambo455/.

88 ■ KWLI-FM - 92.5
1560 Broadway, Ste. 1100
Denver, CO 80202
Phone: (303)832-5665

Format: Country. **Key Personnel:** Bill Gamble, Program Dir., bill.gamble@cbsradio.com. **URL:** http://www.925thewolf.com/.

GREENWOOD VILLAGE

89 ■ KEPN-AM - 1600
7800 E Orchard Rd., Ste. 400
Greenwood Village, CO 80111
Phone: (303)370-1458

Format: Sports. **Key Personnel:** Jay Kisskalt, Local Sales Mgr., jay.kisskalt@lfg.com. **URL:** http://www.espnradio1600.com/.

JULESBURG

90 ■ KJBL-FM - 96.5
205 Elm St.
PO Box 196
Julesburg, CO 80737
Phone: (970)474-0953
E-mail: kjbl@highplainsradio.net

Format: Sports; Oldies. **URL:** http://www.kjblradio.com/.

LONGMONT

91 ■ KSXT-AM - 1570
1270 Boston Ave.
Longmont, CO 80501
Phone: (303)772-7676
Fax: (303)772-0642

Format: Sports. **Key Personnel:** Mark Knudson, General Mgr. **URL:** http://www.1570ksxt.com/.

LOUISVILLE

92 ■ John Lyons' Perfect Horse
Belvoir Media Group, LLC
c/o Pat Eskew, PUB
730 Front St.
Louisville, CO 80027
Phone: (303)661-9282
Fax: (303)661-9298
Publisher E-mail: customer_service@belvoir.com

Magazine featuring horses. **Founded:** 1995. **Freq:** 9/yr. **Key Personnel:** Pat Eskew, Publisher, pat.eskew@horsemediagroup.com. **Subscription Rates:** US$32 individuals; US$3.99 single issue. **Remarks:** Accepts advertising. **URL:** http://www.myhorse.com/magazines/index.aspx. . **Ad Rates:** BW: $2,250, 4C: $3,385. **Circ:** Paid 76,000

93 ■ Spin to Win Rodeo
Belvoir Media Group, LLC
c/o Pat Eskew, PUB
730 Front St.
Louisville, CO 80027
Phone: (303)661-9282
Fax: (303)661-9298
Publisher E-mail: customer_service@belvoir.com

Magazine featuring rodeo. **Founded:** 1997. **Freq:** Monthly. **Key Personnel:** Pat Eskew, Publisher, pat.eskew@horsemediagroup.com. **Subscription Rates:** US$24 individuals; US$3.99 single issue. **Remarks:** Accepts advertising. **URL:** http://www.myhorse.com/magazines/rodeo.aspx. . **Ad Rates:** BW: $1,210, 4C: $2,120. **Circ:** 35,034

LOVELAND

94 ■ Creative Jewelry
Interweave Press L.L.C.
201 E 4th St.
Loveland, CO 80537
Phone: (970)669-7672
Fax: (970)613-4656
Free: (800)272-2193
Publisher E-mail: customerservice@interweave.com

Magazine featuring quick and easy jewelry projects. **Freq:** Annual. **Key Personnel:** Linda Ligon, Founder. **URL:** http://www.interweave.com/magazines/.

95 ■ Knitscene
Interweave Press L.L.C.
201 E 4th St.
Loveland, CO 80537
Phone: (970)669-7672
Fax: (970)613-4656
Free: (800)272-2193
Publisher E-mail: customerservice@interweave.com

Magazine featuring knitting and crochet. **Freq:** Semiannual. **Key Personnel:** Tiffany Ball-Zerges, Display Advertising Rep., phone (970)669-7455, tiffanyb@interweave.com. **Remarks:** Accepts advertising. **URL:** http://www.knitscene.com/; http://www.interweave.com/magazines/. **Circ:** (Not Reported)

96 ■ PieceWork
Interweave Press L.L.C.
201 E 4th St.
Loveland, CO 80537
Phone: (970)669-7672
Fax: (970)613-4656
Free: (800)272-2193
Publication E-mail: piecework@pcspublink.com
Publisher E-mail: customerservice@interweave.com

Magazine featuring jewelry design. **Freq:** 6/yr. **Key Personnel:** Jeane Hutchins, Editor. **Subscription Rates:** US$24 individuals; US$39 two years. **Remarks:** Accepts advertising. **URL:** http://www.interweave.com/needle/. **Circ:** (Not Reported)

PUEBLO

97 ■ The Pomegranate
Equinox Publishing Ltd.
Department of English
Colorado State University-Pueblo
2200 Bonforte Blvd.
Pueblo, CO 81001
Phone: (719)549-2226

Journal covering both ancient and contemporary Pagan religious practices. **Subtitle:** The International Journal of Pagan Studies. **Freq:** 2/yr (May and November). **Key Personnel:** Chas Clifton, Editor, chas.clifton@colostate-pueblo.edu; Steve Barganski, Production Mgr., steve.barganski@virgin.net. **ISSN:** 1528-0268. **Subscription Rates:** 90 institutions; US$165 institutions; 40 institutions developing countries; 40 individuals; US$65 individuals; 18 individuals developing countries; 28 students; US$50 students. **Remarks:** Accepts advertising. **URL:** http://www.equinoxjournals.com/ojs/index.php/POM. **Circ:** (Not Reported)

98 ■ KTPJ-FM - 103.5
PO Box 1018
Laguna Beach, CA 92652
Fax: (469)241-6795
Free: (800)639-5433

Format: Religious. **Owner:** New Life Ministries, at above address. **Key Personnel:** Stephen Arterburn, Founder/Chm. **URL:** http://www.newlife.com.

CONNECTICUT

BRIDGEPORT

99 ■ WSAH-TV - 43
449 Broadway
New York, NY 10013
Phone: (212)966-1059
Fax: (212)966-9580

Owner: Multicultural Radio Broadcasting, Inc., at above address. **Key Personnel:** John Gabel, Contact, phone (212)966-1059, johng@mrbi.net. **URL:** http://www.mrbi.net/tvgroup.htm.

FAIRFIELD

WSUF-FM - See Greenport, New York

HARTFORD

100 ■ WNEZ-AM - 1230
330 Main St.
Hartford, CT 06106
Phone: (860)524-0001

Format: Music of Your Life. **Owner:** Gois Broadcasting of Connecticut, LLC, at above address. **URL:** http://www.latina1230.com/.

WRLI-FM - See Southampton, New York

101 ■ WURH-FM - 104.1
10 Columbus Blvd.
Hartford, CT 06106
Phone: (860)723-6000

Format: Alternative/New Music/Progressive. **URL:** http://www.1041music.com/main.html.

NEW HAVEN

102 ■ Journal of Pediatric Endocrinology & Metabolism
Freund Publishing House Ltd.
Yale Child Health Research Ctr.
464 Congress Ave.
PO Box 208081
New Haven, CT 06520
Publisher E-mail: h_freund@netvision.net.il

Journal covering clinical investigations in pediatric endocrinology and basic research with relevance to clinical pediatric endocrinology and metabolism from all over the world. **Freq:** 12/yr. **Print Method:** 8 1/4 x 10 3/4. **Key Personnel:** Scott A. Rivkees, Editor-in-Chief, rivkees@yale.edu; Fergus Cameron, Editor; Durvan Damiani, Editor; Dennis M. Styne, Editor. **ISSN:** 0334-018X. **Subscription Rates:** US$650 institutions; US$380 individuals. **Remarks:** Accepts advertising. **URL:** http://www.freundpublishing.com/JOURNALS/medicine_and_medical_sciences.htm. . **Ad Rates:** BW: $1,200, 4C: $2,400. **Circ:** (Not Reported)

103 ■ WYBC-AM - 1340
142 Temple St., Ste. 203
New Haven, CT 06510
Phone: (203)776-4118

Format: Full Service. **Owner:** Yale Broadcasting Company Inc., at above address. **Key Personnel:** Jordan Malter, General Mgr.; Julia Galeota, Program Dir.; Alex Civetta, Mktg. Dir. **URL:** http://www.wybc.com/.

NORWALK

104 ■ American Gunsmith
Belvoir Media Group, LLC
800 Connecticut Ave.
Norwalk, CT 06854-1631
Phone: (203)857-3100
Fax: (203)857-3103
Publisher E-mail: customer_service@belvoir.com

Journal focusing on firearms repair and maintenance. **URL:** http://www.belvoir.com/titles/index.html.

105 ■ Arthritis Advisor
Belvoir Media Group, LLC
800 Connecticut Ave.
Norwalk, CT 06854-1631
Phone: (203)857-3100
Fax: (203)857-3103
Publisher E-mail: customer_service@belvoir.com

Magazine featuring information for people with arthritis. **Freq:** 12/yr. **Key Personnel:** Brian Donley, MD, Editor-in-Chief. **Subscription Rates:** US$20 individuals; C$29 Canada; US$42 other countries. **URL:** http://www.arthritis-advisor.com/.

106 ■ Aviation Consumer
Belvoir Media Group, LLC
800 Connecticut Ave.
Norwalk, CT 06854-1631
Phone: (203)857-3100
Fax: (203)857-3103
Publisher E-mail: customer_service@belvoir.com

Journal focusing on the evaluations of aircraft, avionics, accessories, and equipment. **Freq:** Monthly. **Subscription Rates:** US$19.95 individuals 3 months; US$36 individuals 6 months. **URL:** http://www.aviationconsumer.com/.

107 ■ Gun Tests
Belvoir Media Group, LLC
800 Connecticut Ave.
Norwalk, CT 06854-1631
Phone: (203)857-3100
Fax: (203)857-3103
Publisher E-mail: customer_service@belvoir.com

Magazine featuring information and evaluation about handguns, rifles, shotguns, and shooting accessories. **Freq:** Monthly. **Subscription Rates:** US$24 individuals; C$29 Canada; US$42 other countries. **URL:** http://www.gun-tests.com/.

108 ■ IFR Refresher
Belvoir Media Group, LLC
800 Connecticut Ave.
Norwalk, CT 06854-1631
Phone: (203)857-3100
Fax: (203)857-3103
Publisher E-mail: customer_service@belvoir.com

Journal focusing on reviews of IFR rules and procedures. **Freq:** Monthly. **Subscription Rates:** US$24 individuals; C$36 Canada; US$36 other countries. **URL:** http://www.ifr-refresher.com/.

109 ■ Light Plane Maintenance
Belvoir Media Group, LLC
800 Connecticut Ave.
Norwalk, CT 06854-1631
Phone: (203)857-3100
Fax: (203)857-3103
Publisher E-mail: customer_service@belvoir.com

Magazine featuring information about aircraft maintenance. **Freq:** Monthly. **Subscription Rates:** US$19.97 individuals 6 issues; US$29.97 other countries 6 issues. **URL:** http://www.lightplane-maintenance.com/.

110 ■ Living Without
Belvoir Media Group, LLC
800 Connecticut Ave.
Norwalk, CT 06854-1631
Phone: (203)857-3100
Fax: (203)857-3103
Publisher E-mail: customer_service@belvoir.com

Magazine featuring dietary products. **Freq:** 6/yr. **Key Personnel:** Alicia Woodward, Managing Editor. **Subscription Rates:** US$23 individuals; US$42 two years. **Remarks:** Accepts advertising. **URL:** http://www.livingwithout.com/. **Circ:** (Not Reported)

111 ■ Practical Sailor
Belvoir Media Group, LLC
800 Connecticut Ave.
Norwalk, CT 06854-1631
Phone: (203)857-3100
Fax: (203)857-3103
Publisher E-mail: customer_service@belvoir.com

Journal focusing on evaluations of sailing gear and equipment. **Freq:** 7/yr. **Key Personnel:** Darrell Nicholson, Editor. **Subscription Rates:** US$19.97 individuals; C$26 Canada; US$37.97 other countries. **URL:** http://www.practical-sailor.com/.

UNCASVILLE

112 ■ WCSE-FM - 100.1
130 Sharp Hill Rd.
Uncasville, CT 06382
Phone: (860)848-1111
E-mail: info@wcse.org

Format: Contemporary Christian. **URL:** http://wcse.typepad.com/.

DELAWARE

LEWES

WRBG-FM - See Millsboro

MILLSBORO

113 ■ WRBG-FM - 107.9
23136 Prince George Dr.
Lewes, DE 19958-9342
Phone: (302)933-0385

Format: Blues. **Owner:** Joseph D'Alessandro, at above address, (302)945-1554. **URL:** http://wrbg1079fm.com/.

DISTRICT OF COLUMBIA

WASHINGTON

114 ■ Commercial Builder
National Association of Home Builders
Senior Housing Council
1201 15th St. NW
Washington, DC 20005-2800
Phone: (202)266-8200
Fax: (202)266-8400
Free: (800)368-5242
Publisher E-mail: madams@nahb.com

Magazine featuring commercial building industry. **Freq:** Quarterly. **Trim Size:** 10.875 X16.5. **Key Personnel:** Andrew Flank, Contact, aflank@nahb.com. **Subscription Rates:** US$79 individuals. **Remarks:** Accepts advertising. **URL:** http://www.nahb.org/product_details.aspx?forSaleID=2§ionID=155. **Circ:** 10,000

115 ■ 50+ Housing Magazine
National Association of Home Builders
Senior Housing Council
1201 15th St. NW
Washington, DC 20005-2800
Phone: (202)266-8200
Fax: (202)266-8400
Free: (800)368-5242
Publisher E-mail: madams@nahb.com

Magazine featuring housing industry. **Freq:** Quarterly. **Key Personnel:** Harris Floyd, Contact, hfloyd@nahb.com. **Subscription Rates:** US$95 members; US$114 nonmembers. **URL:** http://www.nahb.org/product_details.aspx?forSaleID=13§ionID=155. **Circ:** 20,000

116 ■ Journal of Cardiovascular Computed Tomography
Elsevier
c/o Allen J. Taylor MD, Ed.-in-Ch.
Walter Reed Army Medical Center
6900 Georgia Ave. NW, Bldg. 2, Rm. 4A34
Washington, DC 20307-5001
Publisher E-mail: healthpermissions@elsevier.com

Journal covering the field of cardiovascular CT imaging. **Freq:** 6/yr. **Key Personnel:** Allen J. Taylor, MD, Editor-in-Chief, allen.taylor@na.amedd.army.mil. **ISSN:** 1934-5925. **Subscription Rates:** US$131 individuals; US$66 students; US$148 other countries. **URL:** http://www.cardiacctjournal.com/.

117 ■ Land Development
National Association of Home Builders
Senior Housing Council
1201 15th St. NW
Washington, DC 20005-2800
Phone: (202)266-8200
Fax: (202)266-8400
Free: (800)368-5242
Publisher E-mail: madams@nahb.com

Magazine featuring land development industry. **Freq:** Quarterly. **Trim Size:** 8.25 x 11. **Key Personnel:** Julie Mines, Contact, jmines@nahb.com. **Subscription Rates:** US$40 individuals. **Remarks:** Accepts advertising. **URL:** http://www.nahb.org/generic.aspx?sectionID=1186. **Circ:** 2,000

118 ■ WWWT-FM - 107.7
3400 Idaho Ave. NW
Washington, DC 20016
Phone: (202)895-5000

Format: Talk. **Owner:** Bonneville International, 55 N 300 W, Salt Lake City, UT 84180, (801)575-7500. **Key Personnel:** Joel Oxley, General Mgr., phone (202)895-5012; Greg Tantum, Program Dir., gtantum@3wtradio.com; Steve Goldstein, General Sales Mgr., phone (202)895-5006, sgoldstein@3wtradio.com. **Ad Rates:** Advertising accepted; rates available upon request. **URL:** http://www.3wtradio.com/.

FLORIDA

BOCA RATON

119 ■ Journal of Public Budgeting, Accounting & Financial Management
PrAcademics Press
21760 Mountain Sugar Ln.
Boca Raton, FL 33433
Phone: (561)362-9183
Publisher E-mail: info@pracademicspress.com

Journal covering theories and practices in the fields of public budgeting, governmental accounting, and financial management. **Freq:** Quarterly. **Key Personnel:** Khi V. Thai, Editor, thai@fau.edu; Jack Rabin, Editor. **Subscription Rates:** US$295 individuals and government; US$395 libraries. **URL:** http://www.pracademicspress.com/jpbafm.html.

BONITA SPRINGS

120 ■ WGUF-FM - 98.9
10915 K-Nine Dr.
Bonita Springs, FL 34135
Phone: (239)495-8383
Fax: (239)495-0883

Format: News. **Key Personnel:** David Kaye, Sales Mgr., dkaye@rendabroadcasting.com. **Ad Rates:** Advertising accepted; rates available upon request. **URL:** http://www.wguf989.com/.

COCOA

121 ■ WJFP-FM - 107.1
1150 W King St.
Cocoa, FL 32922
Phone: (321)632-1000

Format: Urban Contemporary. **URL:** http://www.wjfp.com/.

122 ■ WJFP-FM - 91.1
1150 W King St.
Cocoa, FL 32922
Phone: (321)632-1000

Format: Urban Contemporary. **URL:** http://www.wjfp.com/.

123 ■ WJFP-FM - 94.3
1150 W King St.
Cocoa, FL 32922
Phone: (321)632-1000

Format: Urban Contemporary. **URL:** http://www.wjfp.com/.

Circulation: ★ = ABC; △ = BPA; ♦ = CAC; • = CCAB; ❑ = VAC; ⊕ = PO Statement; ‡ = Publisher's Report; Boldface figures = sworn; Light figures = estimated.

124 ■ WJFP-FM - 93.9
1150 W King St.
Cocoa, FL 32922
Phone: (321)632-1000

Format: Urban Contemporary. **URL:** http://www.wjfp.com/.

125 ■ WJFP-FM - 88.5
1150 W King St.
Cocoa, FL 32922
Phone: (321)632-1000

Format: Urban Contemporary. **URL:** http://www.wjfp.com/.

DAYTONA BEACH

126 ■ Advances and Applications in Fluid Mechanics
Pushpa Publishing House
Department of Mathematics
Embry-Riddle Aeronautical University
600 S Clyde Morris Blvd.
Daytona Beach, FL 32114
Publisher E-mail: arun@pphmj.com

Journal covering fields of fluid mechanics. **Freq:** 4/yr (January, April, July and October). **Key Personnel:** Prof. Shahrdad G. Sajjadi, Editor-in-Chief, aaifm@erau.edu; K.K. Azad, Managing Editor. **ISSN:** 0973-4686. **URL:** http://www.pphmj.com/journals/aafm.htm.

127 ■ WHOG-FM - 95.7
126 W International Speedway Blvd.
Daytona Beach, FL 32114
Phone: (386)255-9300

Format: Classic Rock. **Key Personnel:** Donna Fillion, Dir. of Sales, phone (386)238-6071, dfillion@whog.fm. **URL:** http://www.whog.fm/.

DEFUNIAK SPRINGS

128 ■ WAKJ-FM - 91.3
295 Hwy. 90 W
DeFuniak Springs, FL 32435
Phone: (850)892-2107
Fax: (850)892-2507
E-mail: wakj913@embarqmail.com

Format: Contemporary Christian; Religious. **Key Personnel:** John Gradick, Promotions Dir., jgradick@embarqmail.com. **URL:** http://www.wakj.org/.

DESTIN

129 ■ WFFY-FM - 921
743 Harbor Blvd., Ste. 6
Destin, FL 32541
Phone: (850)654-1031

Format: Alternative/New Music/Progressive. **URL:** http://www.fly921online.com/.

FORT LAUDERDALE

130 ■ Boat International USA
Boat International Group
1800 SE 10th Ave., Ste. 440
Fort Lauderdale, FL 33316
Phone: (954)522-2628
Fax: (954)522-2240
Publisher E-mail: info@boatinternational.co.uk

Magazine covering luxury yacht in North America. **Founded:** 1997. **Freq:** 10/yr. **Trim Size:** 223 x 275 mm. **Key Personnel:** Ben Farnborough, Assoc. Publisher, ben@boatinternationalusa.com; Lou Fagas, Advertising Dir., louf@boatinternationalusa.com. **Subscription Rates:** US$24.95 U.S.; C$49.50 Canada; 47 other countries; US$49.90 U.S. 2 years; C$99 Canada 2 years; 94 other countries 2 years. **Remarks:** Accepts advertising. **URL:** http://www.boatinternational.com/mags/mag02.htm. **Circ:** 43,525

131 ■ Home Fort Lauderdale
Bonnier Corporation
445 N Andrews Ave.
Fort Lauderdale, FL 33301
Phone: (954)566-2100
Fax: (954)566-1688

Magazine featuring home design and architecture. **Freq:** 11/yr. **Trim Size:** 9 1/4 x 11 1/8. **Key Personnel:** Jeff Ditmire, Publisher. **Subscription Rates:** US$48 individuals; US$60 two years. **Remarks:** Accepts advertising. **URL:** http://www.homeftl.com/. **Circ:** 21,000

FORT WALTON BEACH

132 ■ WZNS-FM - 96.5
225 NW Hollywood Blvd.
Fort Walton Beach, FL 32548
Phone: (850)243-7676

Format: Alternative/New Music/Progressive. **Owner:** Cumulus Broadcasting, LLC, at above address. **Key Personnel:** Melissa Mack, Contact, melissa.mack@cumulus.com. **URL:** http://www.z96.com/home.shtml.

GAINESVILLE

133 ■ Journal for the Study of Religion, Nature and Culture
Equinox Publishing Ltd.
PO Box 117410
Gainesville, FL 32611-7410
Phone: (352)392-1625
Fax: (352)392-7395

Journal covering relationship between religion and ecology. **Freq:** 4/yr (March, June, September and December). **Key Personnel:** Bron Taylor, Editor, brontaylor@religionandnature.com; Joseph Witt, Managing Editor, joe@religionandnature.com. **ISSN:** 1749-4907. **Subscription Rates:** 165 institutions; US$295 institutions; 50 individuals; US$90 individuals; 35 students; US$63 students. **Remarks:** Accepts advertising. **URL:** http://www.equinoxjournals.com/ojs/index.php/JSRNC. **Circ:** (Not Reported)

HOMESTEAD

134 ■ WWWK-FM - 105.5
27501 S Dixie Hwy., Ste. 208
Homestead, FL 33032

Format: Hispanic. **Key Personnel:** Lilliam M. Sierra, Gen. Mgr., lsierra@myradioexito.com. **Wattage:** 50,000. **URL:** http://myradioexito.com.

JACKSONVILLE

135 ■ WBOB-AM - 1320
4190 Belfort Rd., Ste. 450
Jacksonville, FL 32216
Phone: (904)470-4615

Format: Talk; News; Information. **Owner:** Chesapeake/Portsmouth Broadcasting Inc., at above address. **Key Personnel:** Henry Hoot, General Mgr. **URL:** http://www.1320wbob.com/.

136 ■ WJBC-FM - 91.7
5634 Normandy Blvd.
Jacksonville, FL 32205
Phone: (904)781-4321
Fax: (904)425-1178
E-mail: contact@wjbcfm.com

Format: Gospel. **Ad Rates:** Noncommercial. **URL:** http://www.wjbcfm.com/.

LAKE CITY

137 ■ WJTK-FM - 96.5
229 SW Main Blvd.
Lake City, FL 32025
Phone: (386)758-9696
Fax: (386)754-9650

Format: Talk; News. **Ad Rates:** Advertising accepted; rates available upon request. **URL:** http://965wjtk.com/.

MADISON

138 ■ WMAF-AM - 1230
PO Box 621
Madison, FL 32341
Phone: (850)973-3233
Fax: (850)973-3097
E-mail: countrywmaf@embarqmail.com

Format: Country. **URL:** http://radiowmaf.com/.

MARIANNA

139 ■ WAYP-FM - 88.3
2199 B N Monroe St.
Tallahassee, FL 32303
Phone: (850)422-1929
Fax: (850)297-1888
Free: (888)422-9293

Format: Contemporary Christian. **Owner:** WAY-FM Media Group, PO Box 64500, Colorado Springs, CO 80962, (719)533-0300, Fax: (719)278-4339. **Key Personnel:** Craig Vinson, Operations Dir.; Jill Ruiter, Office Mgr.; Steve Young, General Mgr. **URL:** http://wayp.wayfm.com/.

MIAMI

140 ■ Home Miami
Bonnier Corporation
4040 NE 2nd Ave., Ste. 313
Miami, FL 33137
Phone: (305)673-2112
Publication E-mail: info@homemia.com

Magazine featuring home design and architecture in Miami. **Freq:** 11/yr. **Trim Size:** 9 1/4 x 11 1/8. **Key Personnel:** Beth Dunlop, Editor-in-Chief. **Subscription Rates:** US$48 individuals; US$60 two years. **Remarks:** Accepts advertising. **URL:** http://www.homemia.com/. **Circ:** 21,000

141 ■ WNMA-AM - 1210
7250 NW 58th St.
Miami, FL 33166

Format: Hispanic. **Owner:** Multicultural Radio Broadcasting, Inc., 449 Broadway, New York, NY 10013, (212)966-1059, Fax: (212)966-9580. **Key Personnel:** John Gabel, Contact, johng@mrbi.net. **URL:** http://www.mrbi.net/radiogroup.htm.

PACE

142 ■ WTGF-FM - 90.5
4670 Hwy. 90
Pace, FL 32571
Phone: (850)994-3747
Fax: (850)994-8441

Format: Religious; Gospel. **Operating Hours:** Continuous. **Key Personnel:** John Graves, Gen. Mgr. **Wattage:** 25,000. **URL:** http://www.truthradiofm.org/.

PENSACOLA

143 ■ WDWR-AM - 1230
PO Box 866
Pensacola, FL 32591
Phone: (850)777-1568
E-mail: info@divinewordradio.com

Format: Contemporary Christian. **Owner:** Divine Word Communications, at above address. **URL:** http://www.divinewordradio.com/.

PLANT CITY

144 ■ Paso Fino Horse World
Lionheart Publishing Inc.
101 N Collins St.
Plant City, FL 33563
Phone: (813)719-7777
Fax: (813)719-7872
Publisher E-mail: lpi@lionhrtpub.com

Magazine featuring Paso Fino equine breed and its owners. **Trim Size:** 8 3/8 x 10 7/8. **Key Personnel:** Stephanie Dunkin, Editor. **Subscription Rates:** US$30 individuals; US$45 two years. **Remarks:** Accepts advertising. **URL:** http://pasofinohorseworld.com/. . **Ad Rates:** 4C: $760. **Circ:** (Not Reported)

STUART

145 ■ WHLG-FM - 101.3
1670 NW Federal Hwy.
Stuart, FL 34994
Phone: (772)344-1999
Fax: (772)692-0258

Format: Adult Contemporary. **URL:** http://www.coast1013.com/.

TALLAHASSEE

WAYP-FM - See Marianna

WINTER PARK

146 ■ Destination Weddings & Honeymoons
Bonnier Corporation
460 N Orlando Ave., Ste. 200
Winter Park, FL 32789
Phone: (407)628-4802
Fax: (407)628-7061

Magazine featuring destinations for wedding or honeymoon. **Freq:** Quarterly. **Key Personnel:** Susan Moynihan, Editor-in-Chief. **Subscription Rates:** US$11.97 individuals. **Remarks:** Accepts advertising. **URL:** http://www.destinationweddingmag.com/; http://www.bonniercorp.com/brands/Destination-Weddings-Honeymoons.html. **Circ:** 100,000

147 ■ Florida Travel & Life
Bonnier Corporation
460 N Orlando Ave., Ste. 200
Winter Park, FL 32789
Phone: (407)628-4802
Fax: (407)628-7061

Magazine featuring Florida and its destinations. **Freq:** 7/yr. **Trim Size:** 8 7/8 X 10 7/8. **Key Personnel:** Ana Connery, Editor-in-Chief. **Subscription Rates:** US$11.97 individuals; US$17.97 Canada; US$23.97 other countries. **Remarks:** Accepts advertising. **URL:** http://www.bonniercorp.com/brands/Florida-Travel-Life.html; http://www.floridatravellife.com/. **Circ:** 101,000

148 ■ Kiteboarding
Bonnier Corporation
460 N Orlando Ave., Ste. 200
Winter Park, FL 32789
Phone: (407)628-4802
Fax: (407)628-7061

Magazine featuring kiteboarding. **Freq:** 6/yr. **Key Personnel:** Aaron Sales, Editor. **Subscription Rates:** US$29.97 individuals; US$41.97 Canada; US$65.97 other countries. **Remarks:** Accepts advertising. **URL:** http://www.kiteboardingmag.com/index.jsp; http://www.bonniercorp.com/brands/Kiteboarding.html. . **Ad Rates:** 4C: $6,610. **Circ:** (Not Reported)

ZOLFO SPRINGS

149 ■ WZSP-FM - 105.3
7891 US Hwy. 17 S
Zolfo Springs, FL 33890
Phone: (863)494-1053
Fax: (863)494-4443
Free: (866)357-1053
E-mail: info@lazeta.fm

Format: Hispanic. **Key Personnel:** Hall Kneller, Gen. Mgr., hal@lazeta.fm. **URL:** http://www.lazeta.fm/.

GEORGIA

AMERIBUS

150 ■ WFRP-FM - 88.7
290 Hegenberger Rd.
Oakland, CA 94621
Free: (800)543-1495

Format: Religious. **Owner:** Family Stations Inc., at above address. **URL:** http://www.familyradio.com/.

APPLING

WLGP-FM - See Jacksonville, North Carolina

ATLANTA

151 ■ WCFO-AM - 1160
1100 Spring St., Ste. 610
Atlanta, GA 30309
Phone: (404)681-9307
Fax: (404)870-8859
E-mail: listeners@jwbroadcasting.com

Format: News; Talk. **Owner:** JW Broadcasting, Inc., at above address. **Key Personnel:** Jeff Davis, VP/Gen. Mgr., jeffdavis@jwbroadcasting.com. **URL:** http://www.businessradio1160.com/.

152 ■ WWLG-FM - 96.7
1819 Peachtree Rd., Ste. 700
Atlanta, GA 30309
Phone: (404)875-8080

Format: Country. **Key Personnel:** Shane Collins, Contact, shane@967thelegend.com. **URL:** http://www.967thelegend.com/main.html.

CALHOUN

153 ■ WLOJ-FM - 102.9
PO Box 1018
Laguna Beach, CA 92652
Fax: (469)241-6795
Free: (800)639-5433

Format: Religious. **Owner:** New Life Ministries, at above address. **Key Personnel:** Stephen Arterburn, Founder/Chm. **URL:** http://www.newlife.com/Radio/findradio.asp?id=GA&t=Georgia.

CARROLLTON

154 ■ WGGR-FM - 95.3
PO Box 1018
Laguna Beach, CA 92652
Fax: (469)241-6795
Free: (800)639-5433

Format: Religious. **Owner:** New Life Ministries, at above address. **Key Personnel:** Stephen Arterburn, Founder/Chm. **URL:** http://www.newlife.com/Radio/findradio.asp?id=GA&t=Georgia.

CARTERSVILLE

155 ■ WBHF-AM - 1450
1410 Hwy. 411 NE
Cartersville, GA 30121

Format: Big Band/Nostalgia. **Operating Hours:** Continuous. **Wattage:** 1000. **URL:** http://www.wbhf1450.com/.

COLUMBUS

156 ■ WIOL-FM - 95.7
2203 Wynnton Rd.
Columbus, GA 31906-2531
Phone: (706)576-3565
Fax: (706)576-3683

Format: Alternative/New Music/Progressive. **Key Personnel:** Carl Conner, VP/Mgr., cconner@dbicolumbus.com. **URL:** http://www.mix957fm.com/.

CORDELE

157 ■ WAEF-FM - 90.3
P.O. Drawer 2440
Tupelo, MS 38803
Phone: (662)844-5036
Fax: (662)842-7798

Format: Contemporary Christian. **Owner:** American Family Assoc., at above address. **URL:** http://www.afr.net/newafr/default.asp.

MARIETTA

158 ■ Analytics
Lionheart Publishing Inc.
506 Roswell St., Ste. 220
Marietta, GA 30060
Phone: (770)431-0867
Fax: (770)432-6969
Publisher E-mail: lpi@lionhrtpub.com

Magazine featuring mathematical analysis. **Remarks:** Accepts advertising. **URL:** http://www.lionhrtpub.com/. **Circ:** (Not Reported)

159 ■ Masonry Design
Lionheart Publishing Inc.
506 Roswell St., Ste. 220
Marietta, GA 30060
Phone: (770)431-0867
Fax: (770)432-6969
Publisher E-mail: lpi@lionhrtpub.com

Magazine featuring materials, trends, and technologies for masonry projects. **Freq:** Bimonthly. **Trim Size:** 8 3/8 x 10 7/8. **Key Personnel:** Cory Sekine-Pettite, Editor, cory@lionhrtpub.com. **Remarks:** Accepts advertising. **URL:** http://masonrydesignmagazine.com/index.php. . **Ad Rates:** 4C: $5,655. **Circ:** 15,000

ROME

160 ■ WSRM-FM - 95.3
20 John Davenport Dr.
Rome, GA 30165
Fax: (706)235-7107

Format: News; Talk. **URL:** http://www.wrgarome.com/.

SAVANNAH

161 ■ WSEG-AM - 1400
PO Box 60999
Savannah, GA 31420
Phone: (912)920-4441
Fax: (912)925-7024
E-mail: wsegradio@yahoo.com

Format: Adult Contemporary. **Owner:** Marmac Communications, LLC, at above address. **URL:** http://www.star1400.com/.

162 ■ WSSJ-FM - 100.1
6203 Abercorn St., Ste. 101
Savannah, GA 31405
Phone: (912)691-1934
Fax: (912)691-1936

Format: Adult Contemporary; Jazz. **Wattage:** 50,000. **URL:** http://www.smoothjazz100.com/.

VALDOSTA

163 ■ WVDA-FM - 88.5
2351 Sunset Blvd., Ste. 170-218
Rocklin, CA 95765
Phone: (916)251-1600
Fax: (916)251-1650

Format: Contemporary Christian. **Owner:** Educational Media Foundation, at above address. **URL:** http://www.air1.com/Music/StationList.aspx.

VIDALIA

164 ■ WYUM-FM - 101.7
1501 Mt. Vernon Rd.
PO Box 900
Vidalia, GA 30475
Phone: (912)537-9202

Format: Country. **Owner:** Vidalia Communications, at above address. **Key Personnel:** Collins Knighton, Operation/Sports Dir., collins@vidaliacommunications.com. **URL:** http://www.vidaliacommunications.com/.

WARNER ROBINS

165 ■ WDXQ-FM - 96.7
1350 Radio Loop
Warner Robins, GA 31088
Phone: (478)923-3416

Format: Classic Rock. **Owner:** Georgia Eagle Broadcasting, Inc., at above address. **URL:** http://houstoncountyradio.com/WDXQ.html.

Circulation: ★ = ABC; △ = BPA; ♦ = CAC; • = CCAB; ❑ = VAC; ⊕ = PO Statement; ‡ = Publisher's Report; Boldface figures = sworn; Light figures = estimated.

HAWAII

KAHULUI

166 ■ KJKS-FM - 99.9
311 Ano St.
Kahului, HI 96732-1304
Phone: (808)877-5566

Format: Adult Contemporary. **Key Personnel:** Chuck Bergson, Pres./CEO, bergson@pacificradiogroup.com. **URL:** http://www.kissfmmaui.com/.

167 ■ KUAU-AM - 1570
777 Mokulele Hwy.
Kahului, HI 96732
Fax: (808)871-9708
Free: (888)404-7729
E-mail: info@kingscathedral.com

Format: Talk; Religious. **Owner:** King's Cathedral and Chapels, at above address. **Key Personnel:** Dr. James Marocco, Contact. **URL:** http://www.kingscathedral.com/outreaches.php.

KAMUELA

168 ■ KWYI-FM - 106.9
PO Box 6540
Kamuela, HI 96743
Phone: (808)885-9866

Format: Adult Contemporary; Oldies. **Wattage:** 5500. **Ad Rates:** Advertising accepted; rates available upon request. **URL:** http://www.kwyi.com/.

LIHUE

169 ■ KJMQ-FM - 98.1
4334 Rice St.
Lihue, HI 96766
Phone: (808)246-4444
Fax: (808)246-4405
E-mail: jamz981@hotmail.com

Format: Hip Hop. **Key Personnel:** Danny Hill, Contact, d.hill@hhawaiimedia.com. **URL:** http://jamz981.com/default.aspx.

IDAHO

BOISE

170 ■ KAWO-FM - 104.3
827 E Park Blvd. 201
Boise, ID 83712
Phone: (208)344-6363
Fax: (208)342-0444

Format: Country. **Owner:** Peak Broadcasting, LLC, at above address. **Key Personnel:** Barb McGann, Contact, barb.mcgann@peakbroadcasting.com. **URL:** http://www.wow1043.com/main.html.

171 ■ KDJQ-AM - 890
1050 Clover Dr.
Boise, ID 83703
Phone: (208)424-3689
Fax: (208)433-9318

Format: Oldies. **URL:** http://www.kdjq890.com/.

DRIGGS

172 ■ KCHQ-FM - 102.1
160 N Hwy. 33
PO Box 54
Driggs, ID 83422
Phone: (208)354-4103
Fax: (208)354-4104

Format: Country. **Key Personnel:** Dave Plourde, News Dir., phone (208)354-4101, dave@q102fm.net. **URL:** http://www.q102fm.net/.

MOSCOW

173 ■ MaryJanesFarm
Belvoir Media Group, LLC
1000 Wild Iris Ln.
Moscow, ID 83843
Free: (888)750-6004
Publisher E-mail: customer_service@belvoir.com

Magazine featuring organic farming. **Founded:** Jan. 2002. **URL:** http://www.maryjanesfarm.org/magazine.html.

TWIN FALLS

KJCC-FM - See Carnegie, Oklahoma

KJFT-FM - See Arlee, Montana

174 ■ KTFY-FM - 88.1
131 Grandview Dr.
Twin Falls, ID 83301
Phone: (208)735-0881

Format: Contemporary Christian. **URL:** http://www.881ktfy.org/.

ILLINOIS

BELLEVILLE

175 ■ WXOZ-AM - 1510
6500 W Main St., Ste. 315
Belleville, IL 62223
Phone: (618)394-WXOZ

Format: Talk; Oldies. **Owner:** Insane Broadcasting Company, at above address. **URL:** http://www.wxoz.com/index.php.

CARLINVILLE

WVNL-FM - See Vandalia

CHICAGO

176 ■ Baby & Kids
Talcott Communication Corp.
20 W Kinzie, Ste. 1200
Chicago, IL 60610
Phone: (312)849-2220
Fax: (312)849-2174
Publisher E-mail: talcottpub@talcott.com

Magazine featuring new products for baby and youth market. **Freq:** 4/yr. **Key Personnel:** Daniel von Rabenau, Publisher, dvr@talcott.com. **Subscription Rates:** US$16 individuals. **Remarks:** Accepts advertising. **URL:** http://www.babyandkids.biz/. **Circ:** 8,650

177 ■ Chef Educator Today
Talcott Communication Corp.
20 W Kinzie, Ste. 1200
Chicago, IL 60610
Phone: (312)849-2220
Fax: (312)849-2174
Publisher E-mail: talcottpub@talcott.com

Magazine featuring information for foodservice educators. **Key Personnel:** Daniel von Rabenau, Publisher, dvr@talcott.com. **URL:** http://www.chefedtoday.com.

178 ■ HomeFashion & Furniture Trends
Talcott Communication Corp.
20 W Kinzie, Ste. 1200
Chicago, IL 60610
Phone: (312)849-2220
Fax: (312)849-2174
Publisher E-mail: talcottpub@talcott.com

Magazine featuring products for home decor industry. **Freq:** Bimonthly. **Key Personnel:** Daniel von Rabenau, Publisher, dvr@talcott.com. **Remarks:** Accepts advertising. **URL:** http://www.homefashionmag.biz/. **Circ:** 22,780

179 ■ International Journal of Biomedical Engineering and Technology
Inderscience Publishers
c/o Dr. Nilmini Wickramasinghe, Ed.-in-Ch.
Illinois Institute of Technology
Center for the Management of Medical Technology
565 W Adams St., Ste. 406
Chicago, IL 60661
Publication E-mail: ijbet@inderscience.com
Publisher E-mail: editor@inderscience.com

Journal covering multi-disciplinary area of biomedical engineering & technology. **Founded:** 2007. **Freq:** 4/yr. **Key Personnel:** Dr. Nilmini Wickramasinghe, Editor-in-Chief, nilmini@stuart.iit.edu. **ISSN:** 1752-6418. **Subscription Rates:** EUR470 individuals includes surface mail, print only; EUR640 individuals print & online. **URL:** http://www.inderscience.com/browse/index.php?journalCODE=ijbet.

180 ■ Pizza & Italian Cuisine
Talcott Communication Corp.
20 W Kinzie, Ste. 1200
Chicago, IL 60610
Phone: (312)849-2220
Fax: (312)849-2174
Publisher E-mail: talcottpub@talcott.com

Magazine featuring articles and foodservice recipes for pizza industry. **Freq:** 12/yr. **Key Personnel:** Daniel von Rabenau, Publisher, dvr@talcott.com. **Subscription Rates:** US$16 individuals. **URL:** http://www.pizzazine.com/.

181 ■ WCFS-FM - 105.9
Two Prudential Plz., Ste. 1059
Chicago, IL 60601
Phone: (312)240-7900
E-mail: fresh1059@cbsradio.com

Format: Adult Contemporary. **Owner:** CBS Radio Stations Inc., 1515 Broadway, New York, NY 10036, (212)846-3939. **Ad Rates:** Advertising accepted; rates available upon request. **URL:** http://www.fresh1059.com/.

182 ■ WCPT-AM - 820
6012 S Pulaski Rd.
Chicago, IL 60629
Phone: (773)767-1000

Format: Talk. **Operating Hours:** Continuous. **Key Personnel:** Harvey Wells, General Mgr., hwells@newswebcorporation.com; Gavin Carroll, Program Mgr., gcarroll@wcpt820.com; Melissa Ryzy, Dir. of Sales, mryzy@9chicago.com. **Ad Rates:** Advertising accepted; rates available upon request. **URL:** http://www.wcpt820.com/.

DIXON

183 ■ WRCV-FM - 101.7
1460 S College Ave.
Dixon, IL 61021
Phone: (815)288-3341
Fax: (815)284-1017

Format: Country. **Owner:** NRG Media, 2875 Mt. Vernon Rd. SE, Cedar Rapids, IA 52403, (319)862-0300, Fax: (319)286-9383. **Key Personnel:** Al Knickrehm, Gen. Mgr., aknickrehm@nrgmedia.com. **URL:** http://www.nrgmedia.com/stations.aspx.

EFFINGHAM

184 ■ WEFI-FM - 89.5
P.O. Drawer 2440
Tupelo, MS 38803
Phone: (662)844-5036
Fax: (662)842-7798

Format: Religious. **Owner:** American Family Association, at above address. **URL:** http://www.afr.net/newafr/default.asp.

KANKAKEE

185 ■ WXNU-FM - 106.5
70 Meadowview Ctr.
Kankakee, IL 60901-2047
Phone: (815)935-9555
Fax: (815)935-9593
Free: (877)777-1065

Format: Country. **Key Personnel:** Shaun Kelly, Contact. **URL:** http://www.xcountry1065.com/.

LINCOLNSHIRE

186 ■ WAES-FM - 88.1
One Stevenson Dr.
Lincolnshire, IL 60069
Phone: (847)634-4000

Format: Educational. **URL:** http://www6.district125.k12.il.us/clubs/radiostation/.

MILFORD

187 ■ WJCZ-FM - 91.3
3232 W MacArthur Blvd.
Santa Ana, CA 92707
Phone: (714)825-9663
Fax: (714)825-9660

Format: Religious. **Owner:** CSN International, at above address. **URL:** http://calvaryradionetwork.com/.

PEORIA

188 ■ WIRL-AM - 1290
331 Fulton, Ste. 1200
Peoria, IL 61602
Phone: (309)637-3700

Format: Country. **Key Personnel:** Mike Wild, Mktg. Mgr.; Kevin Cassulo, Sales Mgr.; Randy Rundle, Operations Mgr. **URL:** http://www.1290wirl.com/.

189 ■ WPMJ-FM - 94.3
3641 Meadowbrook Rd.
Peoria, IL 61604
Phone: (309)685-0977
Fax: (309)685-7150

Format: Oldies. **Key Personnel:** Joyce Powell, Contact, jpowell@trueoldies943.com; Lee Malcolm, Program Dir., lmalcolm@trueoldies943.com. **URL:** http://www.trueoldies943.com/.

190 ■ WXMP-FM - 101.1
4234 N Brandywine Dr.
Peoria, IL 61614
Phone: (309)282-4649
Fax: (309)686-0111

Format: Oldies. **Key Personnel:** Jason Stuckwisch, Contact, jasons@impeoria.com. **URL:** http://www.mixpeoria.com/.

191 ■ WZPN-FM - 96.5
4234 N Brandywine Dr., Ste. D
Peoria, IL 61614
Phone: (309)686-0101
Fax: (309)686-0111

Format: Sports. **URL:** http://www.965espn.com/.

RIVERTON

192 ■ WLCE-FM - 97.7
1510 N Third St.
Riverton, IL 62561
Phone: (217)629-9777
Fax: (217)629-7952

Format: Adult Contemporary. **Owner:** Mid-West Family Broadcasting, at above address. **Key Personnel:** Kevan Kavanaugh, Pres./Gen. Mgr. **URL:** http://www.alice.fm/.

193 ■ WYVR-FM - 97.7
1510 N Third St.
Riverton, IL 62561
Phone: (217)629-7077
Fax: (217)629-7952

Format: Adult Contemporary; Top 40. **Owner:** Mid-West Family Broadcasting, at above address. **Key Personnel:** Kevan Kavanaugh, Pres./Gen. Mgr.; Valorie Knight, Program Dir., alice@alice.fm. **URL:** http://www.alice.fm/.

ROCKFORD

194 ■ WRTB-FM - 95.3
2830 Sandy Hollow Rd.
Rockford, IL 61109

Format: Alternative/New Music/Progressive. **Owner:** Maverick Media, at above address. **Key Personnel:** Lisa Chatfield, Contact, phone (815)874-7861, lisachatfield@maverick-media.ws. **Ad Rates:** Advertising accepted; rates available upon request. **URL:** http://www.953bobfm.com/?.

SALEM

195 ■ WSLE-FM - 91.3
P.O. Drawer 2440
Tupelo, MS 38803
Phone: (662)844-5036
Fax: (662)842-7798

Format: Religious. **Owner:** American Family Association, at above address. **URL:** http://www.afr.net/newafr/default.asp.

VANDALIA

196 ■ WVNL-FM - 91.7
PO Box 140
Carlinville, IL 62626
Phone: (217)854-4800
Fax: (217)854-4810
Free: (800)707-9191

Format: Contemporary Christian. **Owner:** Illinois Bible Institute, at above address. **Key Personnel:** Jeremiah Beck, Station Mgr., jeremiahb@wibi.org; Joe Buchanan, Music Dir., joeb@wibi.org; Rob Regal, Program Dir., robr@wibi.org. **URL:** http://www.wibi.org/.

INDIANA

BLOOMINGTON

197 ■ WCLS-FM - 97.7
318 E 3rd St.
Bloomington, IN 47401
Phone: (812)335-0977

Format: Oldies. **Key Personnel:** Ruth Ann Arney, General Mgr.; Charlotte Cook, Traffic/Billing Mgr.; Tony Kale, Program Mgr., wclsfm@smithville.net. **URL:** http://www.wclsfm.com/.

DALEVILLE

198 ■ WURK-FM - 101.7
9821 S County Rd., 800 W
Daleville, IN 47334
Phone: (765)378-2080
Fax: (765)378-2090

Format: Oldies. **Owner:** Backyard Broadcasting, 4237 Salisbury Rd., Ste. 225, Jacksonville, FL 32216, (904)674-0260, Fax: (904)854-4596. **Key Personnel:** Bruce Law, Market Mgr., bruce.law@bybradio.com; Camellia Pflum, Sales Mgr., camellia.pflum@bybradio.com. **URL:** http://www.werkradio.com/.

FORT WAYNE

199 ■ WBOI-FM - 89.1
3204 Clairmont Ct.
PO Box 8459
Fort Wayne, IN 46898-8459
Phone: (260)452-1189
Fax: (260)452-1188
Free: (800)471-9264

Format: Jazz; News. **Owner:** Northeast Indiana Public Radio, at above address. **Key Personnel:** Bruce Haines, General Mgr., bhaines@nipr.fm; Jeanette Dillon, News Dir., jdillon@nipr.fm; Lea Denny, General Mgr., ldenny@nipr.fm. **URL:** http://www.nipr.fm/wboi/index.htm.

200 ■ WQHK-FM - 105.1
2915 Maples Rd.
Fort Wayne, IN 46816
Phone: (260)447-5511

Format: Country. **Key Personnel:** Rob Kelly, Operations Mgr., rkelley@federatedmedia.com; Joel Pyle, General Sales Mgr., jpyle@federatedmedia.com. **URL:** http://www.k105fm.com/.

201 ■ WVBB-FM - 106.3
2100 Goshen Rd.
Fort Wayne, IN 46808
Phone: (260)482-9288
Fax: (260)482-8655

Format: Oldies. **Key Personnel:** Phil Becker, Program Dir., phil@thevibe1063.com; Keith Harris, Production Dir., keith@thevibe1063.com. **URL:** http://thevibe1063.com/default.aspx.

GREENFIELD

202 ■ WJCF-FM - 88.1
PO Box 846
Greenfield, IN 46140
Phone: (317)467-1064
Fax: (317)467-1065
Free: (877)888-5773

Format: Contemporary Christian. **Operating Hours:** Continuous. **URL:** http://www.wjcfradio.com/.

INDIANAPOLIS

203 ■ WRWM-FM - 93.9
6810 N Shadeland Ave.
Indianapolis, IN 46220
Phone: (317)842-9550
E-mail: warm939@indyradio.com

Format: Alternative/New Music/Progressive. **Owner:** Cumulus Media Indianapolis, at above address. **URL:** http://www.warm939.com/.

LOWELL

204 ■ WTMK-FM - 88.5
3232 W MacArthur Blvd.
Santa Ana, CA 92707
Phone: (714)825-9663
Fax: (714)825-9660

Format: Religious. **Owner:** CSN International, at above address. **URL:** http://calvaryradionetwork.com/.

205 ■ WWLO-FM - 89.1
P.O. Drawer 2440
Tupelo, MS 38803
Phone: (662)844-5036
Fax: (662)842-7798

Format: Religious. **Owner:** American Family Association, at above address. **URL:** http://www.afr.net/newafr/default.asp.

NEW ALBANY

WAYI-FM - See Louisville, Kentucky

RICHMOND

206 ■ WVXR-FM - 89.3
PO Box 1018
Laguna Beach, CA 92652
Fax: (469)241-6795
Free: (800)639-5433

Format: Religious. **Owner:** New Life Ministries, at above address. **Key Personnel:** Stephen Arterburn, Founder/Chm. **URL:** http://www.newlife.com/Radio/findradio.asp?id=IN&t=Indiana.

TERRE HAUTE

207 ■ WBOW-FM - 102.7
1301 Ohio St.
Terre Haute, IN 47807
Phone: (812)234-9770
Fax: (812)238-1576

Format: Adult Contemporary. **Key Personnel:** Doug Edge, General Mgr., doug@radioworksforme.com;

Circulation: ★ = ABC; △ = BPA; ♦ = CAC; • = CCAB; ❑ = VAC; ⊕ = PO Statement; ‡ = Publisher's Report; Boldface figures = sworn; Light figures = estimated.

Sketch Brumfield, Program Mgr.; Julie Latta, Webmaster, julie@radioworksforme.com. **URL:** http://www.b1027fm.com/.

IOWA

AMES

KWOI-FM - See Carroll

BOONE

208 ■ KFFF-AM - 1260
900 Eighth St.
Boone, IA 50036
Phone: (515)432-5014
Fax: (515)432-2092
Free: (877)993-1260
E-mail: mail@iowanewstalk.com

Format: News; Talk. **Key Personnel:** Jamie Johnson, Pres./Gen. Mgr. **Ad Rates:** Advertising accepted; rates available upon request. **URL:** http://www.faithandfreedomnetwork.com/.

209 ■ KFFF-FM - 99.3
900 Eighth St.
Boone, IA 50036
Phone: (515)432-5014
Fax: (515)432-2092
Free: (877)993-1260

Format: News; Talk. **Key Personnel:** Jamie Johnson, Pres./Gen. Mgr. **Ad Rates:** Advertising accepted; rates available upon request. **URL:** http://www.faithandfreedomnetwork.com/.

BURLINGTON

210 ■ KHDK-FM - 97.3
2850 Mt. Pleasant St.
Burlington, IA 52601
Phone: (319)752-5402
E-mail: hot973@hot973online.com

Format: Alternative/New Music/Progressive. **Owner:** Pritchard Broadcasting Corp., at above address. **URL:** http://hot973online.com/.

CARROLL

211 ■ KWOI-FM - 90.7
Iowa State University
2022 Communications Bldg.
Ames, IA 50011-3241
Phone: (515)294-2025
Fax: (515)294-1544
Free: (800)861-8000

Format: Public Radio. **URL:** http://www.woi.org/.

CEDAR RAPIDS

212 ■ KGYM-AM - 1600
1110 26th Ave. SW
Cedar Rapids, IA 52404-3430
Phone: (319)363-2061
Fax: (319)363-2948
E-mail: info@1600espn.com

Format: Sports. **Key Personnel:** Scott Unash, Sports and Program Dir., scott.unash@1600espn.com. **URL:** http://www.1600espn.com/.

DAVENPORT

213 ■ KAIP-FM - 88.9
2351 Sunset Blvd., Ste. 170-218
Rocklin, CA 95765
Phone: (916)251-1600
Fax: (916)251-1650

Format: Contemporary Christian. **Owner:** Educational Media Foundation, at above address. **URL:** http://www.air1.com/Music/StationList.aspx.

DES MOINES

214 ■ KJMC-FM - 89.3
1169 25th St.
Des Moines, IA 50311
Phone: (515)279-1811
E-mail: kjmcfm@mchsi.com

Format: Blues; Gospel; Jazz; Hip Hop. **Owner:** Minority Communications, Inc., at above address. **URL:** http://kjmcfm.org/index2.html.

FORT DODGE

215 ■ KEGR-FM - 89.5
290 Hegenberger Rd.
Oakland, CA 94621
Free: (800)543-1495

Format: Religious. **Owner:** Family Stations, Inc., at above address. **URL:** http://www.familyradio.com/.

OKOBOJI

216 ■ KJIA-FM - 88.9
PO Box 738
Okoboji, IA 51355
Phone: (712)332-7184
E-mail: kjia@kjiaradio.com

Format: Contemporary Christian; Religious. **Operating Hours:** Continuous. **Key Personnel:** Matt Dorfner, Exec. Dir. **Wattage:** 50,000. **URL:** http://www.kjiaradio.com/.

SIOUX CITY

217 ■ KOJI-FM - 90.7
4647 Stone Ave.
Sioux City, IA 51106
Phone: (712)274-6406
Fax: (712)274-6411
Free: (800)251-3690

Format: Public Radio; Classical; Jazz. **Key Personnel:** Duane Kraayenbrink, Contact, kraayed@witcc.edu. **Wattage:** 100,000. **URL:** http://www.kwit.org/.

SPENCER

218 ■ KUYY-FM - 100.1
2303 W 18th St.
Spencer, IA 51301
Phone: (712)264-1074
Fax: (712)264-1077

Format: Adult Contemporary. **Owner:** NRG Media, 2875 Mt. Vernon Rd. SE, Cedar Rapids, IA 52403, (319)862-0300, Fax: (319)286-9383. **Key Personnel:** Marty Spies, General Mgr., mspies@nrgmedia.com. **URL:** http://www.nrgmedia.com/stations.aspx.

WEST DES MOINES

219 ■ KNWI-FM - 107.1
3737 Woodland Ave., Ste. 111
West Des Moines, IA 50266
Phone: (515)327-1071
Fax: (515)327-1073
Free: (866)377-1071
E-mail: knwi@desmoines.fm

Format: Contemporary Christian. **Owner:** Northwestern College, 3003 Snelling Ave. N, Saint Paul, MN 55113-1598, (651)631-5100, Free: (800)692-4020. **Key Personnel:** Dave St. John, Program Dir., dave@desmoines.fm; Paul Gurthie, Contact, paul@desmoines.fm; Meridith Foster, Contact, meridith@desmoines.fm; Dick Whitworth, Station Mgr., dick@desmoines.fm; Evelyn Klampe, Admin. Asst., epklampe@nwc.edu; Jerry Chiarmonte, Contact, jachiaramonte@nwc.edu. **URL:** http://knwi.nwc.edu/page.php.

KANSAS

COFFEYVILLE

220 ■ KGGF-AM - 690
PO Box 1087
Coffeyville, KS 67337
Phone: (620)251-3800

Format: Sports; Talk. **Owner:** Radio Results Group, at above address. **URL:** http://www.radioresultsgroup.com/radioresultsgroup4_002.htm.

221 ■ KGGF-FM - 104.1
PO Box 1087
Coffeyville, KS 67337
Phone: (620)251-3800
E-mail: oldies104.1@sbcglobal.net

Format: Oldies. **Owner:** Radio Results Group, at above address. **Key Personnel:** Randy Bale, Program Dir. **URL:** http://www.radioresultsgroup.com/radioresultsgroup4_005.htm.

222 ■ KUSN-FM - 98.1
PO Box 1087
Coffeyville, KS 67337
Phone: (620)251-3800

Format: Country. **Owner:** Radio Results Group, at above address. **URL:** http://www.radioresultsgroup.com/radioresultsgroup4_003.htm.

GARDEN CITY

223 ■ KWKR-FM - 99.9
1402 E Kansas Ave.
Garden City, KS 67846
Phone: (620)276-2366
Fax: (620)276-3568
Free: (800)999-5283

Format: Album-Oriented Rock (AOR). **Key Personnel:** Gil Wohler, General Mgr., gilwohler@amfmradio.biz. **URL:** http://www.wksradio.com/kwkr/index.htm.

LAWRENCE

224 ■ Chelonian Conservation and Biology
Allen Press Inc.
810 E 10th
Lawrence, KS 66044
Phone: (785)843-1234
Fax: (785)843-1244
Free: (800)627-0326
Publisher E-mail: info@allenpress.com

Journal covering turtle and tortoise research. **Subtitle:** International Journal of Turtle and Tortoise Research. **Freq:** Semiannual. **Key Personnel:** Mary Reilly, Publisher, phone (785)843-1234, fax (785)843-1274, mreilly@allenpress.com. **ISSN:** 1071-8443. **Subscription Rates:** US$63 individuals print and online; US$78.75 other countries print and online; US$210 institutions print and online; US$183.75 institutions, other countries online only. **URL:** http://www.acgpublishing.com/dir_Journals/Chelonian2.asp; http://www.chelonianjournals.org/perlserv/?request=get-archive&ct=1.

225 ■ Herpetological Monographs
Allen Press Inc.
810 E 10th
Lawrence, KS 66044
Phone: (785)843-1234
Fax: (785)843-1244
Free: (800)627-0326
Publisher E-mail: info@allenpress.com

Journal covering amphibians and reptiles research. **Founded:** 1982. **Freq:** Annual. **Key Personnel:** Tod Reeder, Editor, treeder@sunstroke.sdsu.edu. **ISSN:** 0733-1347. **URL:** http://www.acgpublishing.com/dir_Journals/HerpetologicalMonographs2.asp; http://www.hljournals.org/perlserv/?request=get-archive&ct=1.

226 ■ Journal of Coastal Research
Allen Press Inc.
810 E 10th
Lawrence, KS 66044
Phone: (785)843-1234
Fax: (785)843-1244
Free: (800)627-0326
Publisher E-mail: info@allenpress.com

Journal covering coastal research. **Freq:** Bimonthly. **ISSN:** 0749-0208. **Subscription Rates:** US$165 individuals print and online; US$185 other countries print and online; US$475 institutions print and online; US$495 institutions, other countries print and online; US$110 individuals online only; US$400 institutions online only. **URL:** http://www.acgpublishing.com/dir_Journals/CoastalResearch2.asp; http://www.jcronline.org/perlserv/?request=index-html&ct=1.

227 ■ The Journal of Wildlife Management
Allen Press Inc.
810 E 10th
Lawrence, KS 66044
Phone: (785)843-1234
Fax: (785)843-1244
Free: (800)627-0326
Publisher E-mail: info@allenpress.com

Journal covering wildlife management. **Freq:** Bimonthly. **ISSN:** 0022-541X. **Subscription Rates:** US$790 institutions print an online; US$840 institutions, other countries print an online; US$665 institutions online only. **URL:** http://www.acgpublishing.com/dir_Journals/TheJournalofWildlifeManagement2.a sp; http://www.wildlifejournals.org/perlserv/?request=index-html&ct=1.

228 ■ Rangelands
Allen Press Inc.
810 E 10th
Lawrence, KS 66044
Phone: (785)843-1234
Fax: (785)843-1244
Free: (800)627-0326
Publisher E-mail: info@allenpress.com

Journal for the Society for Range Management. **Freq:** Bimonthly. **ISSN:** 0190-0528. **Subscription Rates:** US$76 individuals print and online; US$99 other countries print and online; US$160 institutions print and online; US$200 institutions, other countries print and online; US$54 individuals online only; US$165 institutions online only. **URL:** http://www.acgpublishing.com/dir_Journals/Rangelands2.asp; http://www.srmjournals.org/perlserv/?request=index-html&ct=1.

229 ■ Wildlife Monographs
Allen Press Inc.
810 E 10th
Lawrence, KS 66044
Phone: (785)843-1234
Fax: (785)843-1244
Free: (800)627-0326
Publisher E-mail: info@allenpress.com

Journal covering wildlife animals. **ISSN:** 0084-0173. **Subscription Rates:** Included in membership. **URL:** http://www.acgpublishing.com/dir_Journals/WildlifeMonographs2.asp; http://www.wildlifejournals.org/perlserv/?request=index-html.

230 ■ The Wildlife Professional
Allen Press Inc.
810 E 10th
Lawrence, KS 66044
Phone: (785)843-1234
Fax: (785)843-1244
Free: (800)627-0326
Publisher E-mail: info@allenpress.com

Journal covering wildlife professionals. **Freq:** Quarterly. **ISSN:** 1933-2866. **Subscription Rates:** US$125 institutions print and online; US$155 institutions, other countries print and online. **URL:** http://www.acgpublishing.com/dir_Journals/TheWildlifeProfessional.asp.

231 ■ Wildlife Society Bulletin
Allen Press Inc.
810 E 10th
Lawrence, KS 66044
Phone: (785)843-1234
Fax: (785)843-1244
Free: (800)627-0326
Publisher E-mail: info@allenpress.com

Journal covering wildlife society management. **Freq:** 5/yr. **ISSN:** 0091-7648. **URL:** http://www.acgpublishing.com/dir_Journals/WildlifeSocietyBulletin2.asp.

KANV-FM - See Olsburg

232 ■ KMXN-FM - 92.9
3125 W 6th St.
Lawrence, KS 66049
Phone: (785)843-1320
Fax: (785)841-5924

Format: Classic Rock. **Owner:** Jayhawk Broadcasting, at above address. **Key Personnel:** Ron Covert, General Mgr.; Jon Thomas, Program Dir.; Bobby Rock, Music Dir. **URL:** http://www.x929.com/.

OLSBURG

233 ■ KANV-FM - 91.3
1120 W 11th St.
Lawrence, KS 66044
Phone: (785)864-4530
Fax: (785)864-5278
Free: (888)577-5268

Format: Classical; Bluegrass; Jazz; News. **Owner:** Kansas Public Radio, at above address. **Key Personnel:** Janet Campbell, General Mgr., jcampbell@ku.edu. **URL:** http://kansaspublicradio.org/.

PRATT

234 ■ KMMM-AM - 1290
30129 E Hwy. 54
PO Box 486
Pratt, KS 67124
Phone: (620)672-5581
Fax: (620)672-5583

Format: Oldies; Sports. **Key Personnel:** Eric Strobel, General Mgr., estrobel@rockingmradio.com; Carl Raida, Program Dir., craida@rockingmradio.com; Lisa Coss, Bus. Mgr./Traffic Dir., lcoss@rockingmradio.com. **URL:** http://www.superhits1290.com/.

WICHITA

235 ■ KMTW-TV - 36
316 N West St.
Wichita, KS 67203
Phone: (316)942-2424
Fax: (316)942-8927

Key Personnel: Kent Cornish, General Mgr., kentcornish@clearchannel.com; Jeff McCausland, Sales Mgr., jeffmac@clearchannel.com; Ken Whitney, Production Mgr., kenwhitney@clearchannel.com. **URL:** http://www.mytvwichita.com.

236 ■ KYWA-FM - 90.7
110 S Main St., Ste. 1050
Wichita, KS 67202
Phone: (316)831-0907
Fax: (316)831-0910

Format: Contemporary Christian. **Owner:** WAY-FM Media Group, PO Box 64500, Colorado Springs, CO 80962, (719)533-0300, Fax: (719)278-4339. **URL:** http://kywa.wayfm.com/.

KENTUCKY

BOWLING GREEN

237 ■ WAYD-FM - 88.1
1945 Scottsville Rd., B2
PMB 363
Bowling Green, KY 42104-5817
Free: (888)33W-AYFM

Format: Contemporary Christian. **Key Personnel:** Matt Austin, General Mgr. **URL:** http://wayd.wayfm.com/.

DANVILLE

238 ■ WLAI-FM - 107.1
2351 Sunset Blvd., Ste. 170-218
Rocklin, CA 95765
Phone: (916)251-1600
Fax: (916)251-1650

Format: Contemporary Christian. **Owner:** Educational Media Foundation, at above address. **URL:** http://www.air1.com/Music/StationList.aspx.

ELIZABETHTOWN

239 ■ WAKY-FM - 103.5
PO Box 2087
Elizabethtown, KY 42702
Free: (888)766-1035

Format: Country. **Owner:** W & B Broadcasting, at above address. **URL:** http://www.waky1035.com/.

240 ■ WTHX-FM - 107.3
611 W Poplar St., C-2
Elizabethtown, KY 42701
Phone: (270)982-1073
Fax: (270)769-6349

Format: Top 40. **Key Personnel:** Kidd Kraddick, Contact, kidd@kiddlive.com; Cat Michaels, Contact, cat@etownstar.com; Derrick Daniels, Contact, doubled@etownstar.com. **URL:** http://www.etownstar.com/.

HORSE CAVE

241 ■ WHSX-FM - 99.1
1130 S Dixie St.
Horse Cave, KY 42749
Phone: (270)432-7991
Fax: (888)862-4402
Free: (866)991-4677
E-mail: 991@scrtc.com

Format: Blues; Sports. **Wattage:** 6000. **Ad Rates:** Advertising accepted; rates available upon request. **URL:** http://www.thehoss.com/.

LOUISVILLE

242 ■ WAYI-FM - 104.3
3211 Grant Line Rd., Ste. 1
New Albany, IN 47150

Format: Contemporary Christian. **Owner:** WAY-FM Media Group, PO Box 64500, Colorado Springs, CO 80962, (719)533-0300, Fax: (719)278-4339. **Key Personnel:** Matt Hahn, General Mgr.; Bryans Johns, Operations Dir. **URL:** http://wayi.wayfm.com/.

MOREHEAD

243 ■ Softball Youth
Dugout Media Inc.
PO Box 983
Morehead, KY 40351
Fax: (859)201-1107
Free: (866)500-4225
Publisher E-mail: info@dugoutmedia.com

Magazine for 7-14 year-old girls who play fastpitch softball. **Freq:** 4/yr. **Key Personnel:** Scott M. Hacker, Pres./Owner, shacker@dugoutmedia.com. **URL:** http://www.softballyouth.com/; http://www.dugoutmedia.com/publishing.html.

PADUCAH

244 ■ WQQR-FM - 94.7
PO Box 2397
Paducah, KY 42002
Phone: (270)534-9690
Fax: (270)554-5468
Free: (877)947-QQFM

Format: Classic Rock. **Key Personnel:** Jamie Futrell, Contact, phone (270)534-9690. **URL:** http://www.wqqr.com/.

PIKEVILLE

245 ■ WEKB-AM - 1460
1240 Radio Dr.
PO Box 2200
Pikeville, KY 41502
Phone: (606)432-8103
Fax: (606)432-2809

Format: Oldies. **Owner:** East Kentucky Broadcasting Group, at above address. **Wattage:** 5000. **URL:** http://www.myoldiesradio.com/.

Circulation: ★ = ABC; △ = BPA; ♦ = CAC; • = CCAB; ❑ = VAC; ⊕ = PO Statement; ‡ = Publisher's Report; Boldface figures = sworn; Light figures = estimated.

SOMERSET

246 ■ WSGP-FM - 88.3
PO Box 1423
Somerset, KY 42502
Free: (800)408-8888

Format: Gospel. **URL:** http://www.kingofkingsradio.com/.

LOUISIANA

ALEXANDRIA

KBIO-FM - See Natchitoches

WHJM-FM - See Anna, Ohio

BATON ROUGE

247 ■ WPFC-AM - 1550
6943 Titian Ave.
Baton Rouge, LA 70802
Phone: (225)926-6550
Fax: (225)590-3238
Free: (866)439-8026

Format: Gospel. **Key Personnel:** Jacque Griffin, Program Dir., jacquegriffin@wpfc1550.com. **URL:** http://www.wpfc1550.com/.

EUNICE

248 ■ KEUN-FM - 105.5
1237 E Ardion St.
Eunice, LA 70535

Format: Country. **URL:** http://www.keunworldwide.com/.

NATCHITOCHES

249 ■ KBIO-FM - 89.7
601 Washington St.
Alexandria, LA 71301
Free: (888)408-0201

Format: Religious. **Owner:** Radio Maria Inc., at above address. **URL:** http://www.radiomaria.us/.

NEW ORLEANS

250 ■ WNKV-FM - 91.1
2351 Sunset Blvd., Ste. 170-218
Rocklin, CA 95765
Phone: (916)251-1600
Fax: (916)251-1650

Format: Contemporary Christian. **Owner:** Educational Media Foundation, at above address. **Key Personnel:** Chip Bailey, Regional Mgr., phone (225)612-4927. **URL:** http://www.klove.com/.

SHREVEPORT

251 ■ KFLO-FM - 89.1
PO Box 7277
Shreveport, LA 71107
Phone: (318)550-2000
E-mail: info@miracle891.org

Format: Contemporary Christian. **Owner:** Family Life Educational Foundation Inc., at above address. **Key Personnel:** Dan Perkins, Operations Mgr., dperkins@miracle891.org; Joe Miot, Program Dir., jmiot@miracle891.org. **URL:** http://www.miracle891.org/.

MARYLAND

ANNAPOLIS

252 ■ Comparative Islamic Studies
Equinox Publishing Ltd.
Ctr. for Middle East & Islamic Studies
Division of Humanities & Social Sciences
United States Naval Academy
107 Maryland Ave.
Annapolis, MD 21402-5044

Journal covering the study of Islam. **Freq:** 2/yr (June and December). **Key Personnel:** Brannon M. Wheeler, Editor, bwheeler@usna.edu. **ISSN:** 1740-7125. **Subscription Rates:** 90 institutions; US$165 institutions; 40 individuals; US$65 individuals; 28 students; US$50 students; 40 institutions developing countries. **Remarks:** Accepts advertising. **URL:** http://www.equinoxjournals.com/ojs/index.php/CIS. **Circ:** (Not Reported)

BETHESDA

253 ■ American Journal of Physiology - Endocrinology and Metabolism
The American Physiological Society
9650 Rockville Pke.
Bethesda, MD 20814-3991
Phone: (301)634-7164
Fax: (301)634-7241
Publisher E-mail: webmaster@the-aps.org

Journal focusing on the studies of endocrine and metabolic systems. **Freq:** Monthly. **Key Personnel:** Amira Klip, Editor-in-Chief, amira@sickkids.ca. **ISSN:** 0193-1849. **Subscription Rates:** US$135 members; US$175 members Canada, Mexico; US$195 members other countries; US$320 nonmembers; US$360 nonmembers Canada, Mexico; US$380 nonmembers other countries; US$480 institutions; US$520 institutions Canada, Mexico; US$540 nonmembers other countries. **Remarks:** Accepts advertising. **URL:** http://www.the-aps.org/publications/ajpendo/index.htm. **Circ:** (Not Reported)

254 ■ American Journal of Physiology - Gastrointestinal and Liver Physiology
The American Physiological Society
9650 Rockville Pke.
Bethesda, MD 20814-3991
Phone: (301)634-7164
Fax: (301)634-7241
Publisher E-mail: webmaster@the-aps.org

Journal covering all aspects of research involving normal or abnormal function of the gastrointestinal tract, hepatobiliary system, and pancreas. **Freq:** Monthly. **Key Personnel:** Marshall Montrose, Editor-in-Chief, montromh@ucmail.uc.edu. **ISSN:** 0193-1857. **Subscription Rates:** US$135 members; US$175 members Canada, Mexico; US$195 members other countries; US$350 nonmembers; US$390 nonmembers Canada, Mexico; US$410 nonmembers other countries; US$520 institutions; US$560 institutions Canada, Mexico; US$580 nonmembers other countries. **Remarks:** Accepts advertising. **URL:** http://www.the-aps.org/publications/ajpgi/index.htm. **Circ:** 1,650

255 ■ American Journal of Physiology - Heart and Circulatory Physiology
The American Physiological Society
9650 Rockville Pke.
Bethesda, MD 20814-3991
Phone: (301)634-7164
Fax: (301)634-7241
Publisher E-mail: webmaster@the-aps.org

Journal covering the experimental and theoretical studies of cardiovascular function. **Freq:** Monthly. **Key Personnel:** Alberto Nasjletti, Editor-in-Chief, alberto_nasjletti@nymc.edu. **ISSN:** 0363-6135. **Subscription Rates:** US$265 members; US$305 members Canada, Mexico; US$365 members other countries; US$5,646 nonmembers; US$685 nonmembers Canada, Mexico; US$745 nonmembers other countries; US$955 institutions; US$520 Canada and Mexico; US$1,055 nonmembers other countries. **Remarks:** Accepts advertising. **URL:** http://www.the-aps.org/publications/ajpheart/index.htm. **Circ:** 1,750

256 ■ American Journal of Physiology - Lung Cellular and Molecular Physiology
The American Physiological Society
9650 Rockville Pke.
Bethesda, MD 20814-3991
Phone: (301)634-7164
Fax: (301)634-7241
Publisher E-mail: webmaster@the-aps.org

Journal covering the molecular, cellular, and integrative aspects of normal and abnormal function of cells and components of the respiratory system. **Freq:** Monthly. **Key Personnel:** Michael A. Matthay, Editor-in-Chief, michael.matthay@ucsf.edu. **ISSN:** 1040-0605. **Subscription Rates:** US$130 members; US$170 members Canada, Mexico; US$210 members other countries; US$315 nonmembers; US$355 nonmembers Canada, Mexico; US$395 nonmembers other countries; US$470 institutions; US$510 institutions Canada, Mexico; US$550 nonmembers other countries. **Remarks:** Accepts advertising. **URL:** http://www.the-aps.org/publications/ajplung/index.htm. **Circ:** 1,700

257 ■ American Journal of Physiology - Renal Physiology
The American Physiological Society
9650 Rockville Pke.
Bethesda, MD 20814-3991
Phone: (301)634-7164
Fax: (301)634-7241
Publisher E-mail: webmaster@the-aps.org

Journal focusing on the studies of kidney, urinary tract, and their respective cells and vasculature. **Freq:** Monthly. **Key Personnel:** Thomas R. Kleyman, Editor-in-Chief, kleyman@dom.pitt.edu. **ISSN:** 0363-6127. **Subscription Rates:** US$160 members; US$200 members Canada, Mexico; US$220 members other countries; US$320 nonmembers; US$360 nonmembers Canada, Mexico; US$380 nonmembers other countries; US$480 institutions; US$520 institutions Canada, Mexico; US$540 nonmembers other countries. **Remarks:** Accepts advertising. **URL:** http://www.the-aps.org/publications/ajprenal/index.htm. **Circ:** 1,700

BURTONSVILLE

258 ■ Global Digital Business Review
Global Digital Business Association, Inc.
15319 Briarcliff Manor Way
Burtonsville, MD 20866

Journal covering digital business. **Key Personnel:** Dr. Narendra K. Rustagi, Editor, nkrustagi@gdba.us. **URL:** http://www.gdba.us/newJournal.html.

259 ■ Journal of Decision Science and Information Technology
Global Digital Business Association, Inc.
15319 Briarcliff Manor Way
Burtonsville, MD 20866

Journal covering decision science and information technology. **Key Personnel:** Dr. Daniel Okunbar, Editor-in-Chief, diokunbor@uncfsu.edu. **URL:** http://www.gdba.us/newJournal.html.

260 ■ Journal of Global Information Technology
Global Digital Business Association, Inc.
15319 Briarcliff Manor Way
Burtonsville, MD 20866

Journal covering global information technology. **Key Personnel:** Dr. Dinesh Kumar Sharma, Editor, dksharma@gdba.us. **URL:** http://www.gdba.us/newJournal.html.

261 ■ Journal of Innovative Education Strategies
Global Digital Business Association, Inc.
15319 Briarcliff Manor Way
Burtonsville, MD 20866

Journal covering innovative education strategies. **Key Personnel:** Dr. Narendra K. Rustagi, Editor, nkrustagi@gdba.us. **URL:** http://www.gdba.us/newJournal.html.

GLEN BURNIE

262 ■ WFBR-AM - 1590
159 8th Ave. NW
Glen Burnie, MD 21061

Format: Ethnic. **Owner:** Multicultural Radio Broadcasting, Inc., 449 Broadway, New York, NY 10013, (212)966-1059, Fax: (212)966-9580. **Key Personnel:** John Gabel, Contact, johng@mrbi.net. **URL:** http://www.mrbi.net/radiogroup.htm.

ROCKVILLE

263 ■ WLXE-AM - 1600
12216 Parklawn Dr.
Rockville, MD 20852

Format: Hispanic. **Owner:** Multicultural Radio Broadcasting, Inc., 449 Broadway, New York, NY

10013, (212)966-1059, Fax: (212)966-9580. **Key Personnel:** John Gabel, Contact, johng@mrbi.net. **URL:** http://www.mrbi.net/radiogroup.htm.

SILVER SPRING

264 ■ WWXT-FM - 92.7
8121 Georgia Ave., Ste. 1050
Silver Spring, MD 20910
Phone: (301)562-5800

Format: Sports. **Owner:** Red Zebra Broadcasting, at above address. **Operating Hours:** Continuous. **Key Personnel:** Allyson Butler, Mktg. Dir., butlera@redskins.com; Lewis Schreck, Sen. VP, Sales, schreckl@redskins.com. **URL:** http://www.redskinsradioespn.com/.

265 ■ WWXX-FM - 94.3
8121 Georgia Ave., Ste. 1050
Silver Spring, MD 20910
Phone: (301)562-5800

Format: Sports. **Owner:** Red Zebra Broadcasting, at above address. **Operating Hours:** Continuous. **Key Personnel:** Ira Weintraub, Web Ed., weintraubi@redskins.com; Allyson Butler, Mktg. Dir., butlera@redskins.com; Lewis Schreck, Sen. VP, Sales, schreckl@redskins.com. **URL:** http://www.triplexespnradio.com/.

MASSACHUSETTS

CHICOPEE

266 ■ Chicopee Register
Turley Publications
333 Front St.
Chicopee, MA 01013
Phone: (413)592-3599
Fax: (413)592-3568

Community newspaper serving city of Chicopee. **Founded:** Sept. 10, 1998. **Freq:** Weekly (Thurs.). **Key Personnel:** Kathy Mitchell, Editor, kmitchell@turley.com. **Remarks:** Accepts advertising. **URL:** http://www.turley.com/chicopee/; http://www.turley.com/publications/. **Circ:** 13,000

267 ■ The Holyoke Sun
Turley Publications
333 Front St.
Chicopee, MA 01013
Phone: (413)612-2310
Fax: (413)592-3568

Community newspaper serving Greater Holyoke. **Founded:** May 1995. **Freq:** Weekly (Fri.). **Key Personnel:** Aimee Henderson, Editor, ahenderson@turley.com. **Remarks:** Accepts advertising. **URL:** http://www.turley.com/holyoke/; http://www.turley.com/publications/. **Circ:** 7,000

FEEDING HILLS

268 ■ The Agawam Advertiser News
Turley Publications
23 Southwick St.
Feeding Hills, MA 01030
Phone: (413)786-7747
Fax: (413)786-8457

Community newspaper serving Agawam and Feeding Hills. **Founded:** 1965. **Freq:** Weekly (Thurs.). **Key Personnel:** Rick Sardella, General Mgr. **Remarks:** Accepts advertising. **URL:** http://trl.live.mediaspanonline.com/agawam/; http://www.turley.com/publications/. **Circ:** 6,000

269 ■ Southwick Suffield News
Turley Publications
23 Southwick St.
Feeding Hills, MA 01030
Phone: (413)786-7747
Fax: (413)786-8457

Community newspaper serving townspeople of Southwick and Suffield. **Founded:** 1984. **Freq:** Weekly (Fri.). **Key Personnel:** Rich Worth, Editor, rworth@turley.com. **Remarks:** Accepts advertising. **URL:** http://www.turley.com/southwick/; http://www.turley.com/publications/. **Circ:** 6,000

270 ■ WLCQ-FM - 99.7
522 Springfield St.
Feeding Hills, MA 01030
Free: (877)843-7997
E-mail: contact@wlcq.com

Format: Contemporary Christian. **URL:** http://www.wlcq.com/.

GREENFIELD

271 ■ WIZZ-AM - 1520
PO Box 983
Greenfield, MA 01302
Phone: (413)774-5757
Free: (866)844-WIZZ
E-mail: info@wizzradio.com

Format: Big Band/Nostalgia. **Wattage:** 10,000. **Ad Rates:** Advertising accepted; rates available upon request. **URL:** http://www.wizzradio.com/.

272 ■ WLPV-FM - 107.9
450 Davis St.
Greenfield, MA 01301
Phone: (413)773-7300

Format: Contemporary Christian. **Operating Hours:** Continuous. **Wattage:** 100. **URL:** http://www.wlpv1079online.com/.

LOWELL

273 ■ WUML-FM - 91.5
1 University Ave.
Lowell, MA 01854
Phone: (978)934-4969

Format: Information. **Key Personnel:** Andy Bass, General Mgr., gm@wuml.org; Chris Gilroy, Music Dir., md@wuml.org; David Crepeault, Program Dir., pd@wuml.org. **URL:** http://www.wuml.org/.

NEW BEDFORD

274 ■ WFHL-FM - 88.1
71 William St.
New Bedford, MA 02740
Phone: (508)991-7600
E-mail: radio@radiowfhl.com

Format: Religious. **Key Personnel:** Dr. Manuel F.V. Pereira, President. **URL:** http://www.radiowfhl.com/.

NORTH ADAMS

275 ■ WUPE-AM - 1110
466 Curran Hwy.
North Adams, MA 01247
Phone: (413)663-6567
Fax: (413)662-2143

Format: Oldies. **Owner:** Vox Communications, 211 Jason St., Pittsfield, MA 01201-5907, (413)499-3333, Fax: (413)442-1590. **Key Personnel:** Pete Berry, Pres./Market Mgr.; Dick Savage, VP of Sales; Stew Schantz, Operations Mgr. **URL:** http://www.wupe.com/home.php.

NORTH DARTMOUTH

276 ■ International Journal of Indian Culture and Business Management
Inderscience Publishers
c/o Prof. Angappa Gunasekaran, Ed.-in-Ch.
University of Massachusetts-Dartmouth
Department of Decision & Information Sciences, Charlton College of Business
285 Old Westport Rd.
North Dartmouth, MA 02747-2300
Publication E-mail: ijicbm@inderscience.com
Publisher E-mail: editor@inderscience.com

Journal covering field of new developments in Indian culture and their implications on business. **Founded:** 2007. **Freq:** 4/yr. **Key Personnel:** Prof. Angappa Gunasekaran, Editor-in-Chief, agunasekaran@umassd.edu. **ISSN:** 1753-0806. **Subscription Rates:** EUR470 individuals includes surface mail, print only; EUR640 individuals print & online. **URL:** http://www.inderscience.com/browse/index.php?journalCODE=ijicbm.

277 ■ International Journal of Procurement Management
Inderscience Publishers
c/o Prof. Angappa Gunasekaran, Ed.-in-Ch.
University of Massachusetts-Dartmouth
Department of Decision & Information Sciences, Charlton College of Business
285 Old Westport Rd.
North Dartmouth, MA 02747-2300
Publication E-mail: ijpm@inderscience.com
Publisher E-mail: editor@inderscience.com

Journal covering development of procurement resources. **Founded:** 2007. **Freq:** 4/yr. **Key Personnel:** Prof. Angappa Gunasekaran, Editor-in-Chief, agunasekaran@umassd.edu. **ISSN:** 1753-8432. **Subscription Rates:** EUR470 individuals includes surface mail, print only; EUR640 individuals print and online. **URL:** http://www.inderscience.com/browse/index.php?journalCODE=ijpm.

PALMER

278 ■ Tantasqua Town Common
Turley Publications
24 Water St.
Palmer, MA 01069
Fax: (413)289-1977
Free: (800)824-6548

Community newspaper serving Tantasqua. **Freq:** Weekly. **Key Personnel:** Justine Murphy, Editor, jmurphy@turley.com; Tim Kane, Exec. Ed., tkane@turley.com; Dave Anderson, Advertising Mgr., danderson@turley.com. **Remarks:** Accepts advertising. **URL:** http://www.thetantasquatowncommon.com/; http://www.turley.com/publications/. **Circ:** 10,000

PEABODY

279 ■ North Shore Golf
Suburban Publishing Corp.
10 First Ave.
Peabody, MA 01960
Phone: (978)532-5880
Fax: (978)532-4250
Free: (800)221-2078
Publisher E-mail: info@suburbanpublishing.com

Magazine featuring golf in North Shore. **Freq:** Bimonthly. **Trim Size:** 8.25 x 10.625. **Key Personnel:** Richard Ayer, Publisher, rayer@northshoremassgolf.com. **Subscription Rates:** US$12 individuals. **Remarks:** Accepts advertising. **URL:** http://northshoremassgolf.com/. . **Ad Rates:** BW: $1,995. **Circ:** (Not Reported)

280 ■ Suburban Real Estate News
Suburban Publishing Corp.
10 First Ave.
Peabody, MA 01960
Phone: (978)532-5880
Fax: (978)532-4250
Free: (800)221-2078
Publisher E-mail: info@suburbanpublishing.com

Newspaper featuring real estate news. **Freq:** Quarterly. **Key Personnel:** Jim Downey, Circulation Mgr., jdowney@suburbanpublishing.com. **URL:** http://www.suburbanrealestatenews.com/.

PITTSFIELD

281 ■ WBRK-FM - 101.7
100 North St.
Pittsfield, MA 01201
Phone: (413)442-1553
E-mail: wbrk@wbrk.com

Format: Adult Contemporary. **URL:** http://www.wbrk.com/.

Circulation: ★ = ABC; △ = BPA; ♦ = CAC; • = CCAB; ❑ = VAC; ⊕ = PO Statement; ‡ = Publisher's Report; Boldface figures = sworn; Light figures = estimated.

SOUTH HADLEY

282 ■ Town Reminder
Turley Publications
138 College St., Ste. 2
South Hadley, MA 01075
Phone: (413)536-5333
Fax: (413)536-5334
Publication E-mail: townreminder@turley.com

Community newspaper serving South Hadley, Granby, Chicopee, Holyoke, and Ludlow. **Founded:** 1967. **Freq:** Weekly (Fri.). **Key Personnel:** Aimee Henderson, Editor; Jamie Joslyn, Office Mgr. **Remarks:** Accepts advertising. **URL:** http://www.turley.com/tr/; http://www.turley.com/publications/. **Circ:** 11,800

VINEYARD HAVEN

283 ■ WVVY-FM - 93.7
PO Box 1989
Vineyard Haven, MA 02568

Format: News; Talk. **URL:** http://www.wvvy.org/.

WARE

284 ■ Quaboag Current
Turley Publications
80 Main St.
Ware, MA 01082
Phone: (413)967-3505
Fax: (413)967-6009

Community newspaper serving the towns of West Brookfield, East Brookfield, North Brookfield, Warren, West Warren, and New Braintree. **Freq:** Weekly. **Key Personnel:** Douglas Farmer, News Ed., dfarmer@turley.com; Tim Kane, Exec. Ed., tkane@turley.com; Dave Anderson, Advertising Mgr., danderson@turley.com. **Remarks:** Accepts advertising. **URL:** http://www.quaboagcurrent.com/. **Circ:** 8,000

WILBRAHAM

285 ■ Wilbraham-Hampden Times
Turley Publications
2341 Boston Rd.
Wilbraham, MA 01095
Phone: (413)682-0007
Fax: (413)682-0013

Community newspaper serving Wilbraham and Hampden. **Freq:** Weekly (Thurs.). **Key Personnel:** Charlie Bennett, Editor, cbennett@turley.com. **Remarks:** Accepts advertising. **URL:** http://www.turley.com/wht/; http://www.turley.com/publications/. **Circ:** 9,338

WOBURN

286 ■ WAZN-AM - 1470
500 W Cummings Pk.
Woburn, MA 01801
Phone: (212)431-2778

Format: Ethnic. **Owner:** Multicultural Radio Broadcasting, Inc., 449 Broadway, New York, NY 10013, (212)966-1059, Fax: (212)966-9580. **Key Personnel:** Jim Glogowski, Contact, jimg@mrbi.net. **URL:** http://www.mrbi.net/radiogroup.htm.

MICHIGAN

ALPENA

287 ■ WRGZ-FM - 96.7
123 Prentiss St.
Alpena, MI 49707
Phone: (989)354-8400
Fax: (989)354-3436

Format: Information; News. **Key Personnel:** Mike Centala, General Mgr., mcentala@watz.com. **URL:** http://www.watz.com/.

288 ■ WWTH-FM - 100.7
1491 M-32 W
Alpena, MI 49707
Phone: (989)354-4611
Fax: (989)354-4104
Free: (800)743-6424
E-mail: info@1007thundercountry.com

Format: Country. **Wattage:** 20,500. **Ad Rates:** Advertising accepted; rates available upon request. **URL:** http://www.1007thundercountry.com/.

BIG RAPIDS

289 ■ WWBR-FM - 100.9
18720 16 Mile Rd.
Big Rapids, MI 49307
Phone: (231)796-7000

Format: Country. **Owner:** Mentor Partners, Inc., at above address. **Key Personnel:** Jeff Scarpelli, General Mgr. **URL:** http://www.wwbr.com/.

DETROIT

290 ■ American Lift & Handlers
KHL Group
c/o John Wyatt, Ed.
1635 Merrick St.
Detroit, MI 48208
Phone: (313)894-2303
Publisher E-mail: info@khl.com

Magazine for telehandler and aerial work platform industry in North America. **Founded:** Oct. 2006. **Freq:** 6/yr. **Key Personnel:** John Wyatt, Editor, john.wyatt@khl.com. **Subscription Rates:** US$180 individuals; US$325 two years; US$460 individuals 3 years. **Remarks:** Accepts display advertising. **URL:** http://www.khl.com/magazines/information.asp?magazineid=15. . **Ad Rates:** 4C: $1,280. **Circ:** 12,601

291 ■ Creativity
Crain Communications Inc.
1155 Gratiot Ave.
Detroit, MI 48207-2997
Phone: (313)446-6000

Magazine featuring insight and information for creative leaders involved in all aspects of advertising and design. **Freq:** Monthly. **Trim Size:** 10 x 12. **Key Personnel:** Teressa Iezzi, Editor, phone (212)210-0793, tiezzi@crain.com. **Remarks:** Accepts advertising. **URL:** http://www.crain.com/po.html; http://creativity-online.com/. **Circ:** 33,000

292 ■ WDRJ-AM - 1440
2994 E Grand Blvd.
Detroit, MI 48202
Phone: (313)871-1440
Fax: (313)871-6088

Format: Contemporary Christian. **URL:** http://www.1440wdrj.com/.

FLINT

WJIV-FM - See Albany, New York

GAGETOWN

293 ■ WCTP-FM - 88.5
4330 Farver Rd.
Gagetown, MI 48735
Phone: (989)315-8043
Fax: (989)872-3700
E-mail: info@wctpradiofm.com

Format: Contemporary Christian; Gospel. **Owner:** Plonta Broadcasting, at above address. **URL:** http://www.wctpradiofm.com/.

LUPTON

294 ■ WMSD-FM - 90.9
2906 E Heath
Lupton, MI 48635
Phone: (989)473-4616
Free: (866)473-9673

Format: Gospel. **Operating Hours:** Continuous. **Ad Rates:** Noncommercial. **URL:** http://www.wmsdradio.com/.

MARQUETTE

295 ■ WRUP-FM - 98.3
2025 U.S. 41 W
Marquette, MI 49855
Phone: (906)228-6800
Fax: (906)228-8128

Format: Classic Rock. **Owner:** Great Lakes Radio, Inc., at above address. **URL:** http://www.wrup.com/.

MARSHALL

296 ■ WMLY-FM - 93.1
PO Box 1018
Laguna Beach, CA 92652
Fax: (469)241-6795
Free: (800)639-5433

Format: Religious. **Owner:** New Life Ministries, at above address. **Key Personnel:** Stephen Arterburn, Founder/Chm. **URL:** http://www.newlife.com/Radio/findradio.asp?id=MI&t=Michigan.

NEWBERRY

297 ■ WMJT-FM - 96.7
210 W John St.
PO Box 486
Newberry, MI 49868
Phone: (906)293-1400
Fax: (906)293-5161

Format: Top 40; Classical. **Key Personnel:** Kent Smith, General Mgr., kent@radioeagle.com; Teri Petrie, Business Mgr., teri@radioeagle.com; Mike Reling, Operations Mgr., mike@radioeagle.com. **Ad Rates:** Advertising accepted; rates available upon request. **URL:** http://www.radioeagle.com/newberry/index.php.

SOUTHFIELD

KJMU-AM - See Sand Spring, Oklahoma

KTUV-AM - See Little Rock, Arkansas

298 ■ WTOR-AM - 770
21700 Northwestern Hwy.
Tower 14, Ste. 1190
Southfield, MI 48075
Phone: (248)557-3500
Fax: (248)557-2950

Format: Ethnic. **Owner:** Birach Broadcasting Corporation, at above address. **Key Personnel:** Mr. Sima Birach, Operations Mgr., phone (716)754-9514, fax (716)754-9516, sima@birach.com. **URL:** http://www.birach.com/wtor.html.

SPRING ARBOR

299 ■ WJKN-FM - 89.3
106 E Main St.
Spring Arbor, MI 49283
Free: (800)968-9103

Format: Contemporary Christian. **Owner:** Spring Arbor University, at above address. **URL:** http://www.arbor.edu/standarddetail.aspx?id=20219.

WILLIAMSTON

300 ■ WJOJ-FM - 89.7
148 E Grand River
PO Box 388
Williamston, MI 48895
Free: (888)887-7139
E-mail: 411@smile.fm

Format: Contemporary Christian. **Owner:** Superior Communications, at above address. **URL:** http://www.positivehits.com/pages/index.cfm.

MINNESOTA

BRAINERD

301 ■ WWWI-FM - 95.9
PO Box 783
Brainerd, MN 56401
Phone: (218)828-9994
E-mail: info@3wiradio.com

Format: News; Talk. **URL:** http://www.3wiradio.com/.

BROOKLYN CENTER

302 ■ KPNP-AM - 1600
6500 Brooklyn Blvd., Ste. 206
Brooklyn Center, MN 55429
Phone: (612)267-2999
Fax: (763)585-1600
E-mail: info@kpnp1600.com

Format: Ethnic. **URL:** http://www.radiohmong.net/.

CHASKA

303 ■ KQSP-AM - 1530
1107 Hazeltine Blvd., Ste. 301
Chaska, MN 55318
Phone: (952)368-0070
Fax: (952)361-3583

Format: Hispanic. **Owner:** Broadcast One, Inc., at above address. **Key Personnel:** Eddie Cruz, Program Dir., mozkito@broadcastone.us. **URL:** http://www.lapicosa.us/enter.htm.

DULUTH

304 ■ WJRF-FM - 89.5
4604 Airpark Blvd.
Duluth, MN 55811-5751
Phone: (218)722-3017
Free: (866)722-3017
E-mail: airstaff@refugeradio.com

Format: Contemporary Christian. **URL:** http://www.refugeradio.com/.

FARIBAULT

305 ■ KRFO-FM - 104.9
601 Central Ave.
Faribault, MN 55021
Phone: (507)334-0061
Fax: (507)334-7057

Format: Country. **Owner:** Cumulus Media, 3280 Peachtree Rd. NW, Ste. 2300, Atlanta, GA 30305, (404)949-0700, Fax: (404)949-0740. **Key Personnel:** Gary Foss, Market Mgr., gary.foss@cumulus.com. **URL:** http://www.krforadio.com/.

HIBBING

306 ■ WUSZ-FM - 99.9
807 W 37th St.
Hibbing, MN 55746
Phone: (218)263-7531
Fax: (218)263-6112
Free: (866)873-0965
E-mail: info@radiousa.com

Format: Country. **Owner:** Midwest Communications, Inc., 904 Grand Ave., Wausau, WI 54403, (715)842-1437, Fax: (715)842-7061. **URL:** http://www.radiousa.com/.

HINCKLEY

307 ■ WYSG-FM - 96.3
PO Box 1018
Laguna Beach, CA 92652
Fax: (469)241-6795
Free: (800)639-5433

Format: Religious. **Owner:** New Life Ministries, at above address. **Key Personnel:** Stephen Arterburn, Founder/Chm. **URL:** http://www.newlife.com/Radio/findradio.asp?id=MN&t=Minnesota.

MANKATO

308 ■ KRRW-FM - 101.5
PO Box 1420
Mankato, MN 56002

Format: Country. **URL:** http://www.katoinfo.com/linder_radio/krrw/index.php.

MINNEAPOLIS

309 ■ KMNV-AM - 1400
1516 E Lake St., Ste. 200
Minneapolis, MN 55407
Phone: (612)729-5900
E-mail: lainvasora1400@lainvasora1400.com

Format: Hispanic. **URL:** http://www.lainvasora1400.com/.

NORTH MANKATO

310 ■ KQYK-FM - 95.7
1807 Lee Blvd.
North Mankato, MN 56003
Phone: (507)344-0957
E-mail: 957theblaze@gmail.com

Format: Urban Contemporary. **Key Personnel:** Jen Jones, Dir. of Sales, jjones@krbi.threeeagles.com; Jon Schulte, Local Sales Mgr., jschulte@kysm.threeeagles.com; Annie Nicolay, Account Exec., anicolay@kysm.threeeagles.com. **URL:** http://www.957theblaze.com/home/.

OWATONNA

311 ■ KOWZ-FM - 100.9
255 Cedardale
Owatonna, MN 55060
Phone: (507)444-9224

Format: Full Service. **URL:** http://www.kowzonline.com/.

RAMSEY

312 ■ KLCI-FM - 106.1
14443 Armstrong Blvd. NW
Ramsey, MN 55303
Phone: (763)389-1300

Format: Country. **Key Personnel:** Real Howard Johnson, Program Dir., realhojo@realhojo.com. **URL:** http://www.dothebob.com/.

WORTHINGTON

313 ■ KWOA-FM - 95.1
28779 County Hwy. 35
Worthington, MN 56187
Phone: (507)376-6165
Fax: (507)376-5071
E-mail: contactus@951theeagle.com

Format: Adult Contemporary. **Owner:** Three Eagles Communications Company, 3800 Cornhusker Hwy., Lincoln, NE 68504, (402)466-1234, Fax: (402)467-4095. **Key Personnel:** Larry Schultz, General Mgr. **Ad Rates:** Advertising accepted; rates available upon request. **URL:** http://www.951theeagle.com/.

MISSISSIPPI

GREENVILLE

314 ■ WLRK-FM - 91.5
2351 Sunset Blvd., Ste. 170-218
Rocklin, CA 95765
Phone: (916)251-1600
Fax: (916)251-1650

Format: Contemporary Christian. **Owner:** Educational Media Foundation, at above address. **Key Personnel:** Phillip O'Bryan, Regional Mgr., phone (601)992-6988. **URL:** http://www.klove.com/Music/StationList.aspx.

GULFPORT

315 ■ WUJM-FM - 96.7
1909 E Pass Rd., Ste. D-11
Gulfport, MS 39507
Phone: (228)388-2001

Format: Country. **Key Personnel:** Kyle Curley, Program Dir., kcurley@cableone.net; Kevin Harris, Promotions Dir., specialk@datasync.com. **URL:** http://www.967hankfm.net/home.php.

HATTIESBURG

316 ■ International Journal of Intercultural Information Management
Inderscience Publishers
c/o Prof. Chang T. Hsieh, Ed.-in-Ch.
University of Southern Mississippi
118 College Dr., No. 5178
Hattiesburg, MS 39406
Publisher E-mail: editor@inderscience.com

Journal covering intercultural issues related to information management. **Founded:** 2007. **Freq:** 4/yr. **Key Personnel:** Prof. Chang T. Hsieh, Editor-in-Chief, chang-tseh.hsieh@usm.edu. **ISSN:** 1750-0028. **Subscription Rates:** EUR470 individuals includes surface mail, print only; EUR640 individuals print and online. **URL:** http://www.inderscience.com/browse/index.php?journalCODE=ijiim.

317 ■ WLVZ-FM - 107.1
2351 Sunset Blvd., Ste. 170-218
Rocklin, CA 95765
Phone: (916)251-1600
Fax: (916)251-1650

Format: Contemporary Christian. **Owner:** Educational Media Foundation, at above address. **Key Personnel:** Phillip O'Bryan, Regional Mgr., phone (601)992-6988. **URL:** http://www.klove.com/Music/StationList.aspx.

JACKSON

318 ■ WRBJ-FM - 97.7
745 N State St.
Jackson, MS 39202
Phone: (601)974-5700
Fax: (601)974-5711

Format: Urban Contemporary. **Owner:** Roberts Broadcasting Company, 1408 N Kingshighway, Ste. 300, Saint Louis, MO 63113-1420. **Key Personnel:** Keith Smith, General Mgr., phone (601)974-5701, fax (601)974-5719, keiths@roberts-companies.com. **URL:** http://www.cw34jackson.com/news/6082141.html.

KILN

319 ■ WQRZ-FM - 103.5
PO Box 1145
Kiln, MS 39556-1145
Phone: (228)463-1035
E-mail: wqrznews@aol.com

Format: News. **Key Personnel:** Brice L. Phillips, President. **URL:** http://www.wqrz.org/mambo/.

TUPELO

KAKO-FM - See Ada, Oklahoma
KDVI-FM - See Devils Lake, North Dakota
KQPD-FM - See Ardmore, Oklahoma
KXRT-FM - See Idabel, Oklahoma
WAEF-FM - See Cordele, Georgia
WAXU-FM - See Troy, Alabama
WBJV-FM - See Steubenville, Ohio
WEFI-FM - See Effingham, Illinois
WJJE-FM - See Delaware, Ohio
WRAE-FM - See Raeford, North Carolina
WSLE-FM - See Salem, Illinois
WWLO-FM - See Lowell, Indiana

MISSOURI

CAPE GIRARDEAU

320 ■ KSEF-FM - 88.9
One University Plz., MS 0300
Cape Girardeau, MO 63701
Phone: (573)651-5070
Free: (888)651-5070

Format: Public Radio. **Owner:** Southeast Missouri State University, One University Plz., Cape Girardeau, MO 63701, (573)651-2000. **Key Person-

Circulation: ★ = ABC; △ = BPA; ♦ = CAC; • = CCAB; ❑ = VAC; ⊕ = PO Statement; ‡ = Publisher's Report; Boldface figures = sworn; Light figures = estimated.

nel: Dan Woods, General Mgr., djwoods@semo.edu; Jason Brown, Operations Dir., jbrown@semo.edu; Allen Lane, Chief Engineer, aelane@semo.edu. **URL:** http://www.semo.edu/sepr/.

JEFFERSON CITY

321 ■ KZJF-FM - 104.1
3605 Country Club Dr.
Jefferson City, MO 65109
Phone: (573)893-5100
Fax: (573)893-8330

Format: Country. **Key Personnel:** John Walker, Market Mgr.; Brian Pollitt, Program Dir.; Jacqueline Cauthon, Promotions Dir. **Ad Rates:** Advertising accepted; rates available upon request. **URL:** http://www.jeffcountry.com/.

JOPLIN

322 ■ KJML-FM - 107.1
2510 W 20th St.
Joplin, MO 64804
Phone: (417)781-1313
Fax: (417)781-1316

Format: Album-Oriented Rock (AOR). **Key Personnel:** Warren McDonald, Operations Mgr., wmcdonald@crjoplin.com. **Wattage:** 6000. **URL:** http://www.rock105kjml.com/.

323 ■ KZRG-AM - 1310
2702 E 32nd St.
Joplin, MO 64804
Phone: (417)624-1025

Format: News; Talk. **Owner:** Zimmer Radio, Inc., at above address. **Key Personnel:** Rob Hough, Program Dir., robh@zrgmail.com. **URL:** http://www.1310kzrg.com/.

OSAGE BEACH

324 ■ KZWV-FM - 101.9
1081 Osage Beach Rd.
Osage Beach, MO 65065
Fax: (573)746-7874

Format: News. **Key Personnel:** John Caran, Gen. Mgr./Dir. of Sales, jcaran@1019thewave.com; Steve Richards, Programming/Production Dir. **URL:** http://www.1019thewave.com/.

POPLAR BLUFF

325 ■ KYHO-FM - 106.9
PO Box 1018
Laguna Beach, CA 92652
Fax: (469)241-6795
Free: (800)639-5433

Format: Religious. **Owner:** New Life Ministries, at above address. **Key Personnel:** Stephen Arterburn, Founder/Chm. **URL:** http://www.newlife.com.

ROLLA

326 ■ International Journal of Difference Equations
Research India Publications
Department of Mathematics
University of Missouri-Rolla
106 Rolla Bldg.
Rolla, MO 65409-0200
Phone: (573)341-4741
Publisher E-mail: info@ripublication.com

Journal covering areas related to difference equations and dynamical systems. **Freq:** 2/yr. **Key Personnel:** Martin Bohner, Editor-in-Chief, bohner@umr.edu. **ISSN:** 0973-6069. **Subscription Rates:** US$180 libraries; US$180 institutions; US$140 individuals. **URL:** http://www.ripublication.com/ijde.htm.

SAINT CHARLES

327 ■ KHOJ-AM - 1460
2130 Wade Hampton Blvd.
Greenville, SC 29615

Format: Religious. **Owner:** Family Life Center International, at above address. **URL:** http://www.familylifecenter.net/faithfamilystations.asp.

SAINT JOSEPH

328 ■ KSRD-FM - 91.9
1212 Faraon St.
Saint Joseph, MO 64501
Phone: (816)233-5773
Fax: (816)233-5777
Free: (800)681-5773
E-mail: info@ksrdradio.com

Format: Religious. **Owner:** Horizon Broadcast Network, 123-1st St. SW, Ste. 801, Minot, ND 58701, (701)858-1940. **Key Personnel:** Brian KC Jones, General Mgr.; Matt McNeilly, Operations Mgr. **URL:** http://www.ksrdradio.com/.

SAINT LOUIS

329 ■ WHOJ-FM - 91.9
4424 Hampton Ave.
Saint Louis, MO 63109
Phone: (314)752-7000
Free: (877)305-1234

Format: Contemporary Christian. **Owner:** Covenant Network, at above address. **URL:** http://www.covenantnet.net/.

VICKSBURG

330 ■ WVBG-AM - 1490
2221 E Lamar Blvd., Ste. 300
Arlington, TX 76006
Phone: (817)695-1820
Fax: (817)695-0295
Free: (800)288-9227

Format: Information. **Owner:** Midnight Trucking Radio Network, at above address. **URL:** http://www.midnighttrucking.com/sectional.asp?id=24651.

MONTANA

ARLEE

331 ■ KJFT-FM - 90.3
PO Box 391
Twin Falls, ID 83303
Fax: (208)736-1958
Free: (800)357-4226

Format: Religious. **Owner:** CSN International, at above address. **URL:** http://www.csnradio.com/stationsMT.php.

BAKER

332 ■ KFLN-AM - 960
PO Box 790
Baker, MT 59313
Phone: (406)778-3371
E-mail: kfln@midrivers.com

Format: Country; News. **Owner:** Newell Broadcasting Corporation, at above address. **Key Personnel:** Devon Banister, Sales Mgr. **Wattage:** 5000. **URL:** http://www.kfln.info/.

333 ■ KJJM-FM - 100.5
PO Box 790
Baker, MT 59313
Phone: (406)778-3371

Format: Classic Rock; News. **URL:** http://www.kfln.info/.

BILLINGS

334 ■ KLMT-FM - 89.3
PO Box 21888
Carson City, NV 89721
Phone: (775)883-5647
Free: (800)541-5647

Format: Contemporary Christian; Religious. **Owner:** Pilgrim Radio, at above address. **URL:** http://www.pilgrimradio.com/FM_Frequency_List.html.

BLACK EAGLE

335 ■ KFRW-FM - 102.3
290 Hegenberger Rd.
Oakland, CA 94621

Format: Religious. **Owner:** Family Stations Inc., at above address. **URL:** http://www.familyradio.com.

BUTTE

336 ■ KFRD-FM - 88.9
290 Hegenberger Rd.
Oakland, CA 94621

Format: Religious. **Owner:** Family Stations, Inc., at above address. **URL:** http://www.familyradio.com/.

MILES CITY

337 ■ KYUS-FM - 92.3
508 Main St.
Miles City, MT 59301
Phone: (406)234-5626
Fax: (406)874-7000
Free: (888)535-1050

Format: Adult Contemporary. **Owner:** Marks Radio Group, at above address. **Operating Hours:** Continuous. **Key Personnel:** Terry Virag, General Mgr., terry@radiomontana.net; Charice Virag, Traffic Dir., traffic@radiomontana.net. **Wattage:** 100,000. **URL:** http://www.radiomontana.net/kyus.htm.

MISSOULA

338 ■ KDTR-FM - 103.3
2425 W Central Ave., Ste. 203
Missoula, MT 59801
Phone: (406)721-6800

Format: Adult Album Alternative. **Owner:** Simmons Media Group, 515 South 700 E, Ste. 1C, Salt Lake City, UT 84102, (801)524-2600. **URL:** http://trail1033.com/.

339 ■ KHDV-FM - 107.9
725 Strand Ave.
Missoula, MT 59801
Phone: (406)542-1025
Fax: (406)721-1036
E-mail: info@1079valleyfm.com

Format: Country. **Owner:** Mountain Broadcasting, at above address. **Key Personnel:** Chris Brooks, Contact, phone (406)541-1079. **URL:** http://www.moclub.com/valley.php.

340 ■ KMPT-AM - 930
3250 S Reserve, Ste. 200
Missoula, MT 59801
Phone: (406)728-9300

Format: Talk. **Owner:** Gap West Broadcasting, at above address. **URL:** http://www.klcy930.com/main.php.

341 ■ KXGZ-FM - 101.5
3250 S Reserve, Ste. 200
Missoula, MT 59801
E-mail: grizcountry1015@hotmail.com

Format: Country. **URL:** http://www.grizcountry1015.com/main.php.

342 ■ KYJK-FM - 105.9
2425 W Central Ave.
Missoula, MT 59806
Phone: (406)721-6800

Format: Adult Contemporary. **Owner:** Simmons Media Group, 515 S 700 E, Ste. 1C, Salt Lake City, UT 84102, (801)524-2600. **URL:** http://www.1059jackfm.net/.

POLSON

343 ■ KKMT-FM - 99.7
36581 N Reservoir Rd.
Polson, MT 59860
Phone: (406)883-5255
Fax: (406)883-4441

Format: Adult Contemporary. **Owner:** Anderson Radio Broadcasting, Inc., at above address. **URL:** http://www.star99hits.com/.

KQRK-FM - See Ronan

RONAN

344 ■ KQRK-FM - 92.3
36581 N Reservoir Rd.
Polson, MT 59860
Phone: (406)883-9200
Fax: (406)883-4441

Format: Album-Oriented Rock (AOR). **Owner:** Anderson Radio Broadcasting, at above address. **URL:** http://www.power92rocks.com.

NEBRASKA

GRAND ISLAND

345 ■ KKJK-FM - 103.1
3205 W North Front St.
Grand Island, NE 68803
Phone: (308)381-1430
Free: (888)505-1031

Format: Album-Oriented Rock (AOR); Classic Rock. **URL:** http://www.thunder1031.com/.

HASTINGS

346 ■ KICS-AM - 1550
PO Box 726
Hastings, NE 68902
Phone: (402)462-5101

Format: Sports. **Owner:** Platte River Radio Inc., PO Box 130, Kearney, NE 68848-0130, (308)236-9900, Fax: (308)234-6781. **URL:** http://www.espnsuperstation.com/.

NEVADA

CARSON CITY

347 ■ KCMY-AM - 1300
1960 Idaho St.
Carson City, NV 89701
Phone: (775)884-8000

Format: Country. **Owner:** Evans Broadcast Company, at above address. **URL:** http://www.mycountryam1300.com/.

KLMT-FM - See Billings, Montana

ELKO

348 ■ KOYT-FM - 94.5
1750 Manzanita Dr., Ste. 1
Elko, NV 89801
Phone: (775)777-1196
Fax: (775)777-9587

Format: Album-Oriented Rock (AOR); Classic Rock. **Owner:** Ruby Radio Corporation, at above address. **Key Personnel:** Ken Sutherland, President, ken@rubyradio.fm; Alene Sutherland, Vice President, alene@rubyradio.fm. **URL:** http://www.coyote.fm/.

RENO

349 ■ KJFK-AM - 1230
961 Matley Ln., Ste. 120
Reno, NV 89502
Phone: (775)825-7529
Fax: (775)825-3183
E-mail: info@1230kjfk.com

Format: Talk. **Owner:** Americom Broadcasting, at above address. **URL:** http://www.1230kjfk.com/.

350 ■ KLRH-FM - 88.3
PO Box 779002
Rocklin, CA 95677-9972
Free: (800)434-8400

Format: Contemporary Christian. **Owner:** Educational Media Foundation, at above address. **Key Personnel:** Dan Beck, Contact, phone (602)997-4434. **URL:** http://www.klove.com/Music/StationList.aspx.

WINNEMUCCA

351 ■ KKTT-FM - 97.9
PO Box 1018
Laguna Beach, CA 92652
Fax: (469)241-6795
Free: (800)639-5433

Format: Religious. **Owner:** New Life Ministries, at above address. **Key Personnel:** Stephen Arterburn, Founder/Chm. **URL:** http://www.newlife.com.

NEW HAMPSHIRE

CONCORD

WEVJ-FM - See Jackson

HOOKSETT

352 ■ WFNQ-FM - 106.3
11 Kimball Dr., Ste. 114
Hooksett, NH 03106
Phone: (603)889-1063
Fax: (603)882-0688
Free: (877)370-1063

Format: Classic Rock. **Owner:** Nassau Broadcasting Partners, L.P., 619 Alexander Rd., Princeton, NJ 08540, (609)452-9696. **Key Personnel:** A.J. Dukette, Operations Mgr., adukette@nassaubroadcasting.com; Sarah Sullivan, Program Dir., sarah.sullivan@nassaubroadcasting.com; Ryan Muir, Promotions Dir., rmuir@nassaubroadcasting.com. **URL:** http://www.1063frankfm.com/.

JACKSON

353 ■ WEVJ-FM - 99.5
207 N Main St.
Concord, NH 03301-5003
Phone: (603)228-8910
Fax: (603)224-6052
Free: (800)639-4131

Format: Public Radio. **Owner:** New Hampshire Public Radio, at above address. **URL:** http://www.nhpr.org/.

NEW JERSEY

CALIFON

354 ■ Quilter's Home
CK Media LLC
13 Pickle Rd.
Califon, NJ 07830
Publisher E-mail: info@ckmedia.com

Magazine featuring information about quilting. **Freq:** 6/yr. **Subscription Rates:** US$35.94 individuals. **URL:** http://www.quiltershomemag.com/; http://www.ckmedia.com/quilting/quiltershome.php.

CHERRY HILL

355 ■ WKVP-FM - 89.5
2351 Sunset Blvd., Ste. 170-218
Rocklin, CA 95765
Phone: (916)251-1600
Fax: (916)251-1650

Format: Contemporary Christian. **Owner:** Educational Media Foundation, at above address. **Key Personnel:** Glenn Goodwin, Regional Mgr., phone (913)663-9950. **URL:** http://www.klove.com/Music/StationList.aspx.

HOBOKEN

356 ■ Ethnobiology and the Science of Humankind
John Wiley & Sons Inc.
111 River St.
Hoboken, NJ 07030-5774
Phone: (201)748-6000
Fax: (201)748-6088
Free: (800)825-7550
Publisher E-mail: info@wiley.com

Journal covering ethnobiology and anthropological thought. **Subtitle:** Journal of the Royal Anthropological Institute. **Key Personnel:** Roy Ellen, Editor. **Subscription Rates:** US$39.95 individuals. **URL:** http://as.wiley.com/WileyCDA/WileyTitle/productCd-1405145897.html.

357 ■ Evidence-Based Child Health
John Wiley & Sons Inc.
111 River St.
Hoboken, NJ 07030-5774
Phone: (201)748-6000
Fax: (201)748-6088
Free: (800)825-7550
Publisher E-mail: info@wiley.com

Journal covering literature reviews from The Cochrane Library which are relevant to child health. **Subtitle:** A Cochrane Review Journal. **Freq:** 4/yr. **ISSN:** 1557-6272. **URL:** http://as.wiley.com/WileyCDA/WileyTitle/productCd-EBC2.html.

358 ■ International Journal of Cancer
John Wiley & Sons Inc.
111 River St.
Hoboken, NJ 07030-5774
Phone: (201)748-6000
Fax: (201)748-6088
Free: (800)825-7550
Publisher E-mail: info@wiley.com

Journal covering experimental and clinical research. **Freq:** 24/yr. **Trim Size:** 8 1/4 x 11. **Key Personnel:** Prof. H. zur Hausen, Editor-in-Chief. **ISSN:** 0020-7136. **Subscription Rates:** US$295 individuals; US$3,600 institutions. **Remarks:** Accepts advertising. **URL:** http://www3.interscience.wiley.com/journal/29331/home. . **Ad Rates:** BW: $634, 4C: $1,545. **Circ:** (Not Reported)

359 ■ Journal of Field Robotics
John Wiley & Sons Inc.
111 River St.
Hoboken, NJ 07030-5774
Phone: (201)748-6000
Fax: (201)748-6088
Free: (800)825-7550
Publisher E-mail: info@wiley.com

Journal covering the fundamentals of robotics in unstructured and dynamic environment. **Freq:** 11/yr. **Trim Size:** 7 x 10. **Key Personnel:** Sanjiv Singh, Editor-in-Chief. **ISSN:** 1556-4959. **Subscription Rates:** US$385 U.S., Canada, and Mexico; US$457 other countries; US$2,897 institutions; US$3,017 Canada and Mexico; US$3,119 institutions, other countries. **Remarks:** Accepts advertising. **URL:** http://www3.interscience.wiley.com/journal/117946193/grouphome. . **Ad Rates:** BW: $437, 4C: $1,040. **Circ:** (Not Reported)

360 ■ Journal of Leadership Studies
John Wiley & Sons Inc.
111 River St.
Hoboken, NJ 07030-5774
Phone: (201)748-6000
Fax: (201)748-6088
Free: (800)825-7550
Publisher E-mail: info@wiley.com

Journal covering leadership research and theoretical contributions. **ISSN:** 1935-2611. **URL:** http://as.wiley.com/WileyCDA/WileyTitle/productCd-JLS.html.

361 ■ Journal of Peptide Science
John Wiley & Sons Inc.
111 River St.
Hoboken, NJ 07030-5774
Phone: (201)748-6000
Fax: (201)748-6088
Free: (800)825-7550
Publisher E-mail: info@wiley.com

Journal covering peptide science. **Freq:** Monthly. **Trim Size:** 210 x 279. **ISSN:** 1075-2617. **Subscription Rates:** US$985 individuals; US$1,315 institutions. **Remarks:** Accepts advertising. **URL:** http://www3.interscience.wiley.com/journal/6016/home. **Circ:** 1,150

Circulation: ★ = ABC; △ = BPA; ♦ = CAC; • = CCAB; ❑ = VAC; ⊕ = PO Statement; ‡ = Publisher's Report; Boldface figures = sworn; Light figures = estimated.

362 ■ Journal of Tissue Engineering and Regenerative Medicine
John Wiley & Sons Inc.
111 River St.
Hoboken, NJ 07030-5774
Phone: (201)748-6000
Fax: (201)748-6088
Free: (800)825-7550
Publisher E-mail: info@wiley.com

Journal covering tissue engineering and regenerative medicine. **Freq:** 8/yr. **ISSN:** 1932-6254. **URL:** http://as.wiley.com/WileyCDA/WileyTitle/productCd-TERM.html.

363 ■ Journal of Traumatic Stress
John Wiley & Sons Inc.
111 River St.
Hoboken, NJ 07030-5774
Phone: (201)748-6000
Fax: (201)748-6088
Free: (800)825-7550
Publisher E-mail: info@wiley.com

Journal covering biopsychosocial aspects of trauma. **Freq:** 6/yr. **Trim Size:** 7 x 10. **Key Personnel:** Paula P. Schurr, PhD, Editor. **ISSN:** 0894-9867. **Subscription Rates:** US$83 U.S., Canada, and Mexico; US$119 other countries; US$744 institutions; US$804 Canada and Mexico; US$855 institutions, other countries. **Remarks:** Accepts advertising. **URL:** http://www3.interscience.wiley.com/journal/109882595/home. . **Ad Rates:** BW: $634, 4C: $1,545. **Circ:** (Not Reported)

LINDEN

364 ■ Black Men
New Day Associates
PO Box 1041
Linden, NJ 07036
Phone: (908)925-1760
Fax: (908)925-8901
Publisher E-mail: info@lindennewday.com

Magazine featuring men's lifestyle. **Key Personnel:** John Blassingame, Publisher. **URL:** http://www.lindennewday.com/frame.htm.

365 ■ Today's Black Woman
New Day Associates
PO Box 1041
Linden, NJ 07036
Phone: (908)925-1760
Fax: (908)925-8901
Publisher E-mail: info@lindennewday.com

Magazine featuring lifestyles for women. **Key Personnel:** John Blassingame, Publisher. **URL:** http://www.lindennewday.com/frame.htm.

MARLTON

366 ■ WVBV-FM - 90.5
55 E Main St.
Marlton, NJ 08053
Phone: (856)983-1662

Format: Religious. **Owner:** Calvary Chapel of Marlton, at above address. **URL:** http://www.hopefm.net/.

NORTHFIELD

367 ■ WENJ-AM - 1450
950 Tilton Rd., Ste. 200
Northfield, NJ 08225
Phone: (609)645-9797
Fax: (609)272-9228

Format: Talk; Sports. **Key Personnel:** Mike Ruble, General Sales Mgr., mike.ruble@mrgnj.com; Joe Mitchell, Local Sales Mgr., joe.mitchell@mrgnj.com; Tom McNally, Chief Engineer, tom.mcnally@mrgnj.com. **Wattage:** 800 day; 1000 night. **URL:** http://www.1450espn.com/home.php.

NEW MEXICO

ALBUQUERQUE

368 ■ KDLW-FM - 97.7
4125 Carlisle Blvd. NE
Albuquerque, NM 87107
Phone: (505)878-0980
Fax: (505)878-0098
Free: (800)711-5586

Format: Top 40; Hip Hop; Blues. **Owner:** American General Media, at above address. **Key Personnel:** Eddie Go, Program Dir. **URL:** http://www.wild977.net/.

369 ■ KQRI-FM - 90.7
2351 Sunset Blvd., Ste. 170-218
Rocklin, CA 95765
Phone: (916)251-1600
Fax: (916)251-1650

Format: Contemporary Christian. **Owner:** Educational Media Foundation, at above address. **URL:** http://www.air1.com/.

370 ■ KRKE-AM - 1600
1213 San Pedro NE
Albuquerque, NM 87110
Phone: (505)899-5029
E-mail: info@realoldies1600.com

Format: Oldies. **Owner:** Vanguard Media LLC, at above address. **Ad Rates:** Advertising accepted; rates available upon request. **URL:** http://www.joyam.com/index.php/realoldies/.

CLOVIS

371 ■ KKYC-FM - 102.3
1000 W Sycamore
Clovis, NM 88101
Phone: (505)762-6200

Format: Country. **URL:** http://www.1023country.com/.

DULCE

372 ■ KCIE-FM - 90.5
PO Box 603
Dulce, NM 87528
Phone: (505)759-3681
Fax: (505)759-9140

Format: Public Radio. **URL:** http://www.kcie.org/HOME.html.

LAS CRUCES

373 ■ KRUC-FM - 88.9
PO Box 16691
Las Cruces, NM 88004
Phone: (505)521-8053

Format: Hispanic; Religious. **Owner:** World Radio Network, at above address. **Key Personnel:** John Powell, Manager; Paul Salzman, Engr. **Wattage:** 500. **URL:** http://www.wrn-rcm.org/.

SANTA FE

374 ■ KQBA-FM - 107.5
2502 Camino Entrada, Ste. C
Santa Fe, NM 87507
Phone: (505)471-1067
E-mail: info@1075outlawcountry.com

Format: Country. **Owner:** Hutton Broadcasting, at above address. **URL:** http://www.1075outlawcountry.com/.

TAOS

375 ■ KLNN-FM - 103.7
PO Box 1844
Taos, NM 87571
E-mail: connect@luna1037.com

Format: Adult Contemporary. **Key Personnel:** Stephanie Gutz, Contact, phone (575)758-5826. **URL:** http://www.luna1037.com/.

NEW YORK

ALBANY

376 ■ WJIV-FM - 101.9
5210 S Saginaw
Flint, MI 48507
Phone: (810)694-4146
Fax: (810)694-0661

Format: Religious. **Owner:** Christian Broadcasting System Ltd., at above address. **URL:** http://www.cbslradio.com/index.php.

WOSR-FM - See Middletown

ARCADE

377 ■ WCOF-FM - 89.5
PO Box 506
7634 Campbell Creek Rd.
Bath, NY 14810-0506
Phone: (607)776-4151

Format: Religious. **Owner:** Family Life Ministries, Inc., at above address. **URL:** http://www.fln.org/index.php?option=com_wrapper&Itemid=753.

AUBURN

378 ■ WCOV-FM - 93.7
7634 Campbell Creek Rd.
PO Box 506
Bath, NY 14810-0506
Phone: (607)776-4151

Format: Religious. **Owner:** Family Life Ministries, Inc., at above address. **URL:** http://www.fln.org/.

BATH

WCOF-FM - See Arcade

WCOV-FM - See Auburn

BINGHAMTON

379 ■ WBGH-CA - 20
203 Ingraham Hill Rd.
Binghamton, NY 13903
Phone: (607)771-3434

Key Personnel: Jim Ehrnke, News Dir.; Bill Cook, Anchor. **URL:** http://www.newschannel34.com.

380 ■ WIFF-FM - 90.1
111 N Main St.
Elmira, NY 14901
Phone: (607)732-2484
Fax: (607)732-8704

Format: Contemporary Christian. **Owner:** CSN International, PO Box 391, Twin Falls, ID 83303, Fax: (208)736-1958, Free: (800)357-4226. **Key Personnel:** Larry Galletti, Manager. **URL:** http://www.csnradio.com/stationsNY.php.

381 ■ WRRQ-FM - 106.7
495 Court St., 2nd Fl.
Binghamton, NY 13904
Phone: (607)772-1005
Fax: (607)772-2945

Format: Adult Contemporary. **Key Personnel:** George Hawras, General Mgr.; Tom Shiptenko, Station Mgr.; Steve Shimer, Operations Mgr. **URL:** http://www.myq107.com/.

BROCKPORT

382 ■ WRSB-AM - 1310
6675 4th Section Rd.
Brockport, NY 14420
Phone: (716)637-7040

Format: Religious. **Key Personnel:** Dr. David L. Wolfe, General Mgr., docwolfe@hotmail.com. **URL:** http://www.geocities.com/Heartland/Creek/5396/wasb.html.

BUFFALO

383 ■ WNED-FM - 94.5
PO Box 1263
Buffalo, NY 14240-1263
Phone: (716)845-7000
Fax: (716)845-7036

Format: Classical. **Operating Hours:** Continuous. **Key Personnel:** Dick Daly, Sen. VP; Ron Santora, VP/Station Mgr. **URL:** http://www.wned.org/.

WOLN-FM - See Olean

CANTON

WSLJ-FM - See Watertown

384 ■ WXLG-FM - 89.9
80 E Main St.
EJ Noble Medical Bldg., Rm. 201
Canton, NY 13617-1450
Phone: (315)229-5356
Fax: (315)229-5373
Free: (877)388-6277

Format: Public Radio. **URL:** http://www.northcountrypublicradio.org/about/coverage.htmlfreqs.

385 ■ WXLH-FM - 91.3
80 E Main St.
EJ Noble Medical Bldg., Rm. 201
Canton, NY 13617-1450

Format: Public Radio. **URL:** http://www.northcountrypublicradio.org/about/coverage.htmlfreqs.

DIX HILLS

386 ■ WFTU-AM - 1570
305 N Service Rd.
Dix Hills, NY 11746-5857
Phone: (631)424-7000

Format: Full Service. **Owner:** Five Towns Colorado, at above address. **URL:** http://www.ftc.edu/StudentLife/stu_wftu.php.

ELMIRA

WIFF-FM - See Binghamton

GENEVA

387 ■ Art Materials Retailer
Fahy-Williams Publishing Inc.
171 Reed St.
PO Box 1080
Geneva, NY 14456
Phone: (315)789-0458
Fax: (315)789-4263
Free: (800)344-0559
Publisher E-mail: tking@fwpi.com

Magazine featuring art materials supply. **Subtitle:** The Magazine for the Art Supply Industry. **Trim Size:** 8 1/8 x 10 7/8. **Key Personnel:** J. Kevin Fahy, Publisher, kfahy@fwpi.com; Tina Manzer, Editorial Dir.; Mark Stash, Production Mgr., mstash@fwpi.com. **Subscription Rates:** Free. **Remarks:** Accepts advertising. **URL:** http://www.artmaterialsretailer.com/; http://www.fwpi.com/. **Circ:** (Not Reported)

388 ■ The Early Learner
Fahy-Williams Publishing Inc.
171 Reed St.
PO Box 1080
Geneva, NY 14456
Phone: (315)789-0458
Fax: (315)789-4263
Free: (800)344-0559
Publisher E-mail: tking@fwpi.com

Magazine for pre kinder teachers and child care providers. **Subtitle:** Early Childhood Product Review. **Key Personnel:** J. Kevin Fahy, Publisher, kfahy@fwpi.com; Tina Manzer, Editorial Dir.; Mark Stash, Production Mgr., mstash@fwpi.com. **Subscription Rates:** Free. **Remarks:** Accepts advertising. **URL:** http://www.theearlylearner.com/index.php; http://www.fwpi.com/. **Circ:** 50,000

389 ■ edplay
Fahy-Williams Publishing Inc.
171 Reed St.
PO Box 1080
Geneva, NY 14456
Phone: (315)789-0458
Fax: (315)789-4263
Free: (800)344-0559
Publisher E-mail: tking@fwpi.com

Magazine featuring specialty toy industry. **Subtitle:** For specialty toy, game, gift, and museum stores. **Trim Size:** 8 1/8 x 10 7/8. **Key Personnel:** J. Kevin Fahy, Publisher, kfahy@fwpi.com; Tina Manzer, Editorial Dir.; Mark Stash, Production Mgr., mstash@fwpi.com. **Subscription Rates:** Free. **Remarks:** Accepts advertising. **URL:** http://www.edplay.com/; http://www.fwpi.com/. **Circ:** (Not Reported)

390 ■ Life in the Finger Lakes
Fahy-Williams Publishing Inc.
171 Reed St.
PO Box 1080
Geneva, NY 14456
Phone: (315)789-0458
Fax: (315)789-4263
Free: (800)344-0559
Publisher E-mail: tking@fwpi.com

Magazine featuring Finger Lakes each passing season. **Freq:** 4/yr. **Key Personnel:** Mark Stash, Editor, mark@lifeinthefingerlakes.com. **Subscription Rates:** US$12.95 individuals; US$19.95 two years; US$25.95 individuals 3 years; US$27.95 Canada; US$42.95 other countries. **Remarks:** Accepts advertising. **URL:** http://www.lifeinthefingerlakes.com/; http://www.fwpi.com/. **Circ:** (Not Reported)

GREENPORT

391 ■ WSUF-FM - 89.9
5151 Park Ave.
Fairfield, CT 06825
Phone: (203)365-6604

Format: Public Radio. **URL:** http://www.wshu.org/.

KINGSTON

392 ■ WFRH-FM - 91.7
290 Hegenberger Rd.
Oakland, CA 94621
Free: (800)543-1495

Format: Religious. **Owner:** Family Stations, Inc., at above address. **URL:** http://www.familyradio.com/english/connect/broadcast/location-freq.html.

LAKE KATRINE

WGWR-FM - See Liberty

LAKE PLACID

393 ■ WSLP-FM - 93.3
PO Box 368
Lake Placid, NY 12946
Phone: (518)523-4900
Fax: (518)523-4290
E-mail: info@wslpfm.com

Format: Adult Contemporary. **Key Personnel:** Jon Lundin, General Mgr., jonl@wslpfm.com; Jim Williams, General Sales Mgr., jwilliams@wslpfm.com. **Ad Rates:** Advertising accepted; rates available upon request. **URL:** http://www.wslpfm.com/.

LIBERTY

394 ■ WGWR-FM - 88.1
PO Box 777
Lake Katrine, NY 12449
Fax: (845)336-7205
Free: (800)724-8518

Format: Religious. **Owner:** Sound of Life Radio Network, at above address. **URL:** http://www.soundoflife.org/sol/stations.php.

MALONE

395 ■ WMHQ-FM - 90.1
4044 Makyes Rd.
Syracuse, NY 13215
Phone: (315)469-5051
Free: (800)677-1881

Format: Religious. **Owner:** Mars Hill Network, at above address. **URL:** http://www.marshillnetwork.org//coveragemapMHN.html.

MASTIC BEACH

396 ■ WSVV-FM - 100.9
PO Box 1018
Laguna Beach, CA 92652
Fax: (469)241-6795
Free: (800)639-5433

Format: Religious. **Owner:** New Life Ministries, at above address. **Key Personnel:** Stephen Arterburn, Founder/Chm. **URL:** http://www.newlife.com/Radio/findradio.asp?Id=NY&t=New%20York.

MELVILLE

397 ■ Advanced Rescue Technology
Cygnus Business Media Inc.
3 Huntington Quadrangle, Ste. 301N
Melville, NY 11747
Phone: (631)845-2700
Fax: (631)845-7109
Free: (800)308-6397

Magazine covering information on any aspects of rescuing. **Freq:** 6/yr. **Key Personnel:** Harvey Eisner, Editor-in-Chief, phone (631)963-6252, harvey.eisner@cygnuspub.com; Elizabeth Friszell-Neroulas, Managing Editor, phone (631)963-6230, elizabeth.friszell@cygnuspub.com; Jean Rank, Production Mgr., phone (631)963-6237, jean.rank@cygnuspub.com. **URL:** http://www.cygnusb2b.com/PropertyPub.cfm?PropertyID=521. **Circ:** 20,000

398 ■ Concrete Contractor
Cygnus Business Media Inc.
3 Huntington Quadrangle, Ste. 301N
Melville, NY 11747
Phone: (631)845-2700
Fax: (631)845-7109
Free: (800)308-6397

Magazine covering information on concrete equipment and current technology for the concrete contractors. **Founded:** 2002. **Freq:** 8/yr. **Key Personnel:** Rebecca Wasieleski, Editor, phone (920)568-8321, rebecca.wasieleski@cygnuspub.com; Kris Flitcroft, Gp. Publisher, phone (920)563-1646, kris.flitcroft@cygnuspub.com. **Remarks:** Accepts advertising. **URL:** http://www.cygnusb2b.com/PropertyPub.cfm?PropertyID=396; http://www.forconstructionpros.com/cover/Concrete-Contractor/7FCP. **Circ:** △**30,0004**

399 ■ Fuel Advantage
Cygnus Business Media Inc.
3 Huntington Quadrangle, Ste. 301N
Melville, NY 11747
Phone: (631)845-2700
Fax: (631)845-7109
Free: (800)308-6397

Magazine featuring fuel saving products and technologies. **Freq:** Quarterly. **Key Personnel:** Larry M. Greenberger, Publisher, phone (847)454-2722, larry.greenberger@cygnusb2b.com; Mark O'Connell, Editor, phone (920)563-1611, mark.oconnell@cygnusb2b.com; Scott De Laruelle, Asst. Ed., scott@fleetmag.com. **Remarks:** Accepts advertising. **URL:** http://www.cygnusb2b.com/PropertyPub.cfm?PropertyID=520; http://www.fuelpub.com. . **Ad Rates:** 4C: $7,785. **Circ:** 50,000

Circulation: ★ = ABC; △ = BPA; ♦ = CAC; • = CCAB; ☐ = VAC; ⊕ = PO Statement; ‡ = Publisher's Report; Boldface figures = sworn; Light figures = estimated.

400 ■ Light Truck and SUV
Cygnus Business Media Inc.
3 Huntington Quadrangle, Ste. 301N
Melville, NY 11747
Phone: (631)845-2700
Fax: (631)845-7109
Free: (800)308-6397

Magazine featuring light truck and SUV. **Founded:** 1995. **Freq:** 6/yr. **Key Personnel:** Bob Carnahan, Publisher, bob.carnahan@cygnusb2b.com; Dana Nelsen, Editor, dana.nelsen@cygnusb2b.com; Barb Zuehlke, Managing Editor, bzuehlke@lighttruckbiz.com. **Remarks:** Accepts advertising. **URL:** http://www.cygnusb2b.com/PropertyPub.cfm?PropertyID=127; http://www.lighttruckandsuv.com. **Circ:** △**14,005**

401 ■ Snow Pro
Cygnus Business Media Inc.
3 Huntington Quadrangle, Ste. 301N
Melville, NY 11747
Phone: (631)845-2700
Fax: (631)845-7109
Free: (800)308-6397

Magazine featuring equipment, supplies, and services used for winter weather contracting. **Freq:** Semiannual September and November. **Trim Size:** 7 3/8 x 10 3/4. **Key Personnel:** Rick Monogue, Gp. Publisher, rick.monogue@cygnuspub.com; Robert Warde, Editor, robert.warde@cygnusb2b.com; Grant Dunham, Managing Editor, grant.dunham@cygnusb2b.com. **Remarks:** Accepts advertising. **URL:** http://www.cygnusb2b.com/PropertyPub.cfm?PropertyID=433. . **Ad Rates:** 4C: $11,345. **Circ:** 129,000

402 ■ Surface Fabrication
Cygnus Business Media Inc.
3 Huntington Quadrangle, Ste. 301N
Melville, NY 11747
Phone: (631)845-2700
Fax: (631)845-7109
Free: (800)308-6397

Magazine covering solid surface market. **Founded:** 1995. **Freq:** Monthly. **Trim Size:** 9 x 10 7/8. **Key Personnel:** Jay Schneider, Publisher, phone (920)563-1684, jay.schneider@cygnuspub.com; Elizabeth Jackson, Account Exec., phone (847)492-1350, fax (847)492-0085, ejackson@meritdirect.com. **Remarks:** Accepts advertising. **URL:** http://www.cygnusb2b.com/PropertyPub.cfm?PropertyID=86; http://www.surfacefabrication.com. . **Ad Rates:** 4C: $4,975. **Circ:** 10,500

MIDDLETOWN

403 ■ WOSR-FM - 91.7
318 Central Ave.
Albany, NY 12206

Format: Public Radio. **URL:** http://www.wamc.org/station.html.

NEW HARTFORD

404 ■ WUMX-FM - 102.5
39 Kellogg Rd.
New Hartford, NY 13413
Phone: (315)721-0102
Fax: (315)738-1073

Format: Adult Contemporary. **URL:** http://mix1025.com/wp/.

NEW YORK

405 ■ African and Black Diaspora
Routledge Journals
Taylor & Francis Group
270 Madison Ave.
New York, NY 10016
Phone: (212)216-7800
Fax: (212)563-2269

Journal covering the African Diaspora studies. **Freq:** Semiannual. **Key Personnel:** Dr. Fassil Demissie, Editor. **ISSN:** 1752-8631. **Subscription Rates:** US$281 institutions online only; US$70 individuals print and online; US$296 institutions print and online. **URL:** http://www.tandf.co.uk/journals/titles/17528631.asp.

406 ■ Australian Journal of Learning Difficulties
Routledge Journals
Taylor & Francis Group
270 Madison Ave.
New York, NY 10016
Phone: (212)216-7800
Fax: (212)563-2269

Journal featuring theoretical and empirical articles on topics related to the assessment and teaching of students with learning disabilities and learning difficulties. **Freq:** Semiannual. **Key Personnel:** Kevin Wheldall, Editor; Alison Madelaine, Editor. **ISSN:** 1940-4158. **Subscription Rates:** US$171 institutions online only; US$94 individuals print only; US$180 institutions print and online. **URL:** http://www.tandf.co.uk/journals/titles/19404158.asp.

407 ■ China Economic Journal
Routledge Journals
Taylor & Francis Group
270 Madison Ave.
New York, NY 10016
Phone: (212)216-7800
Fax: (212)563-2269

Journal covering China's economic development. **Freq:** 3/yr. **Key Personnel:** Feng Lu, Editor. **ISSN:** 1753-8963. **Subscription Rates:** US$570 institutions online only; US$114 individuals print only; US$601 institutions print and online. **URL:** http://www.tandf.co.uk/journals/titles/17538963.asp.

408 ■ Chinese Journal of Communication
Routledge Journals
Taylor & Francis Group
270 Madison Ave.
New York, NY 10016
Phone: (212)216-7800
Fax: (212)563-2269

Journal focusing on theoretical, empirical, and methodological dimensions of Chinese communication studies. **Freq:** Semiannual. **Key Personnel:** Paul S.N. Lee, Editor. **ISSN:** 1754-4750. **Subscription Rates:** US$259 institutions online only; US$47 individuals print and online; US$273 institutions print and online. **URL:** http://www.tandf.co.uk/journals/titles/17544750.asp.

409 ■ Coaching
Routledge Journals
Taylor & Francis Group
270 Madison Ave.
New York, NY 10016
Phone: (212)216-7800
Fax: (212)563-2269

Journal covering theory, research, and practice of coaching. **Subtitle:** An International Journal of Theory, Research and Practice. **Freq:** Semiannual. **Key Personnel:** Stephen Palmer, PhD, Exec. Ed. **ISSN:** 1752-1882. **Subscription Rates:** US$157 institutions online only; US$49 individuals print and online; US$166 institutions print and online. **URL:** http://www.tandf.co.uk/journals/titles/17521882.asp.

410 ■ Cogeneration & Distributed Generation Journal
Taylor & Francis Group
29 W 35th St.
New York, NY 10001
Phone: (212)216-7800
Fax: (212)244-1563

Journal covering education in energy and management. **Freq:** Quarterly. **ISSN:** 1545-3669. **Subscription Rates:** US$285 institutions online only; US$250 individuals; US$300 institutions print and online. **URL:** http://www.tandf.co.uk/journals/titles/15453669.asp.

411 ■ Contemporary Arab Affairs
Routledge Journals
Taylor & Francis Group
270 Madison Ave.
New York, NY 10016
Phone: (212)216-7800
Fax: (212)563-2269

Journal featuring modern Arab scholarship in the English language. **Freq:** Quarterly. **Key Personnel:** Khair El-Din Haseeb, Editor-in-Chief. **ISSN:** 1755-0920. **Subscription Rates:** US$519 institutions online only; US$88 individuals print only; US$546 institutions print and online. **URL:** http://www.tandf.co.uk/journals/titles/17550912.asp.

412 ■ Criminal Justice Matters
Routledge Journals
Taylor & Francis Group
270 Madison Ave.
New York, NY 10016
Phone: (212)216-7800
Fax: (212)563-2269

Magazine featuring research, analysis, and policy development relating to contemporary social, crime, and justice issues. **Freq:** Quarterly. **Key Personnel:** Enver Solomon, Editor, enver.solomon@kcl.ac.uk; Rebecca Roberts, Editor, rebecca.roberts@kcl.ac.uk. **ISSN:** 0962-7251. **Subscription Rates:** US$181 institutions online only; US$70 individuals print and online; US$191 institutions print and online. **URL:** http://www.tandf.co.uk/journals/titles/09627251.asp.

413 ■ Critical Studies on Terrorism
Routledge Journals
Taylor & Francis Group
270 Madison Ave.
New York, NY 10016
Phone: (212)216-7800
Fax: (212)563-2269

Journal covering research in all aspects of terrorism, counter-terrorism, and state terror. **Freq:** 3/yr. **Key Personnel:** Richard Jackson, Editor. **ISSN:** 1753-9153. **Subscription Rates:** US$324 institutions online only; US$86 individuals print only; US$341 institutions print and online. **URL:** http://www.tandf.co.uk/journals/titles/17539153.asp.

414 ■ Dynamics of Asymmetric Conflict
Routledge Journals
Taylor & Francis Group
270 Madison Ave.
New York, NY 10016
Phone: (212)216-7800
Fax: (212)563-2269

Journal featuring papers and reviews that contribute to understanding and ameliorating conflicts between states and non-state challengers. **Freq:** 3/yr. **Key Personnel:** Clark McCauley, Editor. **ISSN:** 1746-7586. **Subscription Rates:** US$266 institutions online only; US$54 individuals; US$281 institutions print and online. **URL:** http://www.tandf.co.uk/journals/titles/17467586.asp.

415 ■ The Educational Forum
Routledge Journals
Taylor & Francis Group
270 Madison Ave.
New York, NY 10016
Phone: (212)216-7800
Fax: (212)563-2269

Journal covering topics that contributes to the advancement of education. **Freq:** Quarterly. **Key Personnel:** Jan Robertson, Academic Ed. **ISSN:** 0013-1725. **Subscription Rates:** US$119 institutions online only; US$65 individuals print only; US$125 institutions print and online. **URL:** http://www.tandf.co.uk/journals/titles/001317259.asp.

416 ■ Food Additives & Contaminants
Taylor & Francis Group
29 W 35th St.
New York, NY 10001
Phone: (212)216-7800
Fax: (212)244-1563

Journal covering food additives and contaminants. **Subtitle:** Part B - Surveillance Communications.

Freq: Semiannual. **ISSN:** 1939-3210. **URL:** http://www.tandf.co.uk/journals/titles/19393210.asp.

417 ■ The IES Journal Part A
Routledge Journals
Taylor & Francis Group
270 Madison Ave.
New York, NY 10016
Phone: (212)216-7800
Fax: (212)563-2269

Journal covering professional engineering matters. **Subtitle:** Civil & Structural Engineering. **Freq:** Quarterly. **Key Personnel:** Wang Chien Ming, Editor-in-Chief. **ISSN:** 1937-3260. **Subscription Rates:** US$380 institutions online only; US$200 individuals; US$400 institutions print and online. **URL:** http://www.tandf.co.uk/journals/titles/19373260.asp.

418 ■ International Journal of Culture and Mental Health
Routledge Journals
Taylor & Francis Group
270 Madison Ave.
New York, NY 10016
Phone: (212)216-7800
Fax: (212)563-2269

Journal covering topics about cross-cultural issues and mental health. **Freq:** Semiannual. **Key Personnel:** Prof. Dinesh Bhugra, Editor. **ISSN:** 1754-2871. **Subscription Rates:** US$185 institutions online only; US$54 individuals; US$195 institutions print and online. **URL:** http://www.tandf.co.uk/journals/rccm.

419 ■ International Journal of Digital Earth
Taylor & Francis Group
29 W 35th St.
New York, NY 10001
Phone: (212)216-7800
Fax: (212)244-1563

Journal covering the science and technology of Digital Earth and its applications in all major disciplines. **Freq:** Quarterly. **Key Personnel:** Guo Huadong, Editor-in-Chief. **ISSN:** 1753-8947. **Subscription Rates:** US$333 institutions online only; US$176 individuals print only; US$351 institutions print and online. **URL:** http://www.tandf.co.uk/journals/titles/17538947.asp.

420 ■ International Journal of Fashion Design, Technology and Education
Taylor & Francis Group
29 W 35th St.
New York, NY 10001
Phone: (212)216-7800
Fax: (212)244-1563

Journal covering research in fashion design, pattern cutting, apparel production, manufacturing technology, and fashion education. **Freq:** 3/yr. **Key Personnel:** Kristina Shin, Editor-in-Chief. **ISSN:** 1754-3266. **Subscription Rates:** US$296 institutions online only; US$156 individuals print only; US$312 institutions print and online. **URL:** http://www.tandf.co.uk/journals/titles/17543266.asp.

421 ■ International Journal of RF Technologies
Taylor & Francis Group
29 W 35th St.
New York, NY 10001
Phone: (212)216-7800
Fax: (212)244-1563

Journal covering RF technology deployment, data analytics, and business value creation. **Subtitle:** Research and Applications. **Freq:** Quarterly. **Key Personnel:** Bill Hardgrave, Editor-in-Chief, journal@grfla.org. **ISSN:** 1754-5730. **Subscription Rates:** US$355 institutions online only; US$94 individuals; US$374 institutions print and online. **URL:** http://www.tandf.co.uk/journals/titles/17545730.asp.

422 ■ Journal of Asian Public Policy
Routledge Journals
Taylor & Francis Group
270 Madison Ave.
New York, NY 10016
Phone: (212)216-7800
Fax: (212)563-2269

Journal covering the economy of the countries in South East Asia. **Freq:** Semiannual. **Key Personnel:** Ka Ho Mok, Editor. **ISSN:** 1751-6234. **Subscription Rates:** US$333 institutions online only; US$86 individuals print only; US$351 institutions print and online. **URL:** http://www.tandf.co.uk/journals/titles/17516234.asp.

423 ■ Journal of Building Performance Simulation
Taylor & Francis Group
29 W 35th St.
New York, NY 10001
Phone: (212)216-7800
Fax: (212)244-1563

Journal covering topics about the design, construction, operation, and maintenance of new and existing buildings worldwide. **Freq:** Quarterly. **Key Personnel:** Ian Beausoleil-Morrison, Editor, ibeausol@mae.carleton.ca. **ISSN:** 1940-1493. **Subscription Rates:** US$455 institutions online only; US$156 individuals print only; US$468 institutions print and online. **URL:** http://www.tandf.co.uk/journals/titles/19401493.asp.

424 ■ Journal of Cultural Economy
Routledge Journals
Taylor & Francis Group
270 Madison Ave.
New York, NY 10016
Phone: (212)216-7800
Fax: (212)563-2269

Journal covering social sciences and humanities. **Freq:** 3/yr. **Key Personnel:** Tony Bennett, Editor. **ISSN:** 1753-0350. **Subscription Rates:** US$367 institutions online only; US$66 individuals; US$387 institutions print and online. **URL:** http://www.tandf.co.uk/journals/titles/17530350.asp.

425 ■ Journal of International and Intercultural Communication
Routledge Journals
Taylor & Francis Group
270 Madison Ave.
New York, NY 10016
Phone: (212)216-7800
Fax: (212)563-2269

Journal focusing on international, intercultural, and indigenous communication issues. **Freq:** Quarterly. **Key Personnel:** Thomas Nakayama, Editor. **ISSN:** 1751-3057. **Subscription Rates:** US$227 institutions online only; US$61 individuals print only; US$239 institutions print and online. **URL:** http://www.tandf.co.uk/journals/titles/17513057.asp.

426 ■ Journal of Islamic Law and Culture
Routledge Journals
Taylor & Francis Group
270 Madison Ave.
New York, NY 10016
Phone: (212)216-7800
Fax: (212)563-2269

Journal featuring articles and reviews on Islamic law, with an emphasis on the significance of law in the intersection of Western and Muslim legal culture. **Freq:** 3/yr. **Key Personnel:** Aminah Beverly McCloud, Editor-in-Chief. **ISSN:** 1528-817X. **Subscription Rates:** US$222 institutions online only; US$240 institutions print and online. **URL:** http://www.informaworld.com/smpp/title~db=all~content=t781480877~tab=summary.

427 ■ Journal of Mental Health Research in Intellectual Disabilities
Routledge Journals
Taylor & Francis Group
270 Madison Ave.
New York, NY 10016
Phone: (212)216-7800
Fax: (212)563-2269

Journal featuring scientific and scholarly contributions to advance knowledge about mental health issues among persons with intellectual disabilities and related developmental disabilities. **Freq:** Quarterly. **Key Personnel:** Johannes Rojahn, Editor-in-Chief. **ISSN:** 1931-5864. **Subscription Rates:** US$247 institutions online only; US$100 individuals print only; US$260 institutions print and online. **URL:** http://www.tandf.co.uk/journals/titles/19315864.asp.

428 ■ Journal of Research on Educational Effectiveness
Routledge Journals
Taylor & Francis Group
270 Madison Ave.
New York, NY 10016
Phone: (212)216-7800
Fax: (212)563-2269

Journal focusing on the study of educational problems. **Freq:** Quarterly. **Key Personnel:** Barbara R. Foorman, Editor; Larry V. Hedges, Editor. **ISSN:** 1934-5747. **Subscription Rates:** US$261 institutions online only; US$60 individuals print only; US$275 institutions print and online. **URL:** http://www.tandf.co.uk/journals/titles/19345747.asp.

429 ■ Journal of Urbanism
Routledge Journals
Taylor & Francis Group
270 Madison Ave.
New York, NY 10016
Phone: (212)216-7800
Fax: (212)563-2269

Journal focusing on human settlement. **Freq:** 3/yr. **Key Personnel:** Charles C. Bohl, Editor, cbohl@miami.edu. **ISSN:** 1754-9175. **Subscription Rates:** US$396 institutions online only; US$74 individuals print only; US$417 institutions print and online. **URL:** http://www.tandf.co.uk/journals/titles/17549175.asp.

430 ■ Literacy Research & Instruction
Routledge Journals
Taylor & Francis Group
270 Madison Ave.
New York, NY 10016
Phone: (212)216-7800
Fax: (212)563-2269

Journal focusing on reading education and allied literacy fields. **Freq:** Quarterly. **Key Personnel:** Sherry Kragler, Editor; Carolyn Walker, Editor. **ISSN:** 1938-8071. **Subscription Rates:** US$124 institutions online only; US$130 institutions print and online. **URL:** http://www.tandf.co.uk/journals/titles/19388071.asp.

431 ■ Macroeconomics and Finance in Emerging Market Economies
Routledge Journals
Taylor & Francis Group
270 Madison Ave.
New York, NY 10016
Phone: (212)216-7800
Fax: (212)563-2269

Journal covering current economic trends. **Freq:** Semiannual. **Key Personnel:** D.M. Nachane, Editor. **ISSN:** 1752-0843. **Subscription Rates:** US$222 institutions online only; US$68 individuals print only; US$234 institutions print and online. **URL:** http://www.tandf.co.uk/journals/titles/17520843.asp.

Circulation: ★ = ABC; △ = BPA; ♦ = CAC; • = CCAB; ❑ = VAC; ⊕ = PO Statement; ‡ = Publisher's Report; Boldface figures = sworn; Light figures = estimated.

432 ■ Mental Health and Substance Use
Routledge Journals
Taylor & Francis Group
270 Madison Ave.
New York, NY 10016
Phone: (212)216-7800
Fax: (212)563-2269

Journal covering issues on mental health and substance use. **Subtitle:** Dual Diagnosis. **Freq:** 3/yr. **Key Personnel:** Philip D. Cooper, Editor. **ISSN:** 1752-3281. **Subscription Rates:** US$192 institutions online only; US$54 individuals; US$203 institutions print and online. **URL:** http://www.tandf.co.uk/journals/rmhs.

433 ■ Photographies
Routledge Journals
Taylor & Francis Group
270 Madison Ave.
New York, NY 10016
Phone: (212)216-7800
Fax: (212)563-2269

Journal covering photography. **Freq:** Semiannual. **Key Personnel:** David Bate, Editor. **ISSN:** 1754-0763. **Subscription Rates:** US$212 institutions online only; US$58 individuals print only; US$224 institutions print and online. **URL:** http://www.tandf.co.uk/journals/rpho.

434 ■ Planning & Environmental Law
Routledge Journals
Taylor & Francis Group
270 Madison Ave.
New York, NY 10016
Phone: (212)216-7800
Fax: (212)563-2269

Journal covering state judicial decisions and legislative acts that pertain to planning and environmental management. **Freq:** 11/yr. **ISSN:** 1548-0755. **Subscription Rates:** US$371 institutions online only; US$391 institutions print and online. **URL:** http://www.tandf.co.uk/journals/titles/15480755.asp.

435 ■ Postscripts
Equinox Publishing Ltd.
Barnard College
3009 Broadway
New York, NY 10027

Journal covering the study of scriptures around the globe. **Subtitle:** The Journal of Sacred Texts and Contemporary Worlds. **Freq:** 3/yr (April, August and November). **Key Personnel:** Elizabeth Castelli, Editor, ecastell@barnard.edu. **ISSN:** 1743-887X. **Subscription Rates:** 120 institutions; US$195 institutions; 45 individuals; US$80 individuals; 31 students; US$60 students. **Remarks:** Accepts advertising. **URL:** http://www.equinoxjournals.com/ojs/index.php/POST. **Circ:** (Not Reported)

436 ■ SHOT Business
Bonnier Corporation
2 Park Ave.
New York, NY 10016
Phone: (212)779-5000

Magazine featuring shooting sports industry. **Key Personnel:** Slaton L. White, Editor. **URL:** http://www.bonniercorp.com/brands/SHOT-Business.html.

437 ■ The Sixties
Routledge Journals
Taylor & Francis Group
270 Madison Ave.
New York, NY 10016
Phone: (212)216-7800
Fax: (212)563-2269

Journal covering the historical legacy of 1960. **Freq:** Semiannual. **Key Personnel:** Jeremy Varon, Editor. **ISSN:** 1754-1328. **Subscription Rates:** US$244 institutions online only; US$49 individuals print only; US$257 institutions print and online. **URL:** http://www.tandf.co.uk/journals/titles/17541328.asp.

438 ■ Social Dynamics
Routledge Journals
Taylor & Francis Group
270 Madison Ave.
New York, NY 10016
Phone: (212)216-7800
Fax: (212)563-2269

Journal covering articles about humanities and social sciences. **Freq:** Semiannual. **Key Personnel:** S. Levine, Editor. **ISSN:** 0253-3952. **Subscription Rates:** US$234 institutions online only; US$88 individuals print only; US$246 institutions print and online. **URL:** http://www.tandf.co.uk/journals/titles/02533952.asp.

439 ■ Social Sciences in China
Routledge Journals
Taylor & Francis Group
270 Madison Ave.
New York, NY 10016
Phone: (212)216-7800
Fax: (212)563-2269

Journal covering social sciences and humanities in China. **Freq:** Quarterly. **ISSN:** 0252-9203. **Subscription Rates:** US$333 institutions online only; US$98 individuals; US$350 institutions print and online. **URL:** http://www.tandf.co.uk/journals/titles/02529203.asp.

440 ■ South African Journal of International Affairs
Routledge Journals
Taylor & Francis Group
270 Madison Ave.
New York, NY 10016
Phone: (212)216-7800
Fax: (212)563-2269

Journal covering South Africa and its inhabitants. **Freq:** Semiannual. **Key Personnel:** Elizabeth Sidiropoulos, Editor-in-Chief. **ISSN:** 1022-0461. **Subscription Rates:** US$320 institutions online only; US$80 individuals; US$336 institutions print and online. **URL:** http://www.tandf.co.uk/journals/titles/10220461.asp.

441 ■ Studies in Science Education
Routledge Journals
Taylor & Francis Group
270 Madison Ave.
New York, NY 10016
Phone: (212)216-7800
Fax: (212)563-2269

Journal covering science education. **Freq:** Semiannual. **Key Personnel:** Jim Ryder, Editor, j.ryder@education.leeds.ac.uk. **ISSN:** 0305-7267. **Subscription Rates:** US$181 institutions online only; US$125 individuals; US$190 institutions print and online. **URL:** http://www.tandf.co.uk/journals/titles/03057267.asp.

442 ■ Urban Research & Practice
Routledge Journals
Taylor & Francis Group
270 Madison Ave.
New York, NY 10016
Phone: (212)216-7800
Fax: (212)563-2269

Journal covering urban issues. **Freq:** 3/yr. **Key Personnel:** Prof. Rob Atkinson, Editor. **ISSN:** 1753-5069. **Subscription Rates:** US$285 institutions online only; US$78 individuals print and online; US$300 institutions print and online. **URL:** http://www.tandf.co.uk/journals/rurp.

KSCF-FM - See San Diego, California
WRAY-TV - See Wilson, North Carolina
WSAH-TV - See Bridgeport, Connecticut

OGDENSBURG

443 ■ WNCQ-FM - 102.9
1 Bridge Plz., Ste. 204
Ogdensburg, NY 13669
Phone: (315)393-1220
Fax: (315)393-3974

Format: Country. **Key Personnel:** Dave Merz, Program Dir., merz@yesfm.com. **URL:** http://www.q1029.com/index1.html.

OLEAN

444 ■ WOLN-FM - 91.3
3435 Main St.
205 Allen Hall
Buffalo, NY 14214-3003
Phone: (716)829-6000
Fax: (716)829-2277
Free: (888)829-6000

Format: Public Radio. **URL:** http://www.wbfo.org/about/.

OLIVEBRIDGE

445 ■ WFSO-FM - 88.3
PO Box 1520
Olivebridge, NY 12461
Phone: (845)657-6239

Format: Religious. **Owner:** Redeemer Broadcasting, at above address. **URL:** http://www.redeemerbroadcasting.org/pages/faq.php.

OSWEGO

WRVD-FM - See Syracuse
WRVJ-FM - See Watertown
WRVN-FM - See Utica

PLATTSBURGH

446 ■ WGLY-AM - 1070
PO Box 1018
Laguna Beach, CA 92652
Fax: (469)241-6795
Free: (800)639-5433

Format: Religious. **Owner:** New Life Ministries, at above address. **Key Personnel:** Stephen Arterburn, Founder/Chm. **URL:** http://www.newlife.com/Radio/findradio.asp?id=NY&t=New%20York.

POUGHKEEPSIE

447 ■ WKXP-FM - 94.3
PO Box 416
Poughkeepsie, NY 12602
Phone: (845)471-1500
Fax: (845)454-1204

Format: Country. **Key Personnel:** Chuck Benfer, General Mgr., gm@943thewolf.com. **URL:** http://www.943thewolf.com/.

PULASKI

448 ■ WGKV-FM - 101.7
PO Box 779002
Rocklin, CA 95677-9972
Free: (800)434-8400

Format: Religious. **Owner:** Educational Media Foundation, at above address. **Key Personnel:** Philip O'Bryan, Regional Mgr., phone (601)992-6988. **URL:** http://www.klove.com/Music/StationList.aspx.

ROCHESTER

449 ■ WDVI-FM - 100.5
207 Midtown Plz.
Rochester, NY 14604

Format: Adult Album Alternative. **Key Personnel:** Joe Bonacci, Program Dir., joebonacci@clearchannel.com. **URL:** http://www.mydrivefm.com/main.html.

450 ■ WFKL-FM - 93.3
1700 HSBC Plz.
Rochester, NY 14604

Format: Talk; Sports. **Key Personnel:** Joanna Schmidt, Sales Mgr. **URL:** http://wfkl-fm.fimc.net/.

ROME

451 ■ WYFY-AM - 1450
11530 Carmel Commons Blvd.
Charlotte, NC 28226
Phone: (704)523-5555
Free: (800)888-7077

Format: Religious. **Owner:** Bible Broadcasting Network, Inc., at above address. **URL:** http://www.bbnradio.org/wcm4/english/tabid/691/Default.aspx.

ROUSES POINT

452 ■ WKYJ-FM - 88.7
PO Box 779002
Rocklin, CA 95677-9972

Format: Religious. **Owner:** Educational Media Foundation, at above address. **Key Personnel:** Philip O'Bryan, Regional Mgr., phone (601)992-6988. **URL:** http://www.klove.com/Music/StationList.aspx.

SOUTHAMPTON

453 ■ WRLI-FM - 91.3
1049 Asylum Ave.
Hartford, CT 06105
Phone: (860)278-5310

Format: Public Radio. **Owner:** Connecticut Public Broadcasting Network, at above address. **URL:** http://www.cpbn.org/our-television-and-radio-networks.

SYRACUSE

WMHQ-FM - See Malone

454 ■ WRVD-FM - 90.3
7060 State Rte. 104
102 Penfield-SUNY
Oswego, NY 13126
Phone: (315)312-3690
Fax: (315)312-3174
Free: (800)341-3690

Format: Public Radio. **Wattage:** 285. **URL:** http://www.wrvo.fm/thewrvostations.html.

455 ■ WWLF-FM - 100.3
401 W Kirkpatrick St.
Syracuse, NY 13204
Phone: (315)472-0222
E-mail: programming@movin100.com

Format: Urban Contemporary; Adult Contemporary. **Ad Rates:** Advertising accepted; rates available upon request. **URL:** http://www.movin100.com/.

UTICA

456 ■ WRVN-FM - 91.9
7060 State Rte. 104
102 Penfield-SUNY
Oswego, NY 13126
Phone: (315)312-3690
Fax: (315)312-3174
Free: (800)341-3690

Format: Public Radio. **Wattage:** 1900. **URL:** http://www.wrvo.fm/thewrvostations.html.

WATERTOWN

457 ■ WRVJ-FM - 91.7
7060 State Rte. 104
102 Penfield-SUNY
Oswego, NY 13126
Phone: (315)312-3690
Fax: (315)312-3174
Free: (800)341-3690

Format: Public Radio. **Wattage:** 1600. **URL:** http://www.wrvo.fm/thewrvostations.html.

458 ■ WSLJ-FM - 88.9
80 E Main St.
EJ Noble Medical Bldg., Rm. 201
Canton, NY 13617-1450
Phone: (315)229-5356
Fax: (315)229-5373

Format: Public Radio. **Key Personnel:** Ellen Rocco, Station Mgr., ellen@ncpr.org; Shelly Pike, Operations Mgr., shelly@ncpr.org; Jackie Sauter, Program Dir., jackie@ncpr.org. **URL:** http://www.northcountrypublicradio.org/about/coverage.html.

YORKTOWN HEIGHTS

459 ■ International Journal of Business Process Integration & Management
Inderscience Publishers
c/o Liang-Jie Zhang, Ed.-in-Ch.
IBM T.J. Watson Research Center
1101 Kitchawan Rd., Rte. 134
Yorktown Heights, NY 10598
Publisher E-mail: editor@inderscience.com

Journal covering the emerging business process modeling, simulation, integration and management using emerging technologies. **Founded:** 2005. **Freq:** 4/yr. **Key Personnel:** Liang-Jie Zhang, Editor-in-Chief, zhanglj@us.ibm.com; Frank Leymann, Editor-in-Chief, frank.leymann@informatik.uni-stuttgart.de. **ISSN:** 1741-8763. **Subscription Rates:** EUR470 individuals includes surface mail, print only; EUR640 individuals print & online. **URL:** http://www.inderscience.com/browse/index.php?journalCODE=ijbpim.

NORTH CAROLINA

ASHEVILLE

460 ■ WISE-AM - 1310
90 Lookout Rd.
Asheville, NC 28804
Phone: (828)259-9695
Fax: (828)253-5619

Format: Sports. **Key Personnel:** Randy Cable, General Mgr., rcable@saganc.com; Ann Mays, Business Mgr., amays@saganc.com; Chris Hoffman, General Sales Mgr., choffman@saganc.com. **URL:** http://www.1310bigwise.com/.

461 ■ WMXF-AM - 1400
13 Summerlin Rd.
Asheville, NC 28806
Phone: (828)257-2700

Format: Big Band/Nostalgia. **Owner:** Clear Channel Communications, 200 E Basse Rd., San Antonio, TX 78209, (210)822-2929, Fax: (210)832-3428. **URL:** http://www.am1400thepeak.com/main.html.

462 ■ WSKY-AM - 1230
40 Westgate Pky., Ste. F
Asheville, NC 28806
Phone: (828)251-2000

Format: Religious. **Owner:** Wilkins Communications Network, Inc., 292 S Pine St., Spartanburg, SC 29302, (864)585-1885, Fax: (864)597-0687, Free: (888)989-2299. **Key Personnel:** Ruthie Spears, Station Mgr., wsky@wilkinsradio.com. **Wattage:** 1000. **URL:** http://www.wilkinsradio.com/.

463 ■ WTMT-FM - 105.9
1190 Patton Ave.
Asheville, NC 28806
Phone: (828)259-9695
Fax: (828)253-5619

Format: Classic Rock. **Owner:** Saga Communications of NC, LLC, at above address. **Key Personnel:** Matt Stone, Program Dir., mstone@saganc.com; Chris Hoffman, Sales Mgr., choffman@saganc.com. **Wattage:** 50,000. **URL:** http://www.1059themountain.com/.

CARY

464 ■ Adaptation
Oxford University Press
2001 Evans Rd., Ste. 12
Cary, NC 27513
Phone: (919)677-0977
Fax: (919)677-1303
Free: (800)445-9714
Publisher E-mail: custserv.us@oup.com

Journal covering academic articles, film, and book reviews. **Freq:** Semiannual. **Key Personnel:** Deborah Cartmell, Editor; Timothy Corrigan, Editor; Imelda Whelehan, Editor. **ISSN:** 1755-0637. **Subscription Rates:** US$210 institutions print and online; US$200 institutions print only or online only; US$45 individuals print; US$125 institutions single copy; US$28 single issue. **Remarks:** Accepts advertising. **URL:** http://www.oxfordjournals.org/our_journals/adaptation/. **Circ:** (Not Reported)

465 ■ Bioscience Horizons
Oxford University Press
2001 Evans Rd., Ste. 12
Cary, NC 27513
Phone: (919)677-0977
Fax: (919)677-1303
Free: (800)445-9714
Publisher E-mail: custserv.us@oup.com

Journal featuring bioscience research. **Key Personnel:** Dr. Celia Knight, Ch. **Remarks:** Accepts advertising. **URL:** http://www.oxfordjournals.org/our_journals/biohorizons/. **Circ:** (Not Reported)

466 ■ InnovAiT
Oxford University Press
2001 Evans Rd., Ste. 12
Cary, NC 27513
Phone: (919)677-0977
Fax: (919)677-1303
Free: (800)445-9714
Publisher E-mail: custserv.us@oup.com

Journal covering new policies, research, and guidelines affecting the General Practitioners. **Freq:** Monthly. **Key Personnel:** Chantal Simon, Editor. **ISSN:** 1755-7380. **Remarks:** Accepts advertising. **URL:** http://www.oxfordjournals.org/our_journals/innovait/. **Circ:** (Not Reported)

467 ■ International Mathematics Research Notices
Oxford University Press
2001 Evans Rd., Ste. 12
Cary, NC 27513
Phone: (919)677-0977
Fax: (919)677-1303
Free: (800)445-9714
Publisher E-mail: custserv.us@oup.com

Journal covering areas of mathematics. **Freq:** 24/yr. **Key Personnel:** Morris Weisfeld, Editor-in-Chief. **ISSN:** 1073-7928. **Subscription Rates:** US$3,090 institutions print and online; 2,936 institutions online only or print only; US$153 single issue. **Remarks:** Accepts advertising. **URL:** http://imrn.oxfordjournals.org/. **Circ:** (Not Reported)

468 ■ Journal of Plant Ecology
Oxford University Press
2001 Evans Rd., Ste. 12
Cary, NC 27513
Phone: (919)677-0977
Fax: (919)677-1303
Free: (800)445-9714
Publisher E-mail: custserv.us@oup.com

Journal covering plant ecology. **Freq:** Quarterly. **Key Personnel:** Guanghui Lin, Editor-in-Chief; Bernhard Schmid, Editor-in-Chief; Shiqiang Wan, Editor-in-Chief. **ISSN:** 1752-9921. **Subscription Rates:** US$400 institutions; 400 individuals; US$125 single issue. **Remarks:** Accepts advertising. **URL:** http://jpe.oxfordjournals.org/. **Circ:** (Not Reported)

469 ■ Journal of Topology
Oxford University Press
2001 Evans Rd., Ste. 12
Cary, NC 27513
Phone: (919)677-0977
Fax: (919)677-1303
Free: (800)445-9714
Publisher E-mail: custserv.us@oup.com

Journal covering topology, geometry, and adjacent areas of mathematics. **Freq:** Quarterly. **Key Personnel:** Ulrike Tillmann, Managing Editor. **ISSN:** 1753-8416. **Subscription Rates:** US$570 institutions print and online; US$178 institutions single copy. **Remarks:** Accepts advertising. **URL:** http://jtopol.oxfordjournals.org/. **Circ:** (Not Reported)

Circulation: ★ = ABC; △ = BPA; ♦ = CAC; • = CCAB; ❑ = VAC; ⊕ = PO Statement; ‡ = Publisher's Report; Boldface figures = sworn; Light figures = estimated.

470 ■ Journal of World Energy Law & Business
Oxford University Press
2001 Evans Rd., Ste. 12
Cary, NC 27513
Phone: (919)677-0977
Fax: (919)677-1303
Free: (800)445-9714
Publisher E-mail: custserv.us@oup.com

Journal covering legal, business, and policy issues in the international energy industry. **Freq:** 3/yr. **Key Personnel:** Thomas Walde, Editor-in-Chief. **Subscription Rates:** US$475 institutions print and online; US$452 institutions print only or online only; US$181 individuals print only; US$190 members; US$188 institutions single copy; US$75 single issue. **Remarks:** Accepts advertising. **URL:** http://www.oxfordjournals.org/our_journals/jwelb/. **Circ:** (Not Reported)

471 ■ Literary Imagination
Oxford University Press
2001 Evans Rd., Ste. 12
Cary, NC 27513
Phone: (919)677-0977
Fax: (919)677-1303
Free: (800)445-9714
Publisher E-mail: custserv.us@oup.com

Journal covering academic literary study. **Key Personnel:** Peter Campion, Editor. **ISSN:** 1523-9012. **Remarks:** Accepts advertising. **URL:** http://litimag.oxfordjournals.org/. **Circ:** (Not Reported)

472 ■ Molecular Plant
Oxford University Press
2001 Evans Rd., Ste. 12
Cary, NC 27513
Phone: (919)677-0977
Fax: (919)677-1303
Free: (800)445-9714
Publisher E-mail: custserv.us@oup.com

Journal covering plant biology. **Key Personnel:** Sheng Luan, Editor-in-Chief. **ISSN:** 1674-2052. **Remarks:** Accepts advertising. **URL:** http://mplant.oxfordjournals.org/. **Circ:** (Not Reported)

473 ■ NDT PLUS
Oxford University Press
2001 Evans Rd., Ste. 12
Cary, NC 27513
Phone: (919)677-0977
Fax: (919)677-1303
Free: (800)445-9714
Publisher E-mail: custserv.us@oup.com

Journal featuring practical and clinically oriented information for nephrologist. **Freq:** Bimonthly. **Key Personnel:** N. Lameire, Editor-in-Chief. **ISSN:** 1753-0784. **Remarks:** Accepts advertising. **URL:** http://ndtplus.oxfordjournals.org/. **Circ:** (Not Reported)

474 ■ Public Health Ethics
Oxford University Press
2001 Evans Rd., Ste. 12
Cary, NC 27513
Phone: (919)677-0977
Fax: (919)677-1303
Free: (800)445-9714
Publisher E-mail: custserv.us@oup.com

Journal covering issues on public health. **Freq:** 3/yr. **Key Personnel:** Dr. Angus Dawson, Editor; Dr. Marcel Verweij, Editor. **ISSN:** 1072-7928. **Subscription Rates:** US$250 institutions print and online; US$238 institutions print only or online only; US$80 individuals; US$99 institutions single copy; US$33 single issue. **Remarks:** Accepts advertising. **URL:** http://www.oxfordjournals.org/our_journals/phe/. **Circ:** (Not Reported)

CHARLOTTE

475 ■ WBAV-FM - 101.9
1520 South Blvd., Ste. 300
Charlotte, NC 28203
Phone: (704)227-8126

Format: Urban Contemporary. **Owner:** CBS Radio Stations Inc., 1515 Broadway, New York, NY 10036, (212)846-3939. **Key Personnel:** Rob Grossman, General Sales Mgr., phone (704)227-8713, rob.grossman@infinitybroadcasting.com. **Ad Rates:** Advertising accepted; rates available upon request. **URL:** http://www.v1019.com/.

476 ■ WDYT-AM - 1220
131 Providence Rd.
Charlotte, NC 28207
Phone: (704)295-7901
Fax: (704)295-7919

Format: Talk. **Owner:** CRN Communications Network, at above address. **Key Personnel:** Deanna Greco, General Mgr., deanna@1220wdyt.com; Jenifer Roser, General Sales Mgr., phone (704)716-2730, jenifer@1220wdyt.com; Bo Thompson, Program Dir., bo@1220wdyt.com. **URL:** http://www.1220wdyt.com/.

477 ■ WFNA-AM - 1660
1520 South Blvd., Ste. 300
Charlotte, NC 28203
Phone: (704)227-8000

Format: Sports. **Owner:** CBS Radio Stations Inc., 1515 Broadway, New York, NY 10036, (212)846-3939. **Key Personnel:** Anna Lipford, Sales Mgr., anna.lipford@cbsradio.com. **URL:** http://www.wfnz.com/.

WGSP-FM - See Pageland, South Carolina

WOGR-FM - See Salisbury

WXNC-AM - See Monroe

WYFY-AM - See Rome, New York

ELIZABETH CITY

478 ■ WGPS-FM - 88.3
905 Halstead Blvd., Ste. 29
Elizabeth City, NC 27909
Phone: (252)334-1883
Fax: (252)333-1459
Free: (866)444-9477
E-mail: wgpsradio@earthlink.net

Format: Religious. **Owner:** CSN International, PO Box 391, Twin Falls, ID 83303, Fax: (208)736-1958, Free: (800)357-4226. **Key Personnel:** Jeff Ozanne, Station Mgr.; Darla Ozanne, Program Dir.; Leslie Passante, Production Asst. **Wattage:** 50,000. **URL:** http://www.wgpsradio.com/.

ELIZABETHTOWN

479 ■ WDJD-FM - 93.7
PO Box 1018
Laguna Beach, CA 92652
Fax: (469)241-6795
Free: (800)639-5433

Format: Religious. **Owner:** New Life Ministries, at above address. **Key Personnel:** Stephen Arterburn, Founder/Chm. **URL:** http://www.newlife.com/Radio/findradio.asp?id=NC&t=North%20Carolina.

GERMANTON

480 ■ KJBB-FM - 89.1
6704 Hwy. 8 S
Germanton, NC 27019
Phone: (605)868-4567
E-mail: kjbbradio89.1fm@dailypost.com

Format: Southern Gospel. **Operating Hours:** Continuous. **URL:** http://www.kjbbfm.com/.

GOLDSBORO

481 ■ WAGO-FM - 88.7
PO Box 1895
Goldsboro, NC 27533-1895
Phone: (252)747-8887
Fax: (252)747-7888
E-mail: wago@gomixradio.org

Format: Gospel. **URL:** http://www.gomixradio.org/.

482 ■ WZGO-FM - 91.1
PO Box 1895
Goldsboro, NC 27533-1895
Phone: (252)747-8887
Fax: (252)747-7888
E-mail: wago@gomixradio.org

Format: Contemporary Christian. **URL:** http://www.gomixradio.org/.

GREENVILLE

483 ■ WECU-AM - 1570
1413-D Evans St.
PO Box 1534
Greenville, NC 27835
Phone: (252)258-8301
Fax: (252)931-9328

Format: Gospel; Urban Contemporary. **Owner:** CTC Media Group, Inc., 5562 Heron Point Rd., Royal Oak, MD 21662, (410)745-5958. **Ad Rates:** Advertising accepted; rates available upon request. **URL:** http://www.wecu1570.com/.

HIGH POINT

484 ■ WBLO-AM - 790
1607 Country Club Dr.
PO Box 5663
High Point, NC 27260
Phone: (336)887-0983
Fax: (336)887-3055

Format: Sports; Talk. **URL:** http://www.790theball.com/.

JACKSONVILLE

485 ■ WLGP-FM - 100.3
PO Box 510
Appling, GA 30802
Phone: (706)309-9610
Fax: (706)309-9669
Free: (800)926-4669
E-mail: ctbarinowski@comcast.net

Format: Religious. **URL:** http://www.gnnradio.org/.

MONROE

486 ■ WXNC-AM - 1060
4801 E Independence Blvd., Ste. 803
Charlotte, NC 28212
Phone: (704)442-7277
Fax: (704)442-9518
Free: (888)-770-1310

Format: Hispanic. **URL:** http://latremendaradio.com/main.php.

NEW BERN

487 ■ WRHD-FM - 103.7
1307 S Glenburnie Rd.
New Bern, NC 28562
Phone: (252)672-5900

Format: Country. **Owner:** Inner Banks Media, at above address. **Key Personnel:** Henry Hinton, General Mgr., henry@ibxmedia.com; Hank Hinton, General Sales Mgr., hank@ibxmedia.com; Mike Biddle, Program Dir., maddawg@ibxmedia.com. **URL:** http://www.thundercountryonline.com/.

488 ■ WSTK-FM - 104.5
233 Middle St., Ste. 207
New Bern, NC 28560
Free: (877)389-1383

Format: Southern Gospel. **URL:** http://www.1045live.com/.

489 ■ WWNB-FM - 1490
116 S Business Plz.
New Bern, NC 28562
Phone: (252)633-1490
Fax: (252)636-5848
E-mail: info@wwnb1490.com

Format: Sports; Talk. **Owner:** CTC Media Group, Inc., at above address. **URL:** http://www.wwnb1490.com/.

490 ■ WZNB-FM - 88.5
800 Colorado Ct.
New Bern, NC 28562
Fax: (252)638-3538
Free: (800)222-9832

Format: Public Radio. **URL:** http://www.publicradioeast.org/.

NEWTON GROVE

491 ■ WPRZ-FM - 90.7
2351 Sunset Blvd., Ste. 170-218
Rocklin, CA 95765
Phone: (916)251-1600
Fax: (916)251-1650

Format: Contemporary Christian. **Owner:** Educational Media Foundation, at above address. **URL:** http://www.air1.com/Music/StationList.aspx.

RAEFORD

492 ■ WRAE-FM - 88.7
PO Drawer 2440
Tupelo, MS 38803
Phone: (662)844-5036
Fax: (662)842-7798

Format: Contemporary Christian. **Owner:** American Family Association, at above address. **URL:** http://www.afr.net/newafr/stationsbystate/northcarolina.asp.

RALEIGH

493 ■ WDOX-AM - 570
3012 Highwoods Blvd., Ste. 201
Raleigh, NC 27604
Phone: (919)790-9392
Fax: (919)790-8369

Format: News; Talk. **Owner:** Curtis Media Group, at above address. **Key Personnel:** Rick Heilmann, Gen./Sales Mgr., rheilmann@curtismedia.com; Pete Richon, Prog./Oper. Mgr., prichon@curtismedia.com. **Wattage:** 1000 Day; 52 Night. **Ad Rates:** Advertising accepted; rates available upon request. **URL:** http://www.570wdox.com/.

494 ■ WDRU-AM - 1030
6200 Falls of Neuse Rd., Ste. 200
Raleigh, NC 27609
Phone: (919)501-7604
Fax: (919)878-9799

Format: Religious. **Owner:** Truth Radio Network, 4405 Providence Ln., Ste. D, Winston-Salem, NC 27106, (336)759-0363, Fax: (336)759-0366. **URL:** http://www.wtru.com/contact/general-questions/.

495 ■ WLLQ-AM - 1530
150 Fayetteville St. Mall, Ste. 110
Raleigh, NC 27601
Phone: (919)645-1680
Fax: (919)645-1699

Format: Hispanic; Sports. **Operating Hours:** Continuous. **URL:** http://www.quepasamedia.com/web/content/view/13/66/lang,en/.

496 ■ WYMY-FM - 96.9
3012 Highwoods Blvd., Ste. 201
Raleigh, NC 27604
Phone: (919)790-9392
Fax: (919)790-8996

Format: Hispanic. **Owner:** Curtis Media Group, at above address. **Operating Hours:** Continuous. **Key Personnel:** Jon Bloom, Gen./Sales Mgr., jbloom@curtismedia.com; Julie Garza, Program Dir., jgarza@curtismedia.com. **Wattage:** 100,000. **Ad Rates:** Advertising accepted; rates available upon request. **URL:** http://www.curtismedia.com/stations.htm.

REIDSVILLE

497 ■ WCLW-AM - 1130
116 S Franklin St.
Reidsville, NC 27320
Phone: (336)634-1345

Format: Religious. **Owner:** Reidsville Baptist Church, at above address. **Key Personnel:** Dean Lundy, Program Dir. **URL:** http://www.reidsvillebaptist.org/WCLW/index.html.

ROCKY MOUNT

498 ■ WDWG-FM - 98.5
12714 E NC 97
Rocky Mount, NC 27803
Phone: (252)442-8092

Format: Country. **Owner:** First Media Radio, LLC, at above address. **Key Personnel:** Christian Hardy, Dir. of Sales, chardy@firstmedianc.com; Mike Binkley, General Mgr., mbinkley@firstmedianc.com; David Perkins, Operations Mgr., dperkins@firstmedianc.com. **Ad Rates:** Advertising accepted; rates available upon request. **URL:** http://www.bigdawg985.com/.

499 ■ WPWZ-FM - 95.5
12714 E NC 97
Rocky Mount, NC 27803
Phone: (252)442-8092

Format: Urban Contemporary. **Owner:** First Media Radio, LLC, at above address. **URL:** http://www.powerhits95.com/.

500 ■ WZAX-FM - 99.3
12714 E NC 97
Rocky Mount, NC 27803
Phone: (252)442-8092

Format: Adult Contemporary. **Owner:** First Media Radio, LLC, at above address. **Key Personnel:** Christian Hardy, Dir. of Sales, chardy@firstmedianc.com; Mike Binkley, General Mgr., mbinkley@firstmedianc.com; David Perkins, Operations Mgr., dperkins@firstmedianc.com. **Ad Rates:** Advertising accepted; rates available upon request. **URL:** http://www.zaxradio.com/.

SALISBURY

501 ■ WOGR-FM - 93.3
PO Box 16408
Charlotte, NC 28297
Phone: (704)393-1540
Fax: (704)393-1527

Format: Religious. **Owner:** Victory Christian Center, 7728 Old Pineville Rd., Charlotte, NC 28244, (704)602-6010, Fax: (704)602-6046. **URL:** http://www.wordnet.org/.

SANFORD

502 ■ WLHC-FM - 103.1
102 S Steele St., Ste. 301
Sanford, NC 27330
Phone: (919)775-1031
Fax: (919)775-1397
E-mail: wlhc@life1031.com

Format: Adult Contemporary; Middle-of-the-Road (MOR). **Key Personnel:** Al Mangum, Station Mgr., almangum@life1031.com; Mary Button, Contact, mary@life1031.com. **URL:** http://life1031.com/.

WILLIAMSTON

503 ■ WGTI-FM - 97.7
PO Box 590
Williamston, NC 27892

Format: Religious. **Owner:** Johnny Bryant, at above address. **Key Personnel:** Johnny Bryant, Station Mgr./Owner, bryant@opendoorradio.com. **URL:** http://www.opendoorradio.com/.

WILMINGTON

504 ■ WAZO-FM - 107.5
25 N Kerr
Wilmington, NC 28405
Phone: (910)791-3088
Fax: (910)791-0112

Format: Contemporary Hit Radio (CHR). **URL:** http://www.z1075.com/.

505 ■ WBNE-FM - 103.7
122 Cinema Dr.
Wilmington, NC 28403
Phone: (910)772-6300
Fax: (910)772-6310

Format: Classic Rock. **Owner:** Sea-Comm Media, Inc., at above address. **Key Personnel:** Paul Knight, General Mgr., paul.k@sea-comm.com; Max Deutsch, General Sales Mgr., maxdut@sea-comm.com; Zach McHugh, Program Dir., zachm@sea-comm.com. **Ad Rates:** Advertising accepted; rates available upon request. **URL:** http://www.1037thebone.com/.

506 ■ WNTB-FM - 93.7
122 Cinema Dr.
Wilmington, NC 28403
Phone: (910)772-6300
Fax: (910)772-6310

Format: Talk. **Owner:** Sea-Comm Media, at above address. **URL:** http://www.thebigtalkerfm.com/.

507 ■ WUIN-FM - 106.7
122 Cinema Dr.
Wilmington, NC 28403
Phone: (910)332-1067

Format: Adult Album Alternative. **Key Personnel:** Beau Gunn, Program Dir. **URL:** http://www.1067thepenguin.com/.

WILSON

508 ■ WRAY-TV - 30
449 Broadway
New York, NY 10013
Phone: (212)966-1059
Fax: (212)966-9580

Owner: Multicultural Radio Broadcasting, Inc., at above address. **Key Personnel:** John Gabel, Contact, phone (212)966-1059, johng@mrbi.net. **URL:** http://www.mrbi.net/tvgroup.htm.

WINSTON-SALEM

509 ■ Involve
Mathematical Sciences Publishers
c/o Kenneth S. Berenhaut, Mng. Ed.
Department of Mathematics
Wake Forest University
Winston-Salem, NC 27109
Publisher E-mail: contact@mathscipub.org

Journal covering mathematics. **Founded:** 2007. **Freq:** Semiannual. **Key Personnel:** Kenneth S. Berenhaut, Managing Editor, berenhks@wfu.edu; Paulo Ney de Souza, Production Mgr., desouza@math.berkeley.edu. **Subscription Rates:** US$100 individuals. **Remarks:** Accepts advertising. **URL:** http://pjm.math.berkeley.edu/inv/about/cover/cover.html; http://involvemath.org. **Circ:** (Not Reported)

NORTH DAKOTA

BISMARCK

510 ■ KBFR-FM - 91.7
290 Hegenberger Rd.
Oakland, CA 94621
Free: (800)543-1495

Format: Religious. **Owner:** Family Radio Network, at above address. **URL:** http://www.familyradio.com/.

511 ■ KKCT-FM - 97.5
1830 N 11th St.
Bismarck, ND 58501
Phone: (701)250-6602
Fax: (701)250-6632

Format: Top 40. **Owner:** Cumulus Organization, 3280 Peachtree Rd. NW, Ste. 2300, Atlanta, GA 30305, (404)949-0700, Fax: (404)949-0740. **URL:** http://www.hot975fm.com/?pid=8968.

Circulation: ★ = ABC; △ = BPA; ♦ = CAC; • = CCAB; ❑ = VAC; ⊕ = PO Statement; ‡ = Publisher's Report; Boldface figures = sworn; Light figures = estimated.

CASSELTON

512 ■ KVMI-FM - 103.9
4 Langer Ave.
Casselton, ND 58012
Phone: (701)347-5005
Fax: (701)347-5508

Format: Country; Hot Country. **Owner:** Vision Media, at above address. **Key Personnel:** Mike McClain, General Mgr., mike@1039thetruck.com; Jim Babbitt, Owner; Lisa Schaffer, Office Mgr., lisa@1039thetruck.com. **URL:** http://www.1039thetruck.com/.

DEVILS LAKE

513 ■ KDVI-FM - 89.9
PO Drawer 2440
Tupelo, MS 38803
Phone: (662)844-5036
Fax: (662)842-7798

Format: Contemporary Christian. **Owner:** American Family Radio, at above address. **URL:** http://www.afr.net/newafr/stationsbystate/northdakota.asp.

DICKINSON

514 ■ KPAR-FM - 103.3
PO Box 1018
Laguna Beach, CA 92652
Fax: (469)241-6795
Free: (800)639-5433

Format: Religious. **Owner:** New Life Ministries, at above address. **Key Personnel:** Stephen Arterburn, Founder/Chm. **URL:** http://www.newlife.com/Radio/findradio.asp?id=ND&t=North%20Dakota.

FARGO

515 ■ KMJO-FM - 104.7
1020 S 25th St.
Fargo, ND 58103
Phone: (701)237-5346
Fax: (701)237-0980
E-mail: studio@mojo104.com

Format: Classic Rock. **Key Personnel:** Carlos Quintero, Director, phone (701)729-7378, carlosquintero@radiofargomoorhead.com. **Ad Rates:** Advertising accepted; rates available upon request. **URL:** http://www.mojo104.com/.

GRAND FORKS

516 ■ Agripics
Issues Ink
1395-A S Columbia Rd.
PO Box 360
Grand Forks, ND 58201
Free: (877)710-3222
Publisher E-mail: issues@issuesink.com

Magazine featuring image collection of the faces, places, processes, and products of agriculture. **Key Personnel:** Robynne Anderson, President. **Subscription Rates:** C$150 individuals one month; C$500 individuals. **URL:** http://www.issuesink.com/agricpics.html; http://www.agripics.com/.

517 ■ Flavourful
Issues Ink
1395-A S Columbia Rd.
PO Box 360
Grand Forks, ND 58201
Free: (877)710-3222
Publisher E-mail: issues@issuesink.com

Magazine featuring Canada's agri-food products. **Freq:** Semiannual. **Key Personnel:** Robynne Anderson, President. **URL:** http://www.issuesink.com/flavourful.html.

518 ■ Germination
Issues Ink
1395-A S Columbia Rd.
PO Box 360
Grand Forks, ND 58201
Free: (877)710-3222
Publisher E-mail: issues@issuesink.com

Magazine featuring the latest technological developments in Canadian seed industry. **Key Personnel:** Robynne Anderson, President. **URL:** http://www.issuesink.com/germination.html. **Circ:** 5,700

519 ■ Seed Week
Issues Ink
1395-A S Columbia Rd.
PO Box 360
Grand Forks, ND 58201
Free: (877)710-3222
Publisher E-mail: issues@issuesink.com

Magazine featuring latest synopsis of news, views, and breakthroughs about the seed industry. **Freq:** Weekly. **Key Personnel:** Robynne Anderson, President. **URL:** http://www.issuesink.com/seedweek.html; http://www.seedquest.com/hosting/seedworld/seedweek.htm.

520 ■ seed.ab.ca
Issues Ink
1395-A S Columbia Rd.
PO Box 360
Grand Forks, ND 58201
Free: (877)710-3222
Publisher E-mail: issues@issuesink.com

Magazine featuring Alberta seed industry. **Key Personnel:** Robynne Anderson, President. **URL:** http://www.issuesink.com/seedabca.html. **Circ:** 64,000

521 ■ Spud Smart
Issues Ink
1395-A S Columbia Rd.
PO Box 360
Grand Forks, ND 58201
Free: (877)710-3222
Publisher E-mail: issues@issuesink.com

Magazine featuring Canadaian potato industry. **Freq:** 4/yr. **Key Personnel:** Lindsay Hoffman, Contact, lhoffman@issuesink.com. **Subscription Rates:** C$47.70 individuals; US$45 other countries. **URL:** http://www.issuesink.com/spudsmart.html; http://www.spudsmart.com/. **Circ:** 3,400

522 ■ KWTL-AM - 1370
216 Belmont Dr.
Grand Forks, ND 58201
Phone: (701)795-0122
Free: (877)795-0122
E-mail: stationmanager@realpresenceradio.com

Format: Religious. **URL:** http://www.youram1370.com/.

LANGDON

523 ■ KAOC-FM - 105.1
1403 3rd St.
Langdon, ND 58249
Phone: (701)256-1080
Fax: (701)256-1081

Format: Country. **Ad Rates:** Advertising accepted; rates available upon request. **URL:** http://www.maverick105fm.com/.

MANDAN

524 ■ KUSB-FM - 103.3
4303 Memorial Hwy.
Mandan, ND 58554
Phone: (701)663-1033
Free: (800)665-1033

Format: Country. **Owner:** Cumulus Broadcasting, Inc., 3280 Peachtree Rd. NW, Ste. 2300, Atlanta, GA 30305, (404)949-0700, Fax: (404)949-0740. **Key Personnel:** Mike Rose, Program Dir., mike.rose@cumulus.com. **URL:** http://www.uscountryonline.com/.

OHIO

ALLIANCE

525 ■ WDJQ-FM - 92.5
393 Smyth Ave. NE
Alliance, OH 44601
Phone: (330)450-9250
Fax: (330)821-0379

Format: Contemporary Hit Radio (CHR). **Key Personnel:** Don Peterson III, General Mgr., dpiii@q92radio.com. **Wattage:** 50,000. **URL:** http://www.q92radio.com/.

ANNA

526 ■ WHJM-FM - 88.7
601 Washington St.
Alexandria, LA 71301
Free: (888)408-0201

Format: Religious. **Owner:** Radio Maria, Inc., at above address. **URL:** http://www.radiomaria.us/.

ATHENS

WOUH-FM - See Chillicothe

CAMBRIDGE

527 ■ WBIK-FM - 92.1
4988 Skyline Dr.
PO Box 338
Cambridge, OH 43725
Phone: (740)432-5605

Format: Classic Rock. **Owner:** AVC Communications, at above address. **URL:** http://www.yourradioplace.com/WBIK/index.htm.

CANTON

528 ■ WILB-AM - 1060
4365 Fulton Dr.
Canton, OH 44718
Phone: (330)966-2903

Format: Religious. **Owner:** Living Bread Radio, at above address. **Key Personnel:** Dan Clark, Operations Mgr., dmctunes@hotmail.com. **URL:** http://www.livingbreadradio.com/index.php.

CHILLICOTHE

529 ■ WOUH-FM - 91.9
RTVC Bldg.
9 S College St.
Athens, OH 45701
Phone: (740)593-1771
Fax: (740)593-0240
Free: (800)456-2044
E-mail: radio@woub.org

Format: Public Radio; Classical. **Owner:** WOUB Center for Public Media, at above address. **Operating Hours:** Continuous. **URL:** http://www.tcom.ohiou.edu/radio.html.

530 ■ WZRP-FM - 89.3
41 1/2 S Paint St.
Chillicothe, OH 45601
Phone: (740)779-3566
Fax: (740)779-3568
Free: (800)893-1628

Format: Religious. **Key Personnel:** Dan Baughman, General Mgr., dbaughman@promiseradionetwork.com; Scott Thomson, CFO, sthomson@promiseradionetwork.com; Scott Saunders, Program Dir., ssaunders@promiseradionetwork.com. **URL:** http://richmond.promiseradionetwork.com/.

CINCINNATI

531 ■ WCVX-AM - 1050
PO Box 1018
Laguna Beach, CA 92652
Fax: (469)241-6795
Free: (800)639-5433

Format: Religious. **Owner:** New Life Ministries, at above address. **Key Personnel:** Stephen Arterburn, Founder/Chm. **URL:** http://www.newlife.com/Radio/findradio.asp?id=OH&t=Ohio.

532 ■ WNNF-FM - 94.1
8044 Montgomery Rd., Ste. 650
Cincinnati, OH 45236
Phone: (513)686-8300

Format: Adult Contemporary; Alternative/New Music/Progressive. **Key Personnel:** Matt Tobin, Contact, phone (513)686-8431, matttobin@clearchannel.com. **URL:** http://www.radio941.com/main.html.

533 ■ WSWD-FM - 94.9
2060 Reading Rd.
Cincinnati, OH 45202
Phone: (513)749-0949
Fax: (513)699-5000

Format: Alternative/New Music/Progressive. **Key Personnel:** Jay Kruz, Program Dir., phone (513)699-5078, jay@949thesound.com. **URL:** http://www.949thesound.com/.

CLEVELAND

534 ■ Journal of Polymer Engineering
Freund Publishing House Ltd.
Department of Macromolecular Science
Case Western Reserve University
Cleveland, OH 44106
Publisher E-mail: h_freund@netvision.net.il

Journal covering developments in the field of polymer engineering. **Freq:** 9/yr. **Key Personnel:** Ica Manas-Zloczower, Editor-in-Chief, ixm@po.cwru.edu; Robert E. Cohen, Editor, recohen@mit.edu; Tim A. Osswald, Editor, osswald@engr.wisc.edu; Mosto Bousmina, Editor, tim@danu.me.wisc.edu; Nino Grizzuti, Editor, nino.grizzuti@unina.it; Michael S. Silverstein, Editor, nino.grizzuti@unina.it; T.H. Kwon, Editor, thkwon@postech.ac.kr; T. Kanai, Editor, toshitaka.kanai@si.idemitsu.co.jp; Shih-Jung Liu, Editor, shihjung@mail.cgu.edu.tw. **ISSN:** 0250-8079. **Subscription Rates:** US$480 individuals. **URL:** http://www.freundpublishing.com/JOURNALS/materials_science_and_engineering.htm.

COLUMBUS

535 ■ WTDA-FM - 103.9
1458 Dublin Rd.
Columbus, OH 43215

Format: Talk. **Owner:** North American Broadcasting Co. Inc., at above address. **URL:** http://www.tedfm.com/.

DAYTON

536 ■ WDPG-FM - 89.9
126 N Main St.
Dayton, OH 45402
Phone: (937)496-3850
Fax: (937)496-3852

Format: Classical. **Owner:** Dayton Public Radio, Inc., at above address. **Key Personnel:** Georgie M. Woessner, General Mgr., gmw@dpr.org; Shaun Yu, Program Dir., shauny@dpr.org; Larry Coressel, Production Mgr., larryc@dpr.org. **URL:** http://www.dpr.org/.

537 ■ WUDR-FM - 98.1
300 College Pk.
Dayton, OH 45469-2060
Phone: (937)229-2774

Format: Full Service. **Owner:** University of Dayton, at above address. **Key Personnel:** Aaron Moores, President, flyermusic@gmail.com; Casey Drottar, General Mgr., drottaca@notes.udayton.edu; Laura Steffey, General Mgr., steffelm@notes.udayton.edu. **Ad Rates:** Noncommercial. **URL:** http://flyer-radio.udayton.edu/home.htm.

DELAWARE

538 ■ WJJE-FM - 89.1
PO Drawer 2440
Tupelo, MS 38803
Phone: (662)844-5036
Fax: (662)842-7798

Format: Contemporary Christian. **Owner:** American Family Association, at above address. **URL:** http://www.afr.net/newafr/stationsbystate/ohio.asp.

NEWARK

539 ■ WJHE-FM - 98.7
PO Box 1018
Laguna Beach, CA 92652
Fax: (469)241-6795
Free: (800)639-5433

Format: Religious. **Owner:** New Life Ministries, at above address. **Key Personnel:** Stephen Arterburn, Founder/Chm. **URL:** http://www.newlife.com/Radio/findradio.asp?id=OH&t=Ohio.

OBERLIN

540 ■ WDLW-AM - 1380
PO Box 277
Oberlin, OH 44074
Phone: (440)774-1320
Fax: (440)774-1336

Format: Oldies. **Key Personnel:** Doug Wilber, Pres./Gen. Mgr. **URL:** http://www.koolkatwdlw.com/.

RICHFIELD

541 ■ Bath Herald
ScripType Publishing Inc.
4300 W Streetsboro Rd.
Richfield, OH 44286
Phone: (330)659-0303
Fax: (330)659-9488

Magazine featuring township government and local news in Bath. **Founded:** 2005. **Freq:** Monthly. **Key Personnel:** Nancy L. Cushing, Editor, ncushing@scriptype.com; Sue Serdinak, Publisher, sserdinak@scriptype.com. **Remarks:** Accepts advertising. **URL:** http://www.scriptype.com/pages/cities/bath.html. . **Ad Rates:** BW: $425. **Circ:** 8,150

542 ■ Brecksville Magazine
ScripType Publishing Inc.
4300 W Streetsboro Rd.
Richfield, OH 44286
Phone: (330)659-0303
Fax: (330)659-9488

Magazine featuring township government and local news in Brecksville. **Founded:** 1990. **Freq:** Monthly. **Key Personnel:** Marge Palik, Editor, mpalik@scriptype.com; Sue Serdinak, Publisher, sserdinak@scriptype.com. **Remarks:** Accepts advertising. **URL:** http://www.scriptype.com/pages/cities/brecksville.html. . **Ad Rates:** BW: $425. **Circ:** 8,150

543 ■ BroadView Journal
ScripType Publishing Inc.
4300 W Streetsboro Rd.
Richfield, OH 44286
Phone: (330)659-0303
Fax: (330)659-9488

Journal featuring township government and local news in Broadview Heights. **Founded:** 1990. **Freq:** Monthly. **Key Personnel:** Nancy Hudec, Editor, nhudec@sciptype.com; Sue Serdinak, Publisher, sserdinak@scriptype.com. **Remarks:** Accepts advertising. **URL:** http://www.scriptype.com/pages/cities/broadview.html. **Circ:** 8,700

544 ■ Hinckley Record
ScripType Publishing Inc.
4300 W Streetsboro Rd.
Richfield, OH 44286
Phone: (330)659-0303
Fax: (330)659-9488

Magazine featuring government and local news in Hinckley Township. **Founded:** 1992. **Freq:** Monthly. **Key Personnel:** Marge Palik, Editor, mpalik@sciptype.com; Sue Serdinak, Publisher, sserdinak@scriptype.com. **Remarks:** Accepts advertising. **URL:** http://www.scriptype.com/pages/cities/hinckley.html. . **Ad Rates:** BW: $400. **Circ:** 3,450

545 ■ Hudson Life
ScripType Publishing Inc.
4300 W Streetsboro Rd.
Richfield, OH 44286
Phone: (330)659-0303
Fax: (330)659-9488

Magazine featuring township government and local news in Hudson. **Founded:** 1998. **Freq:** Monthly. **Key Personnel:** Linday Sirak, Editor, lsirak@scriptype.com; Sue Serdinak, Publisher, sserdinak@scriptype.com. **Remarks:** Accepts advertising. **URL:** http://www.scriptype.com/pages/cities/hudson.html. . **Ad Rates:** BW: $400. **Circ:** 9,950

546 ■ Richfield Times
ScripType Publishing Inc.
4300 W Streetsboro Rd.
Richfield, OH 44286
Phone: (330)659-0303
Fax: (330)659-9488

Magazine featuring township government and local news in Richfield. **Founded:** 1980. **Freq:** Monthly. **Key Personnel:** Sue Serdinak, Editor, sserdinak@sciptype.com. **Remarks:** Accepts advertising. **URL:** http://www.scriptype.com/pages/cities/richfield.html. . **Ad Rates:** BW: $400. **Circ:** 3,500

RUSSELLS POINT

547 ■ WRPO-FM - 93.5
PO Box 93
Russells Point, OH 43348-0093
Phone: (937)843-6680

Format: Big Band/Nostalgia. **Operating Hours:** Continuous. **Wattage:** 100. **URL:** http://www.wrpo-fm.com/.

SOUTH VIENNA

548 ■ WOAR-FM - 88.3
2351 Sunset Blvd., Ste. 170-218
Rocklin, CA 95765
Phone: (916)251-1600
Fax: (916)251-1650

Format: Contemporary Christian. **Owner:** Educational Media Foundation, at above address. **URL:** http://www.air1.com/common/lowbandwidth.aspx.

STEUBENVILLE

549 ■ WBJV-FM - 88.9
PO Drawer 2440
Tupelo, MS 38803
Phone: (662)844-5036
Fax: (662)842-7798

Format: Contemporary Christian. **Owner:** American Family Association, at above address. **URL:** http://www.afr.net/newafr/stationsbystate/ohio.asp.

WEST UNION

550 ■ WVXW-FM - 89.5
PO Box 1018
Laguna Beach, CA 92652
Fax: (469)241-6795
Free: (800)639-5433

Format: Religious. **Owner:** New Life Ministries, at above address. **Key Personnel:** Stephen Arterburn, Founder/Chm. **URL:** http://www.newlife.com/Radio/findradio.asp?id=OH&t=Ohio.

WESTERVILLE

551 ■ Classroom Connections
National Middle School Association
4151 Executive Pkwy., Ste. 300
Westerville, OH 43081-3871
Phone: (614)895-4730
Fax: (614)895-4750
Free: (800)528-6672
Publisher E-mail: store@nmsa.org

Magazine featuring helpful tips on how to support young adolescents. **Freq:** Quarterly. **URL:** http://www.nmsa.org/Publications/ClassroomConnections/tabid/1118/ Default.aspx.

Circulation: ★ = ABC; △ = BPA; ♦ = CAC; • = CCAB; ❑ = VAC; ⊕ = PO Statement; ‡ = Publisher's Report; Boldface figures = sworn; Light figures = estimated.

OKLAHOMA

ADA

552 ■ KAKO-FM - 91.3
PO Drawer 2440
Tupelo, MS 38803
Phone: (662)844-5036
Fax: (662)842-7798

Format: Contemporary Christian. **Owner:** American Family Association, at above address. **URL:** http://www.afr.net/newafr/stationsbystate/oklahoma.asp.

ARDMORE

553 ■ KQPD-FM - 91.1
PO Drawer 2440
Tupelo, MS 38803
Phone: (662)844-5036
Fax: (662)842-7798

Format: Contemporary Christian. **Owner:** American Family Association, at above address. **URL:** http://www.afr.net/newafr/stationsbystate/oklahoma.asp.

554 ■ KYNZ-FM - 107.1
1205 Northglen
Ardmore, OK 73401
Phone: (580)226-1071
Fax: (580)226-0464

Format: Oldies. **URL:** http://www.kynz.com/.

BARTLESVILLE

555 ■ KRIG-FM - 104.9
1200 SE Frank Phillips Blvd.
Bartlesville, OK 74003
Phone: (918)336-1001
Fax: (918)336-6939
Free: (800)749-5936

Format: Country. **Owner:** KCD Enterprises, Inc., at above address. **URL:** http://www.bartlesvilleradio.com/krigabout.html.

CARNEGIE

556 ■ KJCC-FM - 89.5
PO Box 391
Twin Falls, ID 83303
Fax: (208)736-1958
Free: (800)357-4226

Format: Religious. **Owner:** CSN International, at above address. **URL:** http://www.csnradio.com/stationsOK.php.

COWETA

557 ■ KDIM-FM - 88.1
PO Box 1924
Tulsa, OK 74101
Phone: (918)455-5693

Format: Religious. **Owner:** Oasis Radio Network, at above address. **URL:** http://www.oasisnetwork.org/kdim.asp.

IDABEL

558 ■ KXRT-FM - 90.9
PO Drawer 2440
Tupelo, MS 38803
Phone: (662)844-5036
Fax: (662)842-7798

Format: Contemporary Christian. **Owner:** American Family Association, at above address. **URL:** http://www.afr.net/newafr/stationsbystate/oklahoma.asp.

LAWTON

559 ■ KWKL-FM - 89.9
PO Box 779002
Rocklin, CA 95677-9972
Free: (800)525-LOVE

Format: Religious. **Owner:** Educational Media Foundation, at above address. **URL:** http://www.klove.com/Music/StationList.aspx.

OKLAHOMA CITY

560 ■ KINB-FM - 105.3
4045 NW 64th, Ste. 600
Oklahoma City, OK 73116
Phone: (405)848-0100
Fax: (405)843-5288

Format: Hispanic. **Owner:** Last Bastion Trust, at above address. **URL:** http://www.laindomable.com/.

561 ■ KOCD-FM - 103.7
PO Box 1076
Oklahoma City, OK 73101
Fax: (316)665-6682
Free: (800)516-1037

Format: Jazz. **Key Personnel:** Jason Schlitz, VP/Gen. Mgr., jason@smoothjazzoklahoma.com. **URL:** http://www.smoothjazzoklahoma.com/.

PONCA CITY

KJTH-FM - See Tulsa
KXTH-FM - See Shawnee

SAND SPRING

562 ■ KJMU-AM - 1340
21700 Northwestern Hwy., Tower 14, Ste. 1190
Southfield, MI 48075
Phone: (248)557-3500
Fax: (248)557-2950

Format: Adult Contemporary; Urban Contemporary. **Owner:** Birach Broadcasting Corporation, at above address. **URL:** http://www.birach.com/kjmu.htm.

SHAWNEE

563 ■ KXTH-FM - 89.1
PO Box 14
Ponca City, OK 74602

Format: Contemporary Christian; Religious. **Owner:** Love Station Radio Network, at above address. **URL:** http://www.thehousefm.com/thehouse.asp?ID=about&s=0.

TULSA

564 ■ KBXO-FM - 90.3
PO Box 1924
Tulsa, OK 74101
Phone: (918)455-5693
Free: (888)996-2747

Format: Religious. **URL:** http://www.oasisnetwork.org/.

KDIM-FM - See Coweta

565 ■ KJTH-FM - 89.7
6600 W Hwy. 60
Ponca City, OK 74601
Phone: (580)767-1400
Fax: (580)765-1700
Free: (800)324-8488

Format: Contemporary Christian. **Owner:** Love Station Radio Network, at above address. **Key Personnel:** Doyle Brewer, Station Mgr., doyle@klvv.com; Tony Weir, Prog./Music Dir., tony@klvv.com. **URL:** http://www.thehousefm.com/thehouse.asp.

566 ■ KKCM-FM - 102.3
7136 S Yale, Ste. 500
Tulsa, OK 74136
Phone: (918)493-3434
Fax: (918)493-5383

Format: Contemporary Christian. **Key Personnel:** Chris Kelly, Program Mgr., phone (918)493-7400, chris.kelly3@coxradio.com. **URL:** http://spirit1023.com/.

567 ■ KWTU-FM - 88.7
Kendall Hall 150
600 S College Ave.
Tulsa, OK 74104
Phone: (918)631-2577
Free: (888)594-5947

Format: Classical. **Owner:** University of Tulsa, at above address. **Key Personnel:** Rich Fisher, General Mgr.; Brad Newman, Chief Engineer. **Wattage:** 5000. **Ad Rates:** Noncommercial. **URL:** http://www.kwgs.org/kwtu-answers.html.

OREGON

ASTORIA

568 ■ KKEE-AM - 1230
1006 W Marine Dr.
Astoria, OR 97103
Phone: (503)325-2911

Format: Talk; Public Radio. **Key Personnel:** Paul Mitchell, General Mgr.; Tom Feel, Operations Mgr.; Michael Desmond, News Dir. **URL:** http://www.kkee1230.com/.

BEND

569 ■ KVRA-FM - 89.3
2351 Sunset Blvd., Ste. 170-218
Rocklin, CA 95765
Phone: (916)251-1600
Fax: (916)251-1650

Format: Contemporary Christian. **Owner:** Educational Media Foundation, at above address. **URL:** http://www.air1.com/Music/StationList.aspx.

THE DALLES

570 ■ KMSW-FM - 92.7
502 Washington St., Ste. 203
PO Box 1517
The Dalles, OR 97058
Phone: (541)296-2211
Fax: (541)296-2213
E-mail: info@gorgeradio.com

Format: Classic Rock. **Key Personnel:** Rick Cavagnaro, General Sales Mgr., rick@gorgeradio.com; Gary Grossman, Pres./Gen. Mgr., gary@gorgeradio.com; Mark Bailey, News/Sports Dir., mark@gorgeradio.com. **URL:** http://www.gorgeradio.com/kmsw/main.htm.

EUGENE

571 ■ KOPT-AM - 1600
895 Country Club Rd.
Eugene, OR 97401
Phone: (541)343-4100
Fax: (541)343-0448

Format: Talk. **Owner:** Churchill Media, LLC, at above address. **URL:** http://www.kopt.com/.

572 ■ KXOR-AM - 660
895 Country Club Rd., Ste. A 200
Eugene, OR 97401
Phone: (541)343-4100
Fax: (541)343-0448
Free: (866)599-8746

Format: Hispanic. **URL:** http://www.lax660.com/webphp/index.php.

GRESHAM

573 ■ KHMD-FM - 89.1
26000 SE Stark St.
Gresham, OR 97030
Phone: (503)491-7271

Format: Jazz. **Founded:** 1984. **Operating Hours:** Continuous. **URL:** http://www.kmhd.org/.

HOOD RIVER

574 ■ KBNO-FM - 89.3
PO Box 242
Hood River, OR 97031
Phone: (541)386-8810
E-mail: kbno@lwrn.org

Format: Religious. **Key Personnel:** John Estes, Manager; Mel Rydman, Engr./Prog. Dir. **URL:** http://www.wrn-rcm.org/.

JOHN DAY

575 ■ KGNR-FM - 91.9
166 SE Dayton St.
PO Box 96
John Day, OR 97845
Phone: (541)575-1840
Free: (888)441-8927
E-mail: kgnr@centurytel.net

Format: Religious; Talk. **Owner:** Life Broadcasting, Inc., at above address. **Founded:** June 22, 1993. **Operating Hours:** Continuous. **Key Personnel:** Donn Willey, Manager. **Wattage:** 1500. **URL:** http://www.kgnr.org/.

LA GRANDE

576 ■ KFYL-FM - 94.3
PO Box 1018
Laguna Beach, CA 92652
Fax: (469)241-6795
Free: (800)639-5433

Format: Religious. **Owner:** New Life Ministries, at above address. **Key Personnel:** Stephen Arterburn, Founder/Chm. **URL:** http://www.newlife.com/Radio/findradio.asp?id=OR&t=Oregon.

MADRAS

577 ■ KMAB-FM - 99.3
PO Box 1018
Laguna Beach, CA 92652
Fax: (469)241-6795
Free: (800)639-5433

Format: Religious. **Owner:** New Life Ministries, at above address. **Key Personnel:** Stephen Arterburn, Founder/Chm. **URL:** http://www.newlife.com.

MEDFORD

578 ■ KRVC-FM - 98.9
511 Rossanley Dr.
Medford, OR 97501
Phone: (541)772-0322

Format: Country. **Owner:** Opus Broadcast Systems Inc., at above address. **URL:** http://www.989thewolf.com/.

PORTLAND

579 ■ Research in Middle Level Education Online
National Middle School Association
c/o Micki M. Caskey, PhD, Ed.
Portland State University, Graduate School of Education
Dept. of Curriculum & Instruction
615 SW Harrison
Portland, OR 97201
Phone: (503)725-4749
Fax: (503)725-8475
Publisher E-mail: store@nmsa.org

Journal covering the studies and interpretations of research literature. **Key Personnel:** Micki M. Caskey, PhD, Editor. **ISSN:** 1940-4476. **URL:** http://www.nmsa.org/Publications/RMLEOnline/tabid/426/Default.aspx.

SPRINGFIELD

580 ■ KQFE-FM - 88.9
290 Hegenberger Rd.
Oakland, CA 94621
Free: (800)543-1495

Format: Religious. **Owner:** Family Stations Inc., at above address. **URL:** http://www.familyradio.com/.

TILLAMOOK

581 ■ KAIK-FM - 88.5
2351 Sunset Blvd., Ste. 170-218
Rocklin, CA 95765
Phone: (916)251-1600
Fax: (916)251-1650

Format: Contemporary Christian. **Owner:** Educational Media Foundation, at above address. **URL:** http://www.air1.com/Music/StationList.aspx.

582 ■ KDEP-FM - 105.5
1000 N Main Ave., Ste. 5
Tillamook, OR 97141
Phone: (503)842-3888
Fax: (503)842-5640

Format: Adult Contemporary. **Operating Hours:** Continuous. **Key Personnel:** Shaena Peterson, Station Mgr./Sales Rep., shaena@coast105.com; Tommy Boye, Program Dir., tommy@coast105.com; Kris Ipock, Traffic Mgr., kris@coast105.com. **Ad Rates:** Advertising accepted; rates available upon request. **URL:** http://www.coast105.com/.

583 ■ KIXT-FM - 95.9
1000 N Main Ave., Ste. 5
Tillamook, OR 97141
Phone: (503)842-3888
Fax: (503)842-5640

Format: Country. **Key Personnel:** Tommy Boye, Program Dir., tommy@coast105.com; Rod Strang, Sales Rep., rod@coast105.com; Kris Ipock, Traffic/Office Mgr. **Ad Rates:** Advertising accepted; rates available upon request. **URL:** http://www.kicks96fm.com/.

PENNSYLVANIA

DU BOIS

584 ■ WDSN-FM - 106.5
12 W Long Ave.
Du Bois, PA 15801
Phone: (814)375-5260

Format: Adult Contemporary. **Key Personnel:** Jay Philippone, General Mgr., sunny106@penn.com; Lori Lewis, Station Mgr., lorilewis@sunny1065.fm; Lindsey Schoening, News Dir., news@sunny1065.fm. **URL:** http://www.sunny1065.fm/.

ERIE

585 ■ WEFR-FM - 88.1
290 Hegenberger Rd.
Oakland, CA 94621
Free: (800)543-1495

Format: Religious. **Owner:** Family Station, Inc., at above address. **URL:** http://www.familyradio.com/.

586 ■ WMCE-FM - 88.5
501 E 38th St.
Erie, PA 16546
Phone: (814)824-2261
E-mail: wmce@erieradio.com

Format: Eclectic. **Owner:** Mercyhurst Colorado, at above address. **Key Personnel:** Michael Leal, General Mgr. **URL:** http://wmce.mercyhurst.edu/.

587 ■ WTWF-FM - 93.9
One Boston Store Pl.
Erie, PA 16501
Phone: (814)461-1000
Fax: (814)874-0011

Format: Country. **Key Personnel:** Laurie Thonpson, Sales Mgr. **URL:** http://www.939thewolf.com/.

HARRISBURG

588 ■ WRBT-FM - 94.9
600 Corporate Cir.
Harrisburg, PA 17110
Phone: (717)540-8800
Fax: (717)540-8814
Free: (800)257-9949

Format: Country. **Owner:** Clear Channel Communications, 200 E Basse Rd., San Antonio, TX 78209, (210)822-2828, Fax: (210)832-3428. **URL:** http://www.bobradio.com/main.html.

JERSEY SHORE

589 ■ WJSA-FM - 96.3
262 Allegheny St.
Jersey Shore, PA 17740
Phone: (570)398-7200
Fax: (570)398-7201
E-mail: mail@wjsaradio.com

Format: Religious. **Founded:** Nov. 1, 1984. **Key Personnel:** Liz Brady, News Dir.; John Hogg, Ch. Engr./Gen. Mgr. **Wattage:** 25,000. **URL:** http://www.wjsaradio.com/.

JOHNSTOWN

590 ■ WFGI-FM - 95.5
109 Plaza Dr.
Johnstown, PA 15905
Phone: (814)255-4186
Fax: (814)255-6145
Free: (800)359-5477

Format: Country. **Key Personnel:** Terry Deitz, General Mgr., tdeitz@96key.com; Lara Mosby, Program Dir., lmosby@myfroggy95.com; Shirley Lambert, Business Mgr., slambert@96key.com; Tina Perry, Sales Mgr., tperry@96key.com. **Ad Rates:** Advertising accepted; rates available upon request. **URL:** http://www.myfroggy95.com/.

591 ■ WPRR-AM - 1490
970 Tripoli St.
Johnstown, PA 15902
Phone: (814)534-8975
Fax: (814)534-8979

Format: Sports. **Key Personnel:** Don Bedell, Sales Dir., dbedell@resultsradiopa.com. **Ad Rates:** Advertising accepted; rates available upon request. **URL:** http://espn1490online.com/.

MEADVILLE

592 ■ WUUZ-FM - 107.7
900 Water St., Downtown Mall
Meadville, PA 16335
Phone: (814)724-1111
Fax: (814)333-9628
Free: (800)397-9633
E-mail: radio@zoominternet.net

Format: Classic Rock. **Key Personnel:** Jim Shields, General Mgr. **Ad Rates:** Advertising accepted; rates available upon request. **URL:** http://www.mywuzz.com/.

MIDDLETOWN

593 ■ International Journal of Organization Theory and Behavior
PrAcademics Press
School of Public Affairs
Pennsylvania State University-Harrisburg
777 W Harrisburg Pke.
Middletown, PA 17057
Publisher E-mail: info@pracademicspress.com

Journal covering all private, public, and not-for-profit organizations' theories and behavior. **Freq:** Quarterly. **Key Personnel:** Jack Rabin, Editor, jxr11@psu.edu; Khi V. Thai, Managing Editor, thai@fau.edu. **Subscription Rates:** US$295 individuals and government; US$395 libraries. **URL:** http://www.pracademicspress.com/ijotb.html.

MILLERSVILLE

594 ■ Advanced Materials Research
Trans Tech Publications Inc.
105 Springdale Ln.
Millersville, PA 17551
Publisher E-mail: info@ttp.net

Journal covering areas of materials research. **Freq:** Irregular 8-12/yr. **Key Personnel:** Thomas Wohlbier, Publishing Ed., t.wohlbier@ttp.net; Karen Holford, Editor, holford@cardiff.ac.uk; Alan Kin-tak Lau, Editor, mmktlau@polyu.edu.hk; Xiaozhi Hu, Editor. **ISSN:** 1022-6680. **Subscription Rates:** EUR88

Circulation: ★ = ABC; △ = BPA; ♦ = CAC; • = CCAB; ❑ = VAC; ⊕ = PO Statement; ‡ = Publisher's Report; Boldface figures = sworn; Light figures = estimated.

single issue. **URL:** http://www.scientific.net/AMR/; http://www.ttp.net/1022-6680.html.

PHILADELPHIA

595 ■ Disability and Health Journal
Elsevier
1600 John F. Kennedy Blvd., Ste. 1800
Philadelphia, PA 19103-2899
Phone: (215)239-3900
Fax: (215)238-7883
Publisher E-mail: healthpermissions@elsevier.com

Journal covering reports about disability and health. **Freq:** 4/yr. **Key Personnel:** Suzanne McDermott, PhD, Editor. **ISSN:** 1936-6574. **Subscription Rates:** US$189 U.S. and Canada; US$60 students; US$70 students, other countries; US$199 other countries. **URL:** http://www.disabilityandhealthjnl.com/.

596 ■ JACC
Elsevier
1600 John F. Kennedy Blvd., Ste. 1800
Philadelphia, PA 19103-2899
Phone: (215)239-3900
Fax: (215)238-7883
Publisher E-mail: healthpermissions@elsevier.com

Journal covering all aspects of cardiovascular imaging. **Subtitle:** Cardiovascular Imaging. **Key Personnel:** Jagat Naruda, MD, Editor-in-Chief. **ISSN:** 1936-878X. **Subscription Rates:** US$251 institutions; US$182 individuals; US$80 students; US$320 institutions, other countries; US$242 other countries; US$102 students, other countries. **URL:** http://www.elsevier.com/wps/find/journaldescription.cws_home/712784/descrip tiondescription.

597 ■ Journal of Crohn's and Colitis
Elsevier
1600 John F. Kennedy Blvd., Ste. 1800
Philadelphia, PA 19103-2899
Phone: (215)239-3900
Fax: (215)238-7883
Publisher E-mail: healthpermissions@elsevier.com

Journal covering methods related to inflammatory bowel diseases. **Freq:** 2/yr. **Key Personnel:** Miquel A. Gassull, Editor-in-Chief. **ISSN:** 1873-9946. **Subscription Rates:** US$491 institutions; 59,700¥ institutions Japan; US$365 institutions European countries and Iran; US$284 individuals; 34,300¥ individuals Japan; EUR210 individuals European countries and Iran. **URL:** http://www.elsevier.com/wps/find/journaldescription.cws_home/713442/descrip tiondescription.

598 ■ Journal of the Mechanical Behavior of Biomedical Materials
Elsevier
1600 John F. Kennedy Blvd., Ste. 1800
Philadelphia, PA 19103-2899
Phone: (215)239-3900
Fax: (215)238-7883
Publisher E-mail: healthpermissions@elsevier.com

Journal covering the field of biomaterials. **Freq:** 4/yr. **Key Personnel:** D. Taylor, Editor-in-Chief, dtaylor@tcd.ie. **ISSN:** 1751-6161. **Subscription Rates:** 96,200¥ institutions Japan; EUR769 institutions European countries and Iran; US$895 other countries. **URL:** http://www.elsevier.com/wps/find/journaldescription.cws_home/711005/descrip tiondescription.

PITTSTON

599 ■ WILK-FM - 103.1
305 Hwy. 315
Pittston, PA 18640
Phone: (570)883-9850

Format: News; Talk. **Owner:** Entercom Wilkes-Barre Scranton, LLC, at above address. **Ad Rates:** Advertising accepted; rates available upon request. **URL:** http://wilknetwork.com/pages/611388.php.

SUNBURY

600 ■ WEGH-FM - 107.3
PO Box 1070
Sunbury, PA 17801
Phone: (570)286-5838
Fax: (570)743-7837

Format: Classic Rock. **Owner:** Sunbury Broadcasting Corporation, at above address. **Key Personnel:** Roger S. Haddon, Jr., President, rhaddon@wqkx.com; Kevin Herr, Operations Mgr., kherr@wkok.com; Rob Senter, Program Dir., rsenter@eagle107.com. **URL:** http://www.eagle107.com/Eagle_107/107_HOME.htm.

WHITEHALL

601 ■ WSAN-AM - 1470
1541 Alta Dr., Ste. 400
Whitehall, PA 18052
Phone: (610)434-1742
Fax: (610)434-6288

Format: Sports. **Owner:** Clear Channel Communications, 200 E Basse Rd., San Antonio, TX 78209, (210)822-2828, Fax: (210)832-3428. **Key Personnel:** Connie Uff, Business Mgr., connieuff@clearchannel.com; Craig Stevens, Program Dir., craigstevens@clearchannel.com; Mandy Schnell, Promotions Dir., mandyschnell@clearchannel.com. **Ad Rates:** Advertising accepted; rates available upon request. **URL:** http://www.fox1470.com/main.html.

PUERTO RICO

GUAYNABO

602 ■ WCMA-FM - 96.5
Frances St., Lot 42
Amelia Industrial Pk.
Guaynabo, PR 00968
Phone: (787)622-9700
Fax: (787)622-9477

Format: Eighties. **Owner:** Spanish Broadcasting System, 2601 S Bayshore Dr., PH2, Coconut Grove, FL 33133, (305)441-6901, Fax: (305)446-5148. **URL:** http://www.spanishbroadcasting.com/stations.shtml.

603 ■ WZET-FM - 92.1
Frances St., Lot 42, Amelia Industrial Pk.
Guaynabo, PR 00968
Phone: (787)622-9700
Fax: (787)622-9477

Format: Hispanic. **Owner:** Spanish Broadcasting System, 2601 S Bayshore Dr., PH2, Coconut Grove, FL 33133, (305)441-6901, Fax: (305)446-5148. **URL:** http://www.spanishbroadcasting.com/stations.shtml.

604 ■ WZMT-FM - 93.3
Frances St., Lot 42, Amelia Industrial Pk.
Guaynabo, PR 00968
Phone: (787)622-9700
Fax: (787)622-9477

Format: Hispanic. **Owner:** Spanish Broadcasting System, 2601 S Bayshore Dr., PH2, Coconut Grove, FL 33133, (305)441-6901, Fax: (305)446-5148. **URL:** http://www.spanishbroadcasting.com/stations.shtml.

RHODE ISLAND

JAMESTOWN

605 ■ Boatbuilder
Belvoir Media Group, LLC
PO Box 112
Jamestown, RI 02835
Publisher E-mail: customer_service@belvoir.com

Magazine featuring information about boat kits, boat plans, building materials, and boatbuilding techniques. **Freq:** Bimonthly. **Key Personnel:** Roger Marshall, Publisher/Ed. **Subscription Rates:** US$19.97 individuals; US$24 Canada; US$28.97 other countries. **URL:** http://www.boatbuildermagazine.com/.

SOUTH CAROLINA

AIKEN

606 ■ WASD-FM - 101.9
PO Box 1018
Laguna Beach, CA 92652
Fax: (469)241-6795
Free: (800)639-5433

Format: Religious. **Owner:** New Life Ministries, at above address. **Key Personnel:** Stephen Arterburn, Founder/Chm. **URL:** http://www.newlife.com/Radio/findradio.asp?id=SC&t=South%20Carolina.

607 ■ WLJK-FM - 89.1
1101 George Rogers Blvd.
Columbia, SC 29201-4761

Format: Public Radio. **Owner:** South Carolina Education Television Commission, at above address. **URL:** http://www.myetv.org/about_etv/regional_stations.cfm.

BARNWELL

608 ■ WIIZ-FM - 97.9
8968 Marlboro Ave.
Barnwell, SC 29812
Phone: (803)259-9797
Fax: (803)541-9700
E-mail: thewiz@wiiz979.com

Format: Urban Contemporary. **Owner:** Nicwild Communications, Inc, at above address. **Operating Hours:** Continuous. **Key Personnel:** Bobby Nichols, CEO/Station Mgr.; Cocoa Nichols, Music Dir./Mktg. Dir.; Charlotte Moultrie, Traffic Dir. **Ad Rates:** Advertising accepted; rates available upon request. **URL:** http://www.wiizfm.com/.

CHARLESTON

609 ■ BRAIN STIMULATION
Elsevier
c/o Mark S. George, MD, Ed.-in-Ch.
Medical University of South Carolina
67 President St., 502N IOP
Charleston, SC 29425
Publisher E-mail: healthpermissions@elsevier.com

Journal covering the field of neuromodulation. **Freq:** 4/yr. **Key Personnel:** Mark S. George, MD, Editor-in-Chief. **ISSN:** 1935-861X. **Subscription Rates:** US$550 institutions; US$249 individuals; US$125 students; US$635 institutions, other countries; US$315 other countries; US$175 students, other countries. **URL:** http://www.brainstimjrnl.com/; http://www.elsevier.com/wps/find/journaldescription.cws_home/712317/descriptiondescription.

COLUMBIA

WLJK-FM - See Aiken

610 ■ WNSC-FM - 88.9
1101 George Rogers Blvd.
Columbia, SC 29201-4761
Phone: (803)737-3545

Format: Jazz. **Owner:** South Carolina Education Television Commission, at above address. **URL:** http://www.scetv.org/radio/index.cfm.

GREENVILLE

KHOJ-AM - See Saint Charles, Missouri

611 ■ WOLT-FM - 103.3
225 S Pleasantburg Dr., Ste. B
Greenville, SC 29607
Phone: (864)751-0113

Format: Oldies. **Owner:** Davidson Media Group, at above address. **URL:** http://www.wolt-fm.com/wolt-fm.html.

PAGELAND

612 ■ WGSP-FM - 102.3
4801 E Independence Blvd., Ste. 803
Charlotte, NC 28212
Phone: (704)442-7277
Fax: (704)442-9518
Free: (888)770-1310

Format: Hispanic. **URL:** http://latremendaradio.com/main.php.

SOUTH DAKOTA

EAGLE BUTTE

613 ■ KPSD-FM - 97.1
555 N Dakota St.
PO Box 5000
Vermillion, SD 57069
Free: (800)456-0766

Format: Public Radio. **Owner:** South Dakota Public Broadcasting, at above address. **URL:** http://www.sdpb.org/about/radiochannels.asp.

FREEMAN

614 ■ KVCF-FM - 90.5
3434 W Kilbourn Ave.
Milwaukee, WI 53208
Free: (800)729-9829
E-mail: kvcf@vcyamerica.org

Format: Religious. **Owner:** VCY America, Inc., at above address. **URL:** http://www.vcyamerica.org/.

LOWRY

615 ■ KQSD-FM - 91.9
555 N Dakota St.
PO Box 5000
Vermillion, SD 57069
Free: (800)456-0766

Format: Public Radio. **Owner:** South Dakota Public Broadcasting, at above address. **URL:** http://www.sdpb.org/about/radiochannels.asp.

RAPID CITY

616 ■ KOUT-FM - 98.7
660 Flormann St., Ste. 100
Rapid City, SD 57701
Phone: (605)348-3939
Fax: (605)343-9012

Format: Country. **Owner:** New Rushmore Radio, at above address. **URL:** http://www.katradio.com/.

617 ■ KQRQ-FM - 92.3
PO Box 1760
Rapid City, SD 57709
Phone: (605)342-2000
Fax: (605)342-7305
E-mail: studio@q923.com

Format: Classical. **Owner:** Duhamel Broadcasting, at above address. **URL:** http://www.q923.com/.

618 ■ KZLK-FM - 106.3
518 St. Joe
Rapid City, SD 57709
Phone: (605)721-1063
Fax: (605)721-5732
E-mail: max@maxfmrapidcity.com

Format: Adult Contemporary. **Ad Rates:** Advertising accepted; rates available upon request. **URL:** http://www.maxfmrapidcity.com/.

SPEARFISH

619 ■ KZZI-FM - 95.9
2827 E Colorado Blvd.
Spearfish, SD 57783
Free: (866)495-9963

Format: Country. **URL:** http://www.myeaglecountry.com/.

VERMILLION

KPSD-FM - See Eagle Butte
KQSD-FM - See Lowry

TENNESSEE

ATHENS

620 ■ WKPJ-FM - 104.5
PO Box 1018
Laguna Beach, CA 92652
Fax: (469)241-6795
Free: (800)639-5433

Format: Religious. **Owner:** New Life Ministries, at above address. **Key Personnel:** Stephen Arterburn, Founder/Chm. **URL:** http://www.newlife.com/Radio/findradio.asp?id=TN&t=Tennessee.

CLARKSVILLE

621 ■ WAYQ-FM - 88.3
PO Box 30848
Clarksville, TN 37040
Phone: (931)647-8883

Format: Contemporary Christian. **Owner:** WAY-FM Media Group, PO Box 64500, Colorado Springs, CO 80962, (719)533-0300, Fax: (719)278-4339. **Key Personnel:** Nicole Martin, Bus. Devel. Dir.; Matt Austin, General Mgr.; Tracy Cole, Asst. to Gen. Mgr. **URL:** http://wayq.wayfm.com/.

COOKEVILLE

622 ■ WJNU-FM - 96.9
PO Box 1018
Laguna Beach, CA 92652
Fax: (469)241-6795
Free: (800)639-5433

Format: Religious. **Owner:** New Life Ministries, at above address. **Key Personnel:** Stephen Arterburn, Founder/Chm. **URL:** http://www.newlife.com/Radio/findradio.asp?id=TN&t=Tennessee.

CROSSVILLE

623 ■ WIHG-FM - 105.7
37 South Dr.
Crossville, TN 38555
Phone: (931)484-1057
Fax: (931)707-0580

Format: Classic Rock. **URL:** http://www.1057thehog.com/.

FRANKLIN

WAYM-FM - See Nashville

KNOXVILLE

624 ■ WETR-AM - 760
1621 E Magnolia Ave.
Knoxville, TN 37917
Phone: (865)525-0620
E-mail: info@talkradio760.com

Format: Talk. **Key Personnel:** David Wells, General Mgr., david@talkradio760.com; Becky Mills, Promotions Dir., bmills@talkradio760.com; Bob Bell, Sen. Account Exec., bbell@talkradio760.com. **Ad Rates:** Advertising accepted; rates available upon request. **URL:** http://www.talkradio760.com/.

625 ■ WIFA-AM - 1240
818 Cedar Bluff Rd.
Knoxville, TN 37923
Phone: (865)531-2005
Fax: (865)531-2006

Format: Religious. **URL:** http://www.1240radio.com/1240/index.html.

626 ■ WIJV-FM - 92.7
818 N Cedar Bluff Rd.
Knoxville, TN 37923
Phone: (865)531-2005
Fax: (865)531-2006

Format: Contemporary Christian. **Owner:** Progressive Media, at above address. **URL:** http://www.victory927.com/927/index.html.

627 ■ WQJK-FM - 95.7
1100 Sharps Ridge Rd.
Knoxville, TN 37917
E-mail: hrknox@sccradio.com

Format: Adult Contemporary. **Owner:** South Central Communications Corporation, at above address. **URL:** http://www.jackfmknoxville.com/.

KODAK

628 ■ WSEV-FM - 105.5
196 W Dumplin Valley Rd.
Kodak, TN 37764
Phone: (865)932-6002

Format: Adult Contemporary. **Ad Rates:** Advertising accepted; rates available upon request. **URL:** http://www.easttennesseeradio.com/.

NASHVILLE

629 ■ WAYM-FM - 88.7
1012 McEwen Dr.
Franklin, TN 37067
Phone: (615)261-9293
Fax: (615)261-3967

Format: Contemporary Christian. **Owner:** WAY-FM Media Group, PO Box 64500, Colorado Springs, CO 80962, (719)533-0300, Fax: (719)278-4339. **Key Personnel:** Matt Austin, General Mgr.; Tracy Cole, Asst. to Gen. Mgr.; Jeff Brown, Program Dir. **URL:** http://waym.wayfm.com/.

PARIS

630 ■ WMUF-FM - 104.7
110 India Rd.
Paris, TN 38242
Phone: (731)644-9455
E-mail: mail@wmufradio.com

Format: Country. **Ad Rates:** Advertising accepted; rates available upon request. **URL:** http://www.wmufradio.com/.

631 ■ WTPR-FM - 101.7
206 N Brewer St.
Paris, TN 38242
Phone: (731)642-7100
Fax: (731)644-9367

Format: Oldies. **Key Personnel:** Terry L. Hailey, Pres./Gen. Mgr., thailey@wenkwtpr.com. **Ad Rates:** Advertising accepted; rates available upon request. **URL:** http://www.wenkwtpr.com/w/.

VONORE

632 ■ WTRL-FM - 106.9
PO Box 1018
Laguna Beach, CA 92652
Fax: (469)241-6795
Free: (800)639-5433

Format: Religious. **Owner:** New Life Ministries, at above address. **Key Personnel:** Stephen Arterburn, Founder/Chm. **URL:** http://www.newlife.com/Radio/findradio.asp?id=TN&t=Tennessee.

TEXAS

ARLINGTON

WVBG-AM - See Vicksburg, Missouri

BOERNE

633 ■ Country Lifestyle
Blue Sky Publication, LLC
616 N Main St.
Boerne, TX 78006
Phone: (830)816-7355
Fax: (830)249-7406

Lifestyle magazine about Texas and the Hill Country. **Subtitle:** Fine Living in the Heart of Texas. **Freq:** Bimonthly. **Key Personnel:** Kelley-Jo Glick, Ed./Publisher, phone (210)482-9206, editor@countrylifestyle.net. **Subscription Rates:** US$12.95 individuals; US$24 two years; US$27.95 other

Circulation: ★ = ABC; △ = BPA; ♦ = CAC; • = CCAB; ❑ = VAC; ⊕ = PO Statement; ‡ = Publisher's Report; Boldface figures = sworn; Light figures = estimated.

countries; US$54 other countries 2 years. **URL:** http://www.countrylifestyle.net/.

634 ■ Texas Hills
Blue Sky Publication, LLC
616 N Main St.
Boerne, TX 78006
Phone: (830)816-7355
Fax: (830)249-7406

Magazine featuring Texas real estate industry. **URL:** http://www.texasmags.com/.

635 ■ Texas Magazine
Blue Sky Publication, LLC
616 N Main St.
Boerne, TX 78006
Phone: (830)816-7355
Fax: (830)249-7406

Magazine featuring Texas culture, people, business, education, dining and entertainment. **Freq:** Bimonthly. **URL:** http://www.texas-mag.com/. **Circ:** 50,000

COLLEYVILLE

636 ■ KNOR-FM - 93.7
4201 Pool Rd.
Colleyville, TX 76034
Phone: (817)868-2900
Fax: (817)868-2116

Format: Hispanic. **Owner:** Liberman Broadcasting of Dallas, at above address. **Key Personnel:** Brad Branson, Sales Mgr., bbranson@lbimedia.com; Arturo Buenrostro, Program Dir., abuenrostro@lbimedia.com; Anthony Gutierrez, Promotions Dir., agutierrez@lbimedia.com. **URL:** http://www.laraza937.com/.

DAINGERFIELD

637 ■ KXIV-FM - 94.3
PO Box 1018
Laguna Beach, CA 92652
Fax: (469)241-6795
Free: (800)639-5433

Format: Religious. **Owner:** New Life Ministries, at above address. **Key Personnel:** Stephen Arterburn, Founder/Chm. **URL:** http://www.newlife.com.

GEORGETOWN

638 ■ American Cranes & Transport
KHL Group
c/o D. Ann Slayton Shiffler, Ed.
30325 Oak Tree Dr.
Georgetown, TX 78628
Phone: (512)869-8838
Fax: (512)863-9779
Publisher E-mail: info@khl.com

Magazine featuring North American crane and transport industry. **Freq:** Monthly. **Trim Size:** 7 7/8 x 10 3/4. **Key Personnel:** D. Ann Slayton Shiffler, Editor, d.annshiffler@khl.com. **Subscription Rates:** US$180 individuals; US$325 two years; US$460 individuals 3 years. **Remarks:** Accepts advertising. **URL:** http://www.khl.com/magazines/information.asp?magazineid=8. . **Ad Rates:** 4C: $1,410. **Circ:** △**15,000**

HOUSTON

639 ■ KCHN-AM - 1050
1782 W Sam Houston Pky. N
Houston, TX 77043

Format: Religious. **Owner:** Multicultural Radio Broadcasting, Inc., 449 Broadway, New York, NY 10013, (212)966-1059, Fax: (212)966-9580. **Key Personnel:** John Gabel, Contact, johng@mrbi.net. **URL:** http://www.mrbi.net/radiogroup.htm.

KILLEEN

640 ■ KJHV-FM - 96.3
PO Box 1018
Laguna Beach, CA 92652
Fax: (469)241-6795
Free: (800)639-5433

Format: Religious. **Owner:** New Life Ministries, at above address. **Key Personnel:** Stephen Arterburn, Founder/Chm. **URL:** http://www.newlife.com/Radio/findradio.asp?id=TX&t=Texas.

SAN ANGELO

641 ■ KNAR-FM - 89.3
2351 Sunset Blvd., Ste. 170-128
Rocklin, CA 95765
Phone: (916)251-1600
Fax: (916)251-1650

Format: Contemporary Christian. **Owner:** Educational Media Foundation, at above address. **URL:** http://www.air1.com/.

SAN ANTONIO

642 ■ International Journal of Automation and Control
Inderscience Publishers
c/o Prof. Mo Jamshidi, Ed.-in-Ch.
University of Texas at San Antonio
Department of Electrical & Computer Engineering
BSE 1.544
San Antonio, TX 78249-0665
Publisher E-mail: editor@inderscience.com

Journal covering field of automation & control system technology. **Founded:** 2007. **Freq:** 4/yr. **Key Personnel:** Prof. Mo Jamshidi, Editor-in-Chief, moj@wacong.org; Dr. N. P. Mahalik, Editor-in-Chief, nmahalik@csufresno.edu. **ISSN:** 1740-7516. **Subscription Rates:** EUR470 individuals includes surface mail, print only; EUR640 individuals print & online. **URL:** http://www.inderscience.com/browse/index.php?journalCODE=ijaac.

643 ■ WQSS-FM - 102.5
200 E Basse Rd.
San Antonio, TX 78209
Phone: (210)822-2828
Fax: (210)832-3428

Format: Classical. **Owner:** Clear Channel Communications, at above address. **URL:** http://www.1025thepeak.com/main.html.

TEXARKANA

644 ■ KPGG-FM - 103.9
1323 College Dr.
Texarkana, TX 75503
Phone: (903)793-1039
Fax: (903)794-4717
Free: (888)793-1039

Format: Country. **Key Personnel:** Mike Basso, General Mgr., mikebasso@cableone.net. **Ad Rates:** Advertising accepted; rates available upon request. **URL:** http://www.k-pig.com/.

UTAH

BLUFFDALE

645 ■ Digital Scrapbooking
CK Media LLC
14850 Pony Express Rd.
Bluffdale, UT 84065
Free: (800)815-3538
Publisher E-mail: info@ckmedia.com

Magazine featuring information about scrapbooking through computer. **Key Personnel:** Amy Tanabe, Contact. **URL:** http://www.ckmedia.com/scrapbooking/digitalscrapbooking.php; http://www.digitalscrapbooking.com/info/featured_articles.

646 ■ Sew Simple
CK Media LLC
14850 Pony Express Rd.
Bluffdale, UT 84065
Free: (800)815-3538
Publisher E-mail: info@ckmedia.com

Magazine featuring information about sewing. **Key Personnel:** Amy Stalp, Contact. **Subscription Rates:** US$5.99 single issue. **URL:** http://www.ckmedia.com/sewing/sewsimple.php; http://www.sewnews.com/sewsimple/.

SALT LAKE CITY

647 ■ KEGA-FM - 101.5
515 S 700 E, No. 1C
Salt Lake City, UT 84102
Phone: (801)524-2600

Format: Country. **Owner:** Simmons Media Group, at above address. **Key Personnel:** Kayla Turner, Promotions Dir., kturner@simmonsmedia.com; Allen Hansen, Sales Mgr., ahansen@simmonsmedia.com; Cody Alan, Program Dir., cody@1015theeagle.com. **Ad Rates:** Advertising accepted; rates available upon request. **URL:** http://1015theeagle.com/.

648 ■ KUCW-TV - 30
2175 W 1700 S
Salt Lake City, UT 84104
Phone: (801)975-4444

Owner: Clear Channel Communications, 200 E Basse Rd., San Antonio, TX 78209, (210)822-2828. **Key Personnel:** David D'Antuono, General Mgr. **URL:** http://www.cw30.com.

649 ■ KUDD-FM - 107.9
2835 E 3300 S
Salt Lake City, UT 84109
Phone: (801)412-6040

Format: Adult Contemporary. **Owner:** Millcreek Broadcasting LLC, at above address. **Key Personnel:** Jody Adams, Sales Mgr., phone (801)412-6049. **Ad Rates:** Advertising accepted; rates available upon request. **URL:** http://www.1079themix.com/mixIndex.cfm.

650 ■ KUDE-FM - 103.9
2835 E 3300 S
Salt Lake City, UT 84109
Phone: (801)412-6040
Fax: (801)412-6041

Format: Adult Contemporary. **Owner:** Millcreek Broadcasting LLC, at above address. **Key Personnel:** Jody Adams, Sales Mgr., phone (801)412-6049. **Ad Rates:** Advertising accepted; rates available upon request. **URL:** http://www.1079themix.com/mixindex.cfm?CFID=47066763&CFTOKEN=46215401.

651 ■ KYMV-FM - 100.7
515 S 700 E, Ste. 1C
Salt Lake City, UT 84102
Phone: (801)524-2600

Format: Urban Contemporary. **Owner:** Simmons Media Group, at above address. **Key Personnel:** Steve Johnson, General Mgr., sjohnson@simmonsmedia.com; Randi Wilson, Promotions Dir., randi@x96.com; Tracey Smith, Sales Mgr., tsmith@simmonsmedia.com. **URL:** http://movin1007.com/.

WASHINGTON

652 ■ KZHK-FM - 95.9
204 N Playa Della Rosita
Washington, UT 84780
Phone: (435)674-9959
Free: (888)454-9959

Format: Classic Rock. **Owner:** Canyon Media Broadcasting, at above address. **Ad Rates:** Advertising accepted; rates available upon request. **URL:** http://www.959thehawk.com/index.html.

VERMONT

HARTFORD

653 ■ WVFA-FM - 90.5
PO Box 126
Hartford, VT 05047
Phone: (802)295-9683
Fax: (802)295-9683

Format: Alternative/New Music/Progressive. **Owner:** Green Mountain Educational Fellowship, Inc., at above address. **URL:** http://90.5wvfa.com/.

VIRGINIA

CHANTILLY

654 ■ New Old House
Active Interest Media
4125 Lafayette Center Dr., Ste. 100
Chantilly, VA 20151
Phone: (703)222-9411

Magazine featuring classic American house styles. **Founded:** 2004. **Trim Size:** 9 x 10 7/8. **Key Personnel:** Nancy Berry, Editor, phone (508)362-7007, nberry@homebuyerpubs.com. **Subscription Rates:** US$19.97 individuals 6 issues; US$34.97 individuals 12 issues; C$27.97 Canada 6 issues; C$50.97 Canada 12 issues. **Remarks:** Accepts advertising. **URL:** http://www.newoldhousemag.com/. **Circ:** (Not Reported)

655 ■ Timber Home Living
Active Interest Media
4125 Lafayette Center Dr., Ste. 100
Chantilly, VA 20151
Phone: (703)222-9411
Fax: (703)222-3209

Magazine featuring timber houses. **Key Personnel:** Laurie Sloan, Gp. Publisher, lsloan@homebuyerpubs.com. **Subscription Rates:** US$19.95 individuals; C$29.95 Canada; US$49.95 other countries. **Remarks:** Accepts advertising. **URL:** http://www.aimmedia.com/article_display_36.html. **Circ:** (Not Reported)

CHESAPEAKE

656 ■ WWIP-FM - 89.1
2202 Jolliff Rd.
Chesapeake, VA 23321
Phone: (757)465-1603
Fax: (757)488-7761
E-mail: info@wwip.org

Format: Religious. **Key Personnel:** Colleen Dick, Asst. Station Mgr.; Paul Krismier, Contact; Robert Scott, Contact. **URL:** http://www.wwip.org/.

NORFOLK

657 ■ WYRM-AM - 1110
700 Monticello Ave., Ste. 305
Norfolk, VA 23510
Phone: (757)622-9256
Fax: (757)622-9253
Free: (866)892-9256

Format: Religious. **Key Personnel:** Larry Cobb, General Mgr.; Paula Cobb, Office Mgr. **URL:** http://wyrmradio.com/php/index.php?option=com_frontpage&Itemid=1.

PURCELLVILLE

658 ■ Middle Ground
National Middle School Association
c/o Patricia George, Ed.
37864 Alberts Farm Dr.
Purcellville, VA 20132
Publisher E-mail: store@nmsa.org

Magazine featuring topics about the adolescents and the young generation. **Key Personnel:** Patricia George, Editor, editu2@aol.com. **URL:** http://www.nmsa.org/Publications/MiddleGround/tabid/437/Default.aspx.

SPRINGFIELD

659 ■ Aviation Safety
Belvoir Media Group, LLC
6404 Rivington Rd.
Springfield, VA 22152
Phone: (202)536-5923
Publisher E-mail: customer_service@belvoir.com

Magazine featuring information about risk management and accident prevention. **Freq:** Monthly. **Key Personnel:** Joseph E. Burnside, Editor-in-Chief. **Subscription Rates:** US$16.95 individuals 3 months; US$29.95 individuals 6 months. **URL:** http://www.aviationsafetymagazine.com/.

VIRGINIA BEACH

660 ■ WVBW-FM - 92.9
5589 Greenwich Rd., Ste. 200
Virginia Beach, VA 23462
Phone: (757)671-1000

Format: Adult Contemporary. **Key Personnel:** Nathan James, Promotions Dir., nathan@929thewave.com; Mike Allen, Program Dir., mallen@929thewave.com. **URL:** http://www.929thewave.com/.

WASHINGTON

ANACORTES

661 ■ KWLE-AM - 1340
PO Box 96
Anacortes, WA 98221
Phone: (360)293-3141
Fax: (360)293-9463
E-mail: questions@1340thewhale.com

Format: Adult Contemporary. **Key Personnel:** Rob Uteda, Contact, ruteda@1340thewhale.com; Glen Harris, Contact, gharris@1340thewhale.com. **URL:** http://www.1340thewhale.com/.

LONGVIEW

662 ■ KPPK-FM - 98.3
1130 14th Ave.
Longview, WA 98632
Phone: (360)425-1500
Fax: (360)423-1554

Format: Adult Contemporary. **Owner:** Bicostal Media, 140 N Main St., Lakeport, CA 95453, (707)263-6113, Fax: (707)263-0939. **Wattage:** 6000. **URL:** http://www.bicoastalmedia.com/l_longview.shtml.

SPOKANE

663 ■ American Journal of Media Psychology
Marquette Books LLC
5915 S Regal St., Ste. B-118
Spokane, WA 99223
Phone: (509)443-7057
Fax: (509)448-2191
Publisher E-mail: books@marquettebooks.com

Journal covering the understanding of media effects and processes on individuals in society. **Founded:** Jan. 2008. **Freq:** 2-4/yr. **Key Personnel:** Michael Elasmar, Editor, elasmar@bu.edu. **Subscription Rates:** US$35 individuals; US$85 institutions library. **URL:** http://www.marquettejournals.org/mediapsychology.html.

664 ■ International Journal of Business and Systems Research
Inderscience Publishers
c/o Prof. Jason C.H. Chen, Ed.-in-Ch.
Gonzaga University
Graduate School of Business
E 502 Boone Ave.
Spokane, WA 99258
Publication E-mail: ijbsr@inderscience.com
Publisher E-mail: editor@inderscience.com

Journal covering advances in business & systems research. **Founded:** 2007. **Freq:** 4/yr. **Key Personnel:** Prof. Jason C.H. Chen, Editor-in-Chief, chen@jepson.gonzaga.edu; Dr. Reggie Davidrajuh, Editor, reggie.davidrajuh@uis.no. **ISSN:** 1751-200X. **Subscription Rates:** EUR470 individuals includes surface mail, print only; EUR640 individuals print & online. **URL:** http://www.inderscience.com/browse/index.php?journalCODE=ijbsr.

665 ■ International Journal of Media & Foreign Affairs
Marquette Books LLC
5915 S Regal St., Ste. B-118
Spokane, WA 99223
Phone: (509)443-7057
Fax: (509)448-2191
Publisher E-mail: books@marquettebooks.com

Journal covering roles, functions, and effects of media and foreign affairs. **Founded:** Jan. 2008. **Freq:** 2-4/yr. **Key Personnel:** Peter Gross, Editor, pgross@utk.edu. **Subscription Rates:** US$35 individuals; US$85 institutions library. **URL:** http://www.marquettejournals.org/mediaforeignaffairs.html.

666 ■ Journal of Global Mass Communication
Marquette Books LLC
5915 S Regal St., Ste. B-118
Spokane, WA 99223
Phone: (509)443-7057
Fax: (509)448-2191
Publisher E-mail: books@marquettebooks.com

Journal covering mass communication relations on a global scale. **Founded:** Jan. 2008. **Freq:** 2-4/yr. **Key Personnel:** Prof. Thomas Hanitzsch, Editor, th.hanitzsch@ipmz.uzh.ch; Prof. P. Eric Louw, Editor, e.louw@uq.edu.au; Arnold S. De Beer, Founding Ed., asdebeer@imasa.org. **Subscription Rates:** US$35 individuals; US$85 institutions library. **URL:** http://www.marquettejournals.org/globalmasscommunication.html.

667 ■ Journal of Health & Mass Communication
Marquette Books LLC
5915 S Regal St., Ste. B-118
Spokane, WA 99223
Phone: (509)443-7057
Fax: (509)448-2191
Publisher E-mail: books@marquettebooks.com

Journal covering the understanding of mass media effects or processes with respect to health-related issues. **Founded:** Jan. 2008. **Freq:** 2-4/yr. **Key Personnel:** Prof. Fiona Chew, Editor, cmrfchew@syr.edu. **Subscription Rates:** US$35 individuals; US$85 institutions library. **URL:** http://www.marquettejournals.org/healthmasscomm.html.

668 ■ Journal of Media Law & Ethics
Marquette Books LLC
5915 S Regal St., Ste. B-118
Spokane, WA 99223
Phone: (509)443-7057
Fax: (509)448-2191
Publisher E-mail: books@marquettebooks.com

Journal covering the understanding of media law and ethics and diversity in society. **Founded:** Jan. 2008. **Freq:** 2-4/yr. **Key Personnel:** Eric Easton, Editor, eeaston@ubalt.edu. **Subscription Rates:** US$35 individuals; US$85 institutions library. **URL:** http://www.marquettejournals.org/medialawethics.html.

669 ■ Journal of Media Sociology
Marquette Books LLC
5915 S Regal St., Ste. B-118
Spokane, WA 99223
Phone: (509)443-7057
Fax: (509)448-2191
Publisher E mail: books@marquettebooks.com

Journal covering the understanding of the role and function of mass media and mass communication in society or the world. **Founded:** Jan. 2008. **Freq:** 2-4/yr. **Key Personnel:** Prof. Michael Cheney, Editor, mrcheney@uiuc.edu. **Subscription Rates:** US$35 individuals; US$85 institutions library. **URL:** http://www.marquettejournals.org/mediasociology.html.

Circulation: ★ = ABC; △ = BPA; ♦ = CAC; • = CCAB; ❑ = VAC; ⊕ = PO Statement; ‡ = Publisher's Report; Boldface figures = sworn; Light figures = estimated.

670 ■ Russian Journal of Communication
Marquette Books LLC
5915 S Regal St., Ste. B-118
Spokane, WA 99223
Phone: (509)443-7057
Fax: (509)448-2191
Publisher E-mail: books@marquettebooks.com

Journal covering understanding of communication in Russia. **Founded:** Jan. 2008. **Freq:** 2-4/yr. **Key Personnel:** Igor Klyukanov, Editor, iklyukanov@mail.ewu.edu. **Subscription Rates:** US$35 individuals; US$85 institutions library. **URL:** http://www.marquettejournals.org/russianjournal.html.

WEST VIRGINIA

FAIRMONT

671 ■ WGIE-FM - 92.7
1489 Locust Ave.
Fairmont, WV 26554
Fax: (304)367-1885

Format: Country. **Key Personnel:** David Branham, Contact, david@froggycountry.net. **Ad Rates:** Advertising accepted; rates available upon request. **URL:** http://www.froggycountry.net/.

MORGANTOWN

672 ■ WAJR-FM - 103.3
1251 Earl L. Core Rd.
Morgantown, WV 26505
Phone: (304)296-0029
Fax: (304)296-3876

Format: News; Talk; Sports. **Owner:** West Virginia Radio Corporation, at above address. **URL:** http://www.wajrfm.com/home.php.

MOUNT CLARE

673 ■ WWLW-FM - 106.5
1065 Radio Park Dr.
Mount Clare, WV 26408
Phone: (304)623-6546
Fax: (304)623-6547

Format: Adult Contemporary. **Owner:** West Virginia Corporation, at above address. **Key Personnel:** Max Wolf, Program Dir. **Ad Rates:** Advertising accepted; rates available upon request. **URL:** http://www.wvmagic.com/home.php.

SALEM

674 ■ WVBL-FM - 99.9
PO Box 1018
Laguna Beach, CA 92652
Fax: (469)241-6795
Free: (800)639-5433

Format: Religious. **Owner:** New Life Ministries, at above address. **Key Personnel:** Stephen Arterburn, Founder/Chm. **URL:** http://www.newlife.com/Radio/findradio.asp?id=WV&t=West%20Virginia.

WISCONSIN

FORT ATKINSON

675 ■ EMS Product News
Cygnus Business Media Inc.
1233 Janesville Ave.
PO Box 803
Fort Atkinson, WI 53538-0803
Fax: (920)563-1702
Free: (800)547-7377

Magazine featuring new products information for emergency medical services. **Founded:** 1993. **Freq:** Bimonthly. **Trim Size:** 7 7/8 x 10 3/4. **Key Personnel:** Scott Cravens, Publisher, scott.cravens@cygnusb2b.com; Nancy Perry, Editor, nancy.perry@cygnuspub.com. **Remarks:** Accepts advertising. **URL:** http://www.cygnusb2b.com/PropertyPub.cfm?PropertyID=528; http://www.emsresponder.com/publication/pub.jsp?pubId=2. . **Ad Rates:** BW: $3,835, 4C: $4,385. **Circ:** △25,000

MADISON

676 ■ Occupational Therapy International
John Wiley & Sons Inc.
c/o Dr. Franklin Stein, Ed.
7334 New Washburn Way
Madison, WI 53719-3010
Publisher E-mail: info@wiley.com

Journal covering occupational therapy. **Freq:** Quarterly. **Key Personnel:** Dr. Franklin Stein, Editor. **ISSN:** 0966-7903. **Subscription Rates:** 110 individuals U.K.; US$195 other countries; US$395 institutions. **URL:** http://www3.interscience.wiley.com/journal/112094317/home.

MILWAUKEE

KVCF-FM - See Freeman, South Dakota

WYOMING

CASPER

677 ■ KLWC-FM - 89.1
2351 Sunset Blvd., Ste. 170-128
Rocklin, CA 95765
Phone: (916)251-1600
Fax: (916)251-1650

Format: Contemporary Christian. **Owner:** Educational Media Foundation, at above address. **URL:** http://www.klove.com/Music/StationList.aspx.

CODY

678 ■ American Journal of Lifestyle Medicine
Sage Publishing Co.
PO Box 1090
Cody, WY 82414
Phone: (307)587-2231
Fax: (307)587-5208

Journal covering professional resource for practitioners seeking to incorporate lifestyle practices into clinical medicine. **Freq:** Bimonthly. **Key Personnel:** James M. Rippe, Editor. **ISSN:** 1559-8276. **Subscription Rates:** US$162 institutions print & E-Access; US$146 institutions E-Access; US$159 institutions print only; US$62 individuals print & E-Access. **URL:** http://www.sagepub.com/journalsProdDesc.nav?prodId=Journal201781.

679 ■ Assessment for Effective Intervention
Sage Publishing Co.
PO Box 1090
Cody, WY 82414
Phone: (307)587-2231
Fax: (307)587-5208

Journal featuring articles that describe the relationship between assessment and instruction, introduce innovative assessment strategies; outline diagnostic procedures; analyze relationships between existing instruments; and review assessment techniques, strategies, and instrumentation. **Freq:** Quarterly. **Key Personnel:** Linda K. Elksnin, Editor; Henry N. Elksnin, Editor. **ISSN:** 1534-5084. **Subscription Rates:** US$128 institutions print & E-Access; US$115 institutions E-Access; US$125 institutions print only; US$55 individuals print & E-Access. **URL:** http://www.sagepub.com/journalsProdDesc.nav?prodId=Journal201872.

680 ■ Cultural Sociology
Sage Publishing Co.
PO Box 1090
Cody, WY 82414
Phone: (307)587-2231
Fax: (307)587-5208

Journal covering the sociological comprehension of cultural matters. **Freq:** 3/yr. **Key Personnel:** David Inglis, Editor; Andrew Blaikie, Editor. **ISSN:** 1749-9755. **Subscription Rates:** US$497 institutions print & E-Access; US$447 institutions E-Access; US$487 institutions print only; US$74 individuals print only; US$175 institutions single copy; US$32 single issue. **URL:** http://www.sagepub.com/journalsProdDesc.nav?prodId=Journal201779.

681 ■ Discourse & Communication
Sage Publishing Co.
PO Box 1090
Cody, WY 82414
Phone: (307)587-2231
Fax: (307)587-5208

Journal covering communication research. **Freq:** Quarterly. **Key Personnel:** Teun A. van Dijk, Editor. **ISSN:** 1750-4813. **Subscription Rates:** US$679 institutions print & E-Access; US$611 institutions E-Access; US$665 institutions print only; US$74 individuals print only. **URL:** http://www.sagepub.com/journalsProdDesc.nav?prodId=Journal201784.

EVANSTON

682 ■ KNYN-FM - 99.1
568 Airport Rd.
Evanston, WY 82930
Phone: (307)789-8255
Free: (800)233-9101
E-mail: info@magic99fm.net

Format: Full Service. **URL:** http://www.evanstonradio.net/Magic99Home.html.

GILLETTE

683 ■ KLOF-FM - 88.9
2351 Sunset Blvd., Ste. 170-218
Rocklin, CA
Phone: (916)251-1600
Fax: (916)251-1650

Format: Contemporary Christian. **Owner:** Educational Media Foundation, at above address. **URL:** http://www.klove.com/Music/StationList.aspx.

ALBERTA

BALZAC

684 ■ AirdrieLIFE
Frog Inc.
Site 14 Box 32
Balzac, AB, Canada T0M 0E0
Phone: (403)226-1272
Publisher E-mail: froginc@shaw.ca

Magazine featuring the city of Airdrie. **Subtitle:** The Official Guide to Life in the City of Airdrie. **Trim Size:** 8.25 x 10.75. **Key Personnel:** Sherry Shaw-Froggatt, Contact. **Remarks:** Accepts advertising. **URL:** http://www.airdrielife.com/. **Circ:** (Not Reported)

685 ■ AirdrieWORKS
Frog Inc.
Site 14 Box 32
Balzac, AB, Canada T0M 0E0
Phone: (403)226-1272
Publisher E-mail: froginc@shaw.ca

Magazine featuring successful local businesses in Airdrie. **Trim Size:** 8.21 x 10.75. **Key Personnel:** Sherry Shaw-Froggatt, Contact. **Remarks:** Accepts advertising. **URL:** http://www.airdrielife.com/index.html. **Circ:** (Not Reported)

BLAIRMORE

686 ■ CJPR-FM - 94.9
13213 - 20th Ave.
PO Box 840
Blairmore, AB, Canada T0K 0E0
Phone: (403)562-2807

Format: Country. **Owner:** Newcap Radio, 745 Windmill Rd., Dartmouth, NS, Canada B3B 1C2, (902)468-7557, Fax: (902)468-7558. **Key Personnel:** Barb Kelly, Station Mgr., bkelly@newcap.ca; Randy Spencer, News Dir., rspencer@newcap.ca; Daryl Ferguson, Contact, dferguson@newcap.ca. **Wattage:** 760. **URL:** http://www.mountainradiofm.com/.

BONNYVILLE

687 ■ CJEG-FM - 101.3
PO Box 8251
Bonnyville, AB, Canada T9N 2J5
Free: (877)812-5665

Format: Adult Contemporary. **Owner:** Newcap Radio, 745 Windmill Rd., Dartmouth, NS, Canada B3B 1C2, (902)468-7557, Fax: (902)468-7558. **Key Personnel:** Lise Lacombe, Station Mgr., phone (780)812-3058, fax (780)812-7315, llacombe@newcap.ca; Dave Tymo, Music Dir., dtymo@newcap.ca; Robyn Kettlewell, Mktg. Exec. **Wattage:** 27,000. **URL:** http://www.1013koolfm.com/.

CALGARY

688 ■ The Chronicle of Neurology & Psychiatry
Chronicle Information Resources Ltd.
PO Box 5456, Sta. A
Calgary, AB, Canada T2H 1X8
Phone: (403)229-9544
Fax: (403)229-1661
Publisher E-mail: health@chronicle.org

Newspaper providing news and information on practical therapeutics and clinical progress in cardiovascular medicine. **Freq:** 9/yr. **Key Personnel:** Mitchell Shannon, Publisher, shannon@chronicle.org. **Remarks:** Accepts advertising. **URL:** http://www.chronicle.ca/neuro.htm; http://cns.chronicle.ca/. **Circ:** 6,189

689 ■ Condo Living
Source Media Group
220-5824 2nd St. SW
Calgary, AB, Canada T2H 0H2
Phone: (403)532-3101
Fax: (403)532-3109
Publisher E-mail: info@sourcemediagroup.ca

Magazine featuring the lifestyle of condo living. **Freq:** Bimonthly. **Key Personnel:** Shelley Williamson, Editor. **URL:** http://www.sourcemediagroup.ca/; http://www.condolivingmag.com/index_1.asp.

690 ■ New Home Source
Source Media Group
220-5824 2nd St. SW
Calgary, AB, Canada T2H 0H2
Phone: (403)532-3101
Fax: (403)532-3109
Publisher E-mail: info@sourcemediagroup.ca

Magazine featuring Calgary residential market place. **Freq:** Biweekly. **Key Personnel:** Jim Zang, Assoc. Publisher. **URL:** http://www.sourcemediagroup.ca/; http://www.newhomesource.ca/index_2.asp. **Circ:** 9,000

691 ■ Resorts & Investment Properties
Source Media Group
220-5824 2nd St. SW
Calgary, AB, Canada T2H 0H2
Phone: (403)532-3101
Fax: (403)532-3109
Publisher E-mail: info@sourcemediagroup.ca

Magazine featuring real estate market information, maps, distance charts, and mortgage rates in Alberta and British Columbia. **URL:** http://www.itsaboutliving.com/flash.html; http://resortsmag.com/contents.php.

692 ■ CFEX-FM - 92.9
400, 255 7th Ave. SW
Calgary, AB, Canada T2S 2T8
Phone: (403)670-0210
Fax: (403)212-1399
E-mail: xsales@x929.ca

Format: Alternative/New Music/Progressive. **Owner:** Harvard Broadcasting, Century Plz., 1900 Rose St., Regina, SK, Canada S4P 0A9, (306)546-6200, Fax: (306)781-7338. **Key Personnel:** Christian Hall, Oper. Mgr./Prog. Dir., chall@harvardbroadcasting.com; James Callsen, News Dir., james@x929.ca; Gary Brasil, General Sales Mgr., gbrasil@x929.ca. **URL:** http://www.x929.ca/.

693 ■ CFUL-FM - 90.3
1110 Centre St. NE, Ste. 100
Calgary, AB, Canada T2E 2R2
Phone: (403)271-6366
Fax: (403)278-6772

Format: Adult Album Alternative. **Key Personnel:** Stephen Peck, General Mgr., stephen@fuelcalgary.com; Murray Brookshaw, Operations Mgr., murray@fuelcalgary.com; Michael Godfrey, Mktg. & Promotions Dir., mgodfrey@fuelcalgary.com. **Ad Rates:** Advertising accepted; rates available upon request. **URL:** http://www.fuelcalgary.com/.

694 ■ CIQX-FM - 103.1
1110 Centre St. NE, Ste. 100
Calgary, AB, Canada T2E 2R2
Phone: (403)271-6366
Fax: (403)278-6772
E-mail: feedback@california103.com

Format: Adult Contemporary; Jazz. **Key Personnel:** Hal Gardiner, News Dir., hgardiner@california103.com; Stephen Peck, General Mgr., stephenp@newcap.ca; Murray Brookshaw, Operations Mgr., mbrookshaw@newcap.ca. **URL:** http://www.california103.com/.

695 ■ CJSI-FM - 88.9
4510 MacLeod Trl. S
Calgary, AB, Canada T2G 0A4
Phone: (403)276-1111
Fax: (403)276-1114

Format: Contemporary Christian. **Owner:** Touch Canada Broadcasting, at above address. **Ad Rates:** Advertising accepted; rates available upon request. **URL:** http://www.cjsi.ca/.

CARDSTON

696 ■ CFSO-TV - 32
PO Box 1238
Cardston, AB, Canada T0K 0K0
Phone: (403)653-3792
Fax: (403)653-3792
E-mail: channel32@mac.com

Owner: Logan & Corey McCarthy, at above address. **Founded:** 1983. **Key Personnel:** Logan McCarthy, Station Mgr., Corey McCarthy, Tech. Oper., corey@channel32.ca. **URL:** http://channel32.ca/.

EDSON

697 ■ CFXE-FM - 94.3
422 50th St., 2nd Fl.
Edson, AB, Canada T7E 1T1

Format: Adult Contemporary. **Owner:** Newcap

Circulation: ★ = ABC; △ = BPA; ♦ = CAC; • = CCAB; ❑ = VAC; ⊕ = PO Statement; ‡ = Publisher's Report; Boldface figures = sworn; Light figures = estimated.

Radio, 745 Windmill Rd., Dartmouth, NS, Canada B3B 1C2, (902)468-7557, Fax: (902)468-7558. **Key Personnel:** Dave Schuck, General Mgr., dschuck@newcap.ca; Rob Alexander, Program Dir., ralexander@newcap.ca; Steve Bethge, News Dir., sbethge@newcap.ca. **Wattage:** 11,000. **URL:** http://www.thefoxradio.ca/.

CFXG-AM - See Grande Cache

CFXP-FM - See Jasper

FORT VERMILION

698 ■ CIAM-FM - 92.7
4709 River Rd.
Fort Vermilion, AB, Canada T0H 1N0
Phone: (780)927-2426
Fax: (780)927-2427
Free: (866)927-2426
E-mail: ciam@ciamradio.com

Format: Religious. **Founded:** Jan. 28, 2003. **Operating Hours:** Continuous. **Ad Rates:** Noncommercial. **URL:** http://www.ciam.ciamradio.com/.

GRANDE CACHE

699 ■ CFXG-AM - 1230
422 50th Ave., 2nd Fl.
Edson, AB, Canada T7E 1T1

Format: Full Service. **Owner:** Newcap Radio, 745 Windmill Rd., Dartmouth, NS, Canada B3B 1C2, (902)468-7557, Fax: (902)468-7558. **Wattage:** 50. **URL:** http://www.thefoxradio.ca/default.asp?mn=1.36.21&sfield=content.id&search=1 3.

HINTON

700 ■ CFXH-FM - 97.5
506 Carmichael Ln., No. 102
Hinton, AB, Canada T7V 1S4
Phone: (780)865-8804
Fax: (780)865-7792

Format: Oldies. **Owner:** Newcap Radio, 745 Windmill Rd., Dartmouth, NS, Canada B3B 1C2, (902)468-7557, Fax: (902)468-7558. **Key Personnel:** Terry Nix, Sales Rep. **Wattage:** 1200. **URL:** http://www.thefoxradio.ca/.

JASPER

701 ■ CFXP-FM - 95.5
422 50th St., 2nd Fl.
Edson, AB, Canada T7E 1T1

Format: Oldies. **Owner:** Newcap Radio, 745 Windmill Rd., Dartmouth, NS, Canada B3B 1C2, (902)468-7557, Fax: (902)468-7558. **URL:** http://www.thefoxradio.ca/.

LACOMBE

702 ■ CJUV-FM - 94.1
4725 49B Ave.
Lacombe, AB, Canada T4L 1K1
Phone: (403)786-0194
Fax: (403)786-0199
E-mail: info@sunny94.com

Format: Classical. **Ad Rates:** Advertising accepted; rates available upon request. **URL:** http://www.sunny94.com/.

LETHBRIDGE

703 ■ CJOC-FM - 94.1
220 Third Ave. S, Ste. 400
Lethbridge, AB, Canada T1J 0G9
Phone: (403)388-2910
Fax: (403)388-4648
E-mail: info@loungeradio.ca

Format: Adult Contemporary. **Owner:** Clear Sky Radio, at above address. **Ad Rates:** Advertising accepted; rates available upon request. **URL:** http://www.loungeradio.ca/.

LLOYDMINSTER

704 ■ CILR-FM - 98.9 MHz
5026 - 50th St.
Lloydminster, AB, Canada T9V 1P3
Phone: (780)875-3321
Fax: (780)875-4704

Format: Information. **Owner:** Newcap Radio, 745 Windmill Rd., Dartmouth, NS, Canada B3B 1C2, (902)468-7557, Fax: (902)468-7558. **Key Personnel:** Chad Tabish, Contact. **URL:** http://www.ncc.ca/stationdetails.asp?id=21.

MEDICINE HAT

705 ■ CJLT-FM - 99.5
901 3rd Ave. SW
Medicine Hat, AB, Canada T1A 4Z2
Phone: (403)529-9599
Fax: (403)529-9282
E-mail: alive995@telus.net

Format: Contemporary Christian. **Operating Hours:** Continuous. **URL:** http://alivefm.com/.

RED DEER

706 ■ CFDV-FM - 106.7
2840 Bremmer Ave.
Red Deer, AB, Canada T4R 1M9
Phone: (403)343-7105
Fax: (403)343-2573
E-mail: rock@1067thedrive.fm

Format: Classic Rock. **Key Personnel:** Joyce Ross, Creative Dir., thinktank2@big105.fm. **Ad Rates:** Advertising accepted; rates available upon request. **URL:** http://www.1067thedrive.fm/.

REDCLIFF

707 ■ CFMY-FM - 96.1
Media Ctr.
10 Boundary Rd.
Redcliff, AB, Canada T0J 2P0
Phone: (403)548-8282
Fax: (403)548-8270

Format: Adult Contemporary. **Owner:** Jim Pattison Broadcast Group, 460 Pemberton Ter., Kamloops, BC, Canada V2C 1T5, (250)372-3322, Fax: (250)374-0445. **URL:** http://www.my96fm.com/.

WAINWRIGHT

708 ■ CKWY-FM - 93.7
1037 2nd Ave., 2nd Fl.
Wainwright, AB, Canada T9W 1K7
Phone: (780)842-4311
Fax: (780)842-4636

Format: Adult Contemporary. **Owner:** Newcap Radio, 745 Windmill Rd., Dartmouth, NS, Canada B3B 1C2, (902)468-7557, Fax: (902)468-7558. **Key Personnel:** Hugh MacDonald, Station Mgr., hmacdonald@newcap.ca; Kathy Romanowicz, Traffic Coord., kromanowicz@newcap.ca. **Wattage:** 100,000. **Ad Rates:** Advertising accepted; rates available upon request. **URL:** http://www.waynefm.com/.

WETASKIWIN

709 ■ CIHS-FM - 93.5
5206 - 50 Ave.
Wetaskiwin, AB, Canada T9A 0S8
Phone: (780)361-0245
Free: (866)409-2797
E-mail: mail@cihsfm.com

Format: Gospel; Country. **Operating Hours:** Continuous. **Ad Rates:** Advertising accepted; rates available upon request. **URL:** http://www.cihsfm.com/.

WHITECOURT

710 ■ CIXM-FM - 105.3
4912A 50th Ave.
PO Box 1050
Whitecourt, AB, Canada T7S 1N9
Phone: (780)706-1053
Fax: (780)706-1017
Free: (866)706-1053

Format: Country. **Key Personnel:** Tyran Ault, News Dir./Reporter; Gene Fabro, President; Andrew Joseph, Program Dir. **Ad Rates:** Advertising accepted; rates available upon request. **URL:** http://www.xm105.com/.

BRITISH COLUMBIA

ABBOTSFORD

711 ■ CFSR-FM - 107.1
318-31935 S Fraser Way
Abbotsford, BC, Canada V2T 5N7
Phone: (604)853-4756
Fax: (604)853-1071

Format: Country. **Ad Rates:** Advertising accepted; rates available upon request. **URL:** http://www.country1071.com/.

BURNABY

712 ■ CFML-FM - 107.9
3700 Willingdon Ave.
Burnaby, BC, Canada V5G 3H2
Phone: (604)432-8510
E-mail: contests@evolution1079.com

Format: Adult Album Alternative. **Key Personnel:** Anjee Gill, Contact; Anita Rai, Contact. **URL:** http://www.evolution1079.com/.

CANOE

713 ■ CHBC-TV Canoe - 6
342 Leon Ave.
Kelowna, BC, Canada V1Y 6J2
Phone: (250)762-4535
Fax: (250)868-0662
Free: (888)762-4535
E-mail: comments@chbcnews.ca

Key Personnel: Keith Williams, General Mgr. **Ad Rates:** Advertising accepted; rates available upon request. **URL:** http://www.chbcnews.ca/modules.php?name=Content&pa=showpage&pid=17.

CRESTON

714 ■ CIDO-FM - 97.7
PO Box 8
Creston, BC, Canada V0B 1G0
Phone: (250)402-6772
E-mail: info@crestonradio.ca

Format: Information. **Owner:** Creston Community Radio Society, at above address. **URL:** http://www.crestonradio.ca/.

DAWSON CREEK

715 ■ CJDC-TV - 5
901 102nd Ave.
Dawson Creek, BC, Canada V1G 2B6
Phone: (250)782-3341
Fax: (250)782-3154

Owner: Astral Media, 2100, rue Ste.-Catherine Ouest, Bureau 1000, Montreal, QC, Canada H3H 2T3, (514)939-5000, Fax: (514)939-1515. **Key Personnel:** Ed Ylanen, General Mgr.; Dave McConnell, Operations Mgr.; Brian Hill, News Dir. **Ad Rates:** Advertising accepted; rates available upon request. **URL:** http://www.cjdctv.com.

DUNCAN

716 ■ CJSU-FM - 89.7
130 Trans Canada Hwy.
Duncan, BC, Canada V9L 3P7
Phone: (250)746-0897
Free: (877)722-0897
E-mail: onair@897sunfm.com

Format: Adult Album Alternative. **URL:** http://www.897sunfm.com/.

ENDERBY

717 ■ CHBC-TV Enderby - 16
342 Leon Ave.
Kelowna, BC, Canada V1Y 6J2
Phone: (250)762-4535
Fax: (250)868-0662
Free: (888)762-4535
E-mail: comments@chbcnews.ca

Key Personnel: Keith Williams, General Mgr. **Ad Rates:** Advertising accepted; rates available upon request. **URL:** http://www.chbcnews.ca/modules.php?name=Content&pa=showpage&pid=17.

FORT SAINT JOHN

718 ■ CKFU-FM - 100.1
10423 101st Ave.
Fort Saint John, BC, Canada V1J 2B7
Phone: (250)787-7100
Fax: (250)263-9749

Format: Classical. **Key Personnel:** Russ Beerling, Gen. Mgr./Owner, rbeerling@moosefm.ca; Adam Reaburn, Prog. Dir./Owner, areaburn@moosefm.ca; Dwight Ford, News Dir., news@moosefm.ca. **URL:** http://www.moosefm.ca/.

KELOWNA

CHBC-TV Canoe - See Canoe
CHBC-TV Enderby - See Enderby
CHBC-TV Naramata - See Naramata
CHBC-TV Osoyoos - See Osoyoos
CHBC-TV Princeton - See Princeton
CHBC-TV Revelstoke - See Revelstoke
CHBC-TV Salmon Arm - See Salmon Arm
CHBC-TV Summerland - See Summerland
CHBC-TV Vernon - See Vernon

NANAIMO

719 ■ Canadian Powersport Trade
Point One Media Inc.
3-2232 Wilgress Rd.
Nanaimo, BC, Canada V9S 4N4
Fax: (250)758-8665
Free: (877)755-2762
Publisher E-mail: info@pointonemedia.com

Magazine featuring the motorcycle, ATV, PWC, and snowmobile industries in Canada. **Trim Size:** 8 x 10.5. **Key Personnel:** Joe Perraton, Publisher, jperraton@pointonemedia.com. **Remarks:** Accepts advertising. **URL:** http://powersporttrade.com/. **Circ:** 3,200

720 ■ Canadian Wall & Ceiling Journal
Point One Media Inc.
3-2232 Wilgress Rd.
Nanaimo, BC, Canada V9S 4N4
Fax: (250)758-8665
Free: (877)755-2762
Publisher E-mail: info@pointonemedia.com

Journal covering all aspects of the Eastern Canadian wall and ceiling industry. **Key Personnel:** Lara Perraton, Gp. Publisher, fax (250)758-8665, lperraton@pointonemedia.com. **Remarks:** Accepts advertising. **URL:** http://www.wallandceiling.ca/. **Circ:** 4,500

721 ■ Sheet Metal Journal
Point One Media Inc.
3-2232 Wilgress Rd.
Nanaimo, BC, Canada V9S 4N4
Fax: (250)758-8665
Free: (877)755-2762
Publisher E-mail: info@pointonemedia.com

Magazine featuring information about sheet metal industry. **Freq:** Bimonthly. **Key Personnel:** Lara Perraton, Gp. Publisher, lperraton@pointonemedia.com. **Remarks:** Accepts advertising. **URL:** http://www.sheetmetaljournal.com/. **Circ:** 2,500

722 ■ The Trowel
Point One Media Inc.
3-2232 Wilgress Rd.
Nanaimo, BC, Canada V9S 4N4
Fax: (250)758-8665
Free: (877)755-2762
Publisher E-mail: info@pointonemedia.com

Magazine featuring wall and ceiling industry of Western Canada. **Founded:** 1953. **Key Personnel:** Jessica Krippendorf, Editor, jkrippendorf@pointonemedia.com. **Remarks:** Accepts advertising. **URL:** http://www.thetrowel.ca/. **Circ:** 3,000

723 ■ Western Woodland
Point One Media Inc.
3-2232 Wilgress Rd.
Nanaimo, BC, Canada V9S 4N4
Fax: (250)758-8665
Free: (877)755-2762
Publisher E-mail: info@pointonemedia.com

Magazine featuring communities, businesses, and people in the forest industry of Canada. **Freq:** Bimonthly. **Key Personnel:** Lara Perraton, Editor, lperraton@pointonemedia.com. **Remarks:** Accepts advertising. **URL:** http://www.westernwoodland.com/. **Circ:** (Not Reported)

NARAMATA

724 ■ CHBC-TV Naramata - 7
342 Leon Ave.
Kelowna, BC, Canada V1Y 6J2
Phone: (250)762-4535
Fax: (250)868-0662
Free: (888)762-4535
E-mail: comments@chbcnews.ca

Key Personnel: Keith Williams, General Mgr. **Ad Rates:** Advertising accepted; rates available upon request. **URL:** http://www.chbcnews.ca/modules.php?name=Content&pa=showpage&pid=17.

OSOYOOS

725 ■ CHBC-TV Osoyoos - 8
342 Leon Ave.
Kelowna, BC, Canada V1Y 6J2
Phone: (250)762-4535
Fax: (250)868-0662
Free: (888)762-4535
E-mail: comments@chbcnews.ca

Key Personnel: Keith Williams, General Mgr. **Ad Rates:** Advertising accepted; rates available upon request. **URL:** http://www.chbcnews.ca/modules.php?name=Content&pa=showpage&pid=17.

PARKSVILLE

726 ■ CHPQ-FM - 99.9
141 Memorial Ave.
PO Box 1370
Parksville, BC, Canada V9P 2H3
Phone: (250)248-4211
Fax: (250)248-4210
E-mail: info@thelounge999.com

Format: Easy Listening; Middle-of-the-Road (MOR). **Owner:** Jim Pattison Broadcast Group, 460 Pemberton Ter., Kamloops, BC, Canada V2C 1T5, (250)372-3322, Fax: (250)374-0445. **Key Personnel:** Mr. Rob Bye, General Mgr.; Mr. Kent Wilson, Program Dir.; Mr. Ray Evans, Music Dir. **Ad Rates:** Advertising accepted; rates available upon request. **URL:** http://www.thelounge999.com/home.php.

PRINCETON

727 ■ CHBC-TV Princeton - 27
342 Leon Ave.
Kelowna, BC, Canada V1Y 6J2
Phone: (250)762-4535
Fax: (250)868-0662
Free: (888)762-4535
E-mail: comments@chbcnews.ca

Key Personnel: Keith Williams, General Mgr. **Ad Rates:** Advertising accepted; rates available upon request. **URL:** http://www.chbcnews.ca/modules.php?name=Content&pa=showpage&pid=17.

REVELSTOKE

728 ■ CHBC-TV Revelstoke - 9
342 Leon Ave.
Kelowna, BC, Canada V1Y 6J2
Phone: (250)762-4535
Fax: (250)868-0662
Free: (888)762-4535
E-mail: comments@chbcnews.ca

Key Personnel: Keith Williams, General Mgr. **Ad Rates:** Advertising accepted; rates available upon request. **URL:** http://www.chbcnews.ca/modules.php?name=Content&pa=showpage&pid=17.

SALMON ARM

729 ■ CHBC-TV Salmon Arm - 9
342 Leon Ave.
Kelowna, BC, Canada V1Y 6J2
Phone: (250)762-4535
Fax: (250)868-0662
Free: (888)762-4535
E-mail: comments@chbcnews.ca

Key Personnel: Keith Williams, General Mgr. **Ad Rates:** Advertising accepted; rates available upon request. **URL:** http://www.chbcnews.ca/modules.php?name=Content&pa=showpage&pid=17.

SUMMERLAND

730 ■ CHBC-TV Summerland - 13
342 Leon Ave.
Kelowna, BC, Canada V1Y 6J2
Phone: (250)762-4535
Fax: (250)868-0662
Free: (888)762-4535
E-mail: comments@chbcnews.ca

Key Personnel: Keith Williams, General Mgr. **Wattage:** 300. **Ad Rates:** Advertising accepted; rates available upon request. **URL:** http://www.chbcnews.ca/modules.php?name=Content&pa=showpage&pid=17.

VALEMOUNT

731 ■ CHVC-TV - 7
Box 922
Valemount, BC, Canada V0E 2Z0
Phone: (250)566-8288
Fax: (250)566-4645
E-mail: tv@vctv.ca

Owner: Valemount Entertainment Society, at above address. **Key Personnel:** Andru McCracken, Station Mgr. **URL:** http://www.vctv.ca/.

VANCOUVER

732 ■ CNHW-FM - 89.3
1033 Haro St., Ste. 100
Vancouver, BC, Canada V6E 1C8
Phone: (778)737-1910
Fax: (778)737-1592
E-mail: info@vancouvertouristradio.com

Format: Information. **URL:** http://vancouvertouristradio.com/.

Circulation: ★ = ABC; △ = BPA; ♦ = CAC; • = CCAB; ❑ = VAC; ⊕ = PO Statement; ‡ = Publisher's Report; Boldface figures = sworn; Light figures = estimated.

VERNON

733 ■ CHBC-TV Vernon - 7
342 Leon Ave.
Kelowna, BC, Canada V1Y 6J2
Phone: (250)762-4535
Fax: (250)868-0662
Free: (888)762-4535
E-mail: comments@chbcnews.ca

Key Personnel: Keith Williams, General Mgr. **Wattage:** 310. **Ad Rates:** Advertising accepted; rates available upon request. **URL:** http://www.chbcnews.ca/modules.php?name=Content&pa=showpage&pid=17.

VICTORIA

734 ■ Douglas
Page One Publishing Inc.
1322A Government St.
Victoria, BC, Canada V8W 1Y8
Phone: (250)595-7243
Fax: (250)595-1626
Free: (866)595-7243
Publisher E-mail: info@pageonepublishing.ca

Magazine featuring issues and articles for Greater Victoria business community. **Subtitle:** Victoria's Business Magazine. **Freq:** Bimonthly. **Key Personnel:** Norman Gidney, Editor, phone (250)474-6145, ngidney@douglasmagazine.com. **Subscription Rates:** C$15 individuals; C$26 two years. **Remarks:** Accepts advertising. **URL:** http://www.douglasmagazine.com/. **Circ:** 20,000

735 ■ CHTT-FM - 103.1
817 Fort St.
Victoria, BC, Canada V8W 1H6
Phone: (250)382-0900
Fax: (250)382-4358

Format: Adult Contemporary. **Key Personnel:** Don Landels, Sales Mgr., phone (250)414-4501, fax (250)414-4516, don.landels@rci.rogers.com. **Ad Rates:** Advertising accepted; rates available upon request. **URL:** http://www.1031jackfm.ca/.

736 ■ CILS-FM - 107.9
200 - 535 Yates St.
Victoria, BC, Canada V8W 2Z6
Phone: (250)220-4139
Fax: (250)388-6280
E-mail: radio@francocentre.com

Format: Ethnic. **Key Personnel:** Jules Desjarlais, Program Dir. **URL:** http://cilsfm.ca/en/.

MANITOBA

THOMPSON

737 ■ CINC-FM - 96.3
1507 Inkster Blvd.
Winnipeg, MB, Canada R2X 1R2
Phone: (204)772-8255
Fax: (204)779-5628

Format: Ethnic. **Ad Rates:** Advertising accepted; rates available upon request. **URL:** http://www.ncifm.com/.

WINNIPEG

CINC-FM - See Thompson

NEW BRUNSWICK

BALMORAL

738 ■ CIMS-FM - 103.9
1991, ave. des Pionniers
C.P. 2561
Balmoral, NB, Canada E8E 2W7
Phone: (506)826-1040
Fax: (506)826-2400
E-mail: info@cimsfm.ca

Format: Ethnic. **Owner:** Cooperative Radio Restigouche, at above address. **Key Personnel:** Jean-Guy Landry, Musical & Stimulating Dir., jean-guy@cimsfm.ca. **URL:** http://www.cimsfm.ca/cims/index.cfm.

BEAVER DAM

739 ■ Cosmetology Association Magazine
Partners Publishing
2289 Rte. 101
Beaver Dam, NB, Canada E3B 7T9
Phone: (506)450-9768
Fax: (506)450-2546
Publisher E-mail: info@partnerspublishing.com

Magazine featuring hair salons, beauty salons, skin care salons, esthetic & electrolysis salons, day spas, educational & technical organizations and suppliers around the Atlantic Provinces. **Key Personnel:** Peter Cole, Pres./Ed. in Ch., pcole@partnerspublishing.com. **Subscription Rates:** Free. **Remarks:** Accepts advertising. **URL:** http://www.partnerspublishing.com/cosmo/cam_association_nb.htm. . **Ad Rates:** BW: $700, 4C: $850. **Circ:** 6,000

740 ■ New Brunswick Road Builder
Partners Publishing
2289 Rte. 101
Beaver Dam, NB, Canada E3B 7T9
Phone: (506)450-9768
Fax: (506)450-2546
Publisher E-mail: info@partnerspublishing.com

Magazine featuring complete information on the Road Builders Association Of New Brunswick. **Freq:** Annual. **Key Personnel:** Peter Cole, Pres./Ed. in Ch., pcole@partnerspublishing.com. **Subscription Rates:** Free. **Remarks:** Accepts advertising. **URL:** http://www.partnerspublishing.com/road/road_builder.htm. . **Ad Rates:** BW: $750, 4C: $900. **Circ:** (Not Reported)

741 ■ Spa Canada Magazine
Partners Publishing
2289 Rte. 101
Beaver Dam, NB, Canada E3B 7T9
Phone: (506)450-9768
Fax: (506)450-2546
Publisher E-mail: info@partnerspublishing.com

Magazine featuring various aspects of the aesthetic, fitness, and spa industries. **Freq:** 6/yr. **Print Method:** offset. **Trim Size:** 8 1/2 x 11. **Key Personnel:** Peter Cole, Pres./Ed. in Ch., pcole@partnerspublishing.com. **Remarks:** Accepts advertising. **URL:** http://www.partnerspublishing.com/spa/spa_canada.htm. **Circ:** 12,000

FREDERICTON

742 ■ CFRK-FM - 92.3
77 Westmorland St., Ste. 400
Fredericton, NB, Canada E3B 6Z3
Phone: (506)455-0923
Fax: (506)455-3602
E-mail: fred@fredfm.ca

Format: Classic Rock. **Owner:** Newcap Radio, 745 Windmill Rd., Dartmouth, NS, Canada B3B 1C2, (902)468-7557, Fax: (902)468-7558. **Key Personnel:** Hilary Montbourquette, General Mgr., hilary@fredfm.ca; Brad Muir, Oper. Mgr./Prog. Dir., bmuir@fredfm.ca; John Knox, Promotions/Mktg. Dir., jknox@fredfm.ca. **URL:** http://fredfm.ca/.

GRAND FALLS

743 ■ CIKX-FM - 93.5
399 Boul. Broadway Blvd.
Grand Falls, NB, Canada E3Z 2K5
Phone: (506)473-9393
Fax: (506)473-3893
E-mail: k93@radioatl.ca

Format: Adult Contemporary. **Owner:** Astral Media Inc., 2100, rue Ste.-Catherine Ouest, Bureau 1000, Montreal, QC, Canada H3H 2T3, (514)939-5000, Fax: (514)939-1515. **Key Personnel:** Jacques LaFrance, Sales Mgr., phone (506)473-5732, jlfrance@radioatl.ca; Janice Paradis, Advertising Sales Rep., phone (506)473-6438, jparadis@radioatl.ca; Rick McGuire, Program Dir., phone (506)325-3035, rmcguire@radioatl.ca. **URL:** http://www.k93.ca/.

MONCTON

744 ■ CKUM-FM - 93.5
Ctr. Entudiant
University de Moncton
Moncton, NB, Canada E1A 3E9
Phone: (506)858-3750
Fax: (506)858-4524

Format: Educational. **Key Personnel:** Jean-Sebastien Levesque, Gen. Dir., dgckum@umoncton.ca; Pierre-Luc Larocque, Program Dir., phone (506)858-3777, progckum935@yahoo.ca; Carolynn McNally, Music Dir., phone (506)856-5772, musiqueradioj935@yahoo.ca. **URL:** http://www.umoncton.ca/ckum/.

SAINT JOHN

745 ■ CINB-FM - 96.1
PO Box 96
Saint John, NB, Canada E2L 3X1
Phone: (506)657-9600
E-mail: staff@newsongfm.com

Format: Religious. **URL:** http://www.newsongfm.com/.

746 ■ CJEF-FM - 103.5
87 Lansdowne Ave.
Saint John, NB, Canada E2K 3A1
Phone: (506)657-2533
Fax: (506)642-7408
E-mail: onair@thepirate.ca

Format: Urban Contemporary. **Key Personnel:** Geoff Rivett, CEO/Gen. Mgr., geoff@thepirate.ca; Steph Downey, Promotions Dir., steph@thepirate.ca; Marc Henwood, Music Dir. **URL:** http://www.thepirate.ca/.

747 ■ CKLT-TV - 9
12 Smythe St., Ste. 126
Saint John, NB, Canada E2L 5G5
Phone: (506)658-1010
Fax: (506)658-1208

Owner: CTVglobemedia Inc., 9 Channel Nine Ct., Scarborough, ON, Canada M1S 4B5, (416)332-5000, Fax: (416)332-5283. **Ad Rates:** Advertising accepted; rates available upon request. **URL:** http://www.ctv.ca/servlet/ArticleNews/show/CTVShows/1065812845544_61211386.

NOVA SCOTIA

ANTIGONISH

748 ■ International Journal of Embedded Systems
Inderscience Publishers
c/o Prof. Laurence T. Yang, Ed.-in-Ch.
St. Francis Xavier University
Department of Computer Science
Antigonish, NS, Canada B2G 2W5
Publication E-mail: ijes@inderscience.com
Publisher E-mail: editor@inderscience.com

Journal covering all aspects of embedded computing systems. **Founded:** 2005. **Freq:** 4/yr. **Key Personnel:** Prof. Laurence T. Yang, Editor-in-Chief, lyang@stfx.ca; Prof. Minyi Guo, Editor-in-Chief. **ISSN:** 1741-1068. **Subscription Rates:** EUR470 individuals includes surface mail, print only; EUR640 individuals print & online. **URL:** http://www.inderscience.com/browse/index.php?journalCODE=ijes.

ARICHAT

749 ■ CIMC-TV - 10
705 Lower Rd.
PO Box 87
Arichat, NS, Canada B0E 1A0
Phone: (902)226-1928
Fax: (902)226-1331
E-mail: telile@telile.tv

Founded: 1994. **Key Personnel:** Gloria Hill, General Mgr., gloria@telile.tv; Rhonda LeBlanc, Community Prog., rhonda@telile.tv; Julie Boudreau, Admin. Asst., julie@telile.tv. **Ad Rates:** Advertising accepted; rates available upon request. **URL:** http://www.telile.tv/.

HALIFAX

750 ■ CBHT-TV - 3
PO Box 3000
Halifax, NS, Canada B3J 3E9
Phone: (902)420-8311

Owner: Canadian Broadcasting Corporation, PO Box 500, Sta. A, Toronto, ON, Canada M5W 1E6, Free: (866)306-4636. **Key Personnel:** Mike Linder, Regional Dir., phone (780)468-7505; Judy Piercey, Managing Editor, phone (780)468-7526; Kelly Walter, Sales Mgr., phone (780)468-2344. **Ad Rates:** Advertising accepted; rates available upon request. **URL:** http://www.cbc.ca/ns/.

ONTARIO

BARRIE

751 ■ CKMB-FM - 107.5
431 Huronia Rd., Unit 10
Barrie, ON, Canada L4N 9B3
Phone: (705)725-7304
Fax: (705)721-7842

Format: Adult Contemporary. **Founded:** Jan. 1, 2006. **Wattage:** 40,000. **Ad Rates:** Advertising accepted; rates available upon request. **URL:** http://1075koolfm.com/home.php.

BELLEVILLE

752 ■ CHCQ-FM - 100.1
354 Pinnacle St.
Belleville, ON, Canada K8N 3B4
Phone: (613)966-0955
Fax: (613)967-2565

Format: Country. **Key Personnel:** Darren Matassa, Sales Mgr., darrenm@classichits955.fm; Paul Martin, News Dir., news@aclassichits955.fm; John Sherratt, Owner/Mgr., johns@cool100.fm. **URL:** http://cool100.fm/.

753 ■ CKJJ-FM - 102.3
PO Box 23095
Belleville, ON, Canada K8P 5J3
Phone: (613)966-4822
Fax: (613)966-3211

Format: Contemporary Christian. **Owner:** United Christian Broadcasters Canada, at above address. **Key Personnel:** James Hunt, CEO; Garry Quinn, General Mgr.; Melanie Linn, Music Dir. **URL:** http://ucbcanada.com/.

BRACEBRIDGE

754 ■ Summer in Muskoka
Osprey Media
Box 180
Bracebridge, ON, Canada P1L 1T6
Phone: (705)646-1314
Fax: (705)645-6424

Magazine featuring Muskoka District. **Key Personnel:** Don Smith, Contact. **Remarks:** Accepts advertising. **URL:** http://ospreymedialp.com/corporate/Default.asp?section=publications&paper=SummerinMuskoka. **Circ:** (Not Reported)

BRANTFORD

755 ■ VIBRANT
Osprey Media
53 Dalhousie St.
PO Box 965
Brantford, ON, Canada N3T 5S8
Phone: (519)756-2020
Fax: (519)756-4911

Magazine covering different topics like lifestyles and fashion for the Brant County residents. **Key Personnel:** Michael Pearce, Publisher. **Remarks:** Accepts advertising. **URL:** http://ospreymedialp.com/corporate/Default.asp?section=publications&paper=V IBRANT. **Circ:** (Not Reported)

BURLINGTON

756 ■ GolfStyle
Osprey Media
1074 Cooke Blvd.
Burlington, ON, Canada L7T 4A8
Phone: (905)634-8003
Fax: (905)634-7661

Magazine covering golf and lifestyle. **Freq:** 3/yr. **Print Method:** Web offset. **Trim Size:** 9 x 10.875. **Key Personnel:** Christina Trembelas, Contact, ctrembelas@townmedia.ca; Wayne Narciso, Contact, wayne@townmedia.ca; Ted McIntyre, Contact. **Remarks:** Accepts advertising. **URL:** http://www.golfstylemagazine.ca; http://ospreymedialp.com/corporate/Default.asp?section=publications&paper=GolfStyle. **Circ:** Combined 50,000

757 ■ Ontario Golf
Osprey Media
1074 Cooke Blvd.
Burlington, ON, Canada L7T 4A8

Magazine featuring golf in Ontario. **Freq:** 5/yr. **Trim Size:** 8.125 x 10.875. **Key Personnel:** Ted McIntyre, Editor, tedbits@golfontario.ca; Wayne Narciso, Publisher, wayne@golfontario.ca; Kate Sharrow, Art Dir., ksharrow@townmedia.ca. **Subscription Rates:** C$12 individuals; C$20 two years; C$27 individuals 3 years. **Remarks:** Accepts advertising. **URL:** http://www.golfontario.ca/sitepages; http://ospreymedialp.com/corporate/Default.asp?section=publications&paper=O ntarioGolf. **Circ:** Combined 55,500

758 ■ Toronto Golf
Osprey Media
1074 Cooke Blvd.
Burlington, ON, Canada L7T 4A8
Phone: (905)634-8003
Fax: (905)634-7661

Magazine covering golf in Toronto. **Key Personnel:** Wayne Narciso, Contact, wayne@townmedia.ca; Christina Trembelas, Contact, ctrembelas@townmedia.ca; Ted McIntyre, Contact. **Remarks:** Accepts advertising. **URL:** http://ospreymedialp.com/corporate/Default.asp?section=publications&paper=TorontoGolf. **Circ:** (Not Reported)

759 ■ Vines
Osprey Media
1074 Cooke Blvd.
Burlington, ON, Canada L7T 4A8
Phone: (905)634-8003

Magazine covering wine industry. **Key Personnel:** Wayne Narciso, Contact, wayne@townmedia.ca; Christina Trembelas, Contact, ctrembelas@townmedia.ca; Christopher Waters, Contact, chris@townmedia.ca. **Remarks:** Accepts advertising. **URL:** http://ospreymedialp.com/corporate/Default.asp?section=publications&paper=V ines. **Circ:** (Not Reported)

760 ■ Visitors
Osprey Media
1074 Cooke Blvd.
Burlington, ON, Canada L7T 4A8
Phone: (905)634-8003
Fax: (905)634-7661

Magazine featuring Hamilton/Burlington region. **Key Personnel:** Wayne Narciso, Contact, wayne@townmedia.ca; Joanne Bouchard, Contact, jbouchard@townmedia.ca; David Young, Contact, david@townmedia.ca. **Remarks:** Accepts advertising. **URL:** http://ospreymedialp.com/corporate/Default.asp?section=publications&paper=V isitors. **Circ:** (Not Reported)

761 ■ CITS-TV - 36
1295 N Service Rd.
Burlington, ON, Canada L7R 4X5
Phone: (905)331-7333
Fax: (905)332-6005

Key Personnel: Glenn Stewart, Dir., Sales & Mktg., gstewart@ctstv.com; Rob Sheppard, Program Mgr., rsheppard@ctstv.com; Michelle Gillies, Promotions Dir., mgillies@ctstv.com. **Ad Rates:** Advertising accepted; rates available upon request. **URL:** http://www.ctstv.com/index.asp.

DRYDEN

762 ■ CJIV-FM - 97.3
PO Box 112
Dryden, ON, Canada P8N 2Y7
Phone: (807)937-9731

Format: Religious. **Key Personnel:** Adrian Rogers, Contact. **URL:** http://www.cjiv973.net/.

DUNDAS

763 ■ Canadian Hearing Report
Andrew John Publishing Inc.
115 King St. W, Ste. 220
Dundas, ON, Canada L9H 1V1
Phone: (905)628-4309
Fax: (905)628-6847
Publisher E-mail: info@andrewjohnpublishing.com

Magazine featuring information for hearing health care professionals. **Trim Size:** 8 1/8 x 10 3/4. **Key Personnel:** Brenda Robinson, Contact, phone (905)628-4309, fax (905)628-6847, brobinson@andrewjohnpublishing.com. **ISSN:** 1718-1860. **Remarks:** Accepts advertising. **URL:** http://www.andrewjohnpublishing.com/pub.html. . **Ad Rates:** 4C: $2,113. **Circ:** (Not Reported)

764 ■ Canadian Journal of General Internal Medicine
Andrew John Publishing Inc.
115 King St. W, Ste. 220
Dundas, ON, Canada L9H 1V1
Phone: (905)628-4309
Fax: (905)628-6847
Publisher E-mail: info@andrewjohnpublishing.com

Magazine featuring topics and issues about general internal medicine. **Freq:** 4/yr. **Trim Size:** 8 1/8 x 10 3/4. **Key Personnel:** John Birkby, Advertising Sales Rep., phone (905)628-4309, fax (905)628-6847, jbirkby@andrewjohnpublishing.com. **ISSN:** 1911-1606. **Remarks:** Accepts advertising. **URL:** http://www.andrewjohnpublishing.com/pub.html. . **Ad Rates:** BW: $775. **Circ:** 4,100

765 ■ CASLPO Today
Andrew John Publishing Inc.
115 King St. W, Ste. 220
Dundas, ON, Canada L9H 1V1
Phone: (905)628-4309
Fax: (905)628-6847
Publisher E-mail: info@andrewjohnpublishing.com

Magazine featuring information for audiologist and speech language pathologist. **Freq:** 5/yr. **Trim Size:** 8 1/8 x 10 3/4. **Key Personnel:** John Birkby, Advertising Sales Rep., phone (905)628-4309, fax (905)628-6847, jbirkby@andrewjohnpublishing.com. **ISSN:** 1713-8922. **Remarks:** Accepts advertising. **URL:** http://www.andrewjohnpublishing.com/pub.html. . **Ad Rates:** 4C: $1,890. **Circ:** 3,000

766 ■ College Contact
Andrew John Publishing Inc.
115 King St. W, Ste. 220
Dundas, ON, Canada L9H 1V1
Phone: (905)628-4309
Fax: (905)628-6847
Publisher E-mail: info@andrewjohnpublishing.com

Magazine featuring products and procedures affecting dentistry. **Trim Size:** 8 1/8 x 10 3/4. **Key Personnel:** John Birkby, Advertising Sales Rep., phone (905)628-4309, fax (905)628-6847, jbirkby@andrewjohnpublishing.com. **ISSN:** 1913-424X. **Remarks:** Accepts advertising. **URL:** http://www.andrewjohnpublishing.com/pub.html. . **Ad Rates:** 4C: $750. **Circ:** 1,000

Circulation: ★ = ABC; △ = BPA; ♦ = CAC; • = CCAB; ❑ = VAC; ⊕ = PO Statement; ‡ = Publisher's Report; Boldface figures = sworn; Light figures = estimated.

767 ■ Listen/Ecoute
Andrew John Publishing Inc.
115 King St. W, Ste. 220
Dundas, ON, Canada L9H 1V1
Phone: (905)628-4309
Fax: (905)628-6847
Publisher E-mail: info@andrewjohnpublishing.com

Magazine featuring information about hearing aids. **Trim Size:** 8 1/8 x 10 3/4. **Key Personnel:** John Birkby, Advertising Sales Rep., phone (905)628-4309, fax (905)628-6847, jbirkby@andrewjohnpublishing.com. **ISSN:** 1813-1820. **Remarks:** Accepts advertising. **URL:** http://www.andrewjohnpublishing.com/pub.html. . **Ad Rates:** 4C: $850. **Circ:** (Not Reported)

768 ■ Wavelength
Andrew John Publishing Inc.
115 King St. W, Ste. 220
Dundas, ON, Canada L9H 1V1
Phone: (905)628-4309
Fax: (905)628-6847
Publisher E-mail: info@andrewjohnpublishing.com

Magazine featuring information about and for public safety communications professionals. **Trim Size:** 8 1/2 x 11. **Key Personnel:** Brenda Robinson, Contact, phone (905)628-4309, fax (905)628-6847, brobinson@andrewjohnpublishing.com. **Remarks:** Accepts advertising. **URL:** http://www.andrewjohnpublishing.com/pub.html. . **Ad Rates:** BW: $320, 4C: $1,320. **Circ:** 2,000

GODERICH

769 ■ CHWC-FM - 104.9
300 Suncoast Dr., Unit E
Goderich, ON, Canada N7A 4N7
Phone: (519)612-1149
Fax: (519)612-1050
E-mail: thebeach@1049thebeach.ca

Format: Adult Contemporary. **Key Personnel:** Dale Roth, Asst. Prog. Dir.; Kevin MacDonald, Music Dir. **Ad Rates:** Advertising accepted; rates available upon request. **URL:** http://www.1049thebeach.ca/.

HANOVER

770 ■ CFBW-FM - 91.3
267 10th St.
Hanover, ON, Canada N4N 1P1
Phone: (519)364-0200
Fax: (519)364-5175
E-mail: bluewaterradio@on.aibn.com

Format: Adult Contemporary. **Owner:** Bluewater Radio, at above address. **Operating Hours:** Continuous. **Key Personnel:** Andrew McBride, Station Mgr./Prog. Dir. **Ad Rates:** Advertising accepted; rates available upon request. **URL:** http://www.bluewaterradio.ca/.

HUNTSVILLE

771 ■ CFBK-FM - 105.5
15-2 Main St. E
Huntsville, ON, Canada P1H 2C9
Phone: (705)789-4461
Fax: (705)789-1269

Format: Adult Contemporary. **URL:** http://www.zeuter.com/more/.

KANATA

772 ■ Canadian Wildlife
Canadian Wildlife Federation
350 Michael Cowpland Dr.
Kanata, ON, Canada K2M 2W1
Phone: (613)599-9594
Fax: (613)599-4428
Free: (800)563-WILD
Publisher E-mail: info@cwf-fcf.org

Magazine featuring stories about Canadian and international wildlife for teenagers and adults. **Freq:** 6/yr. **URL:** http://28005.vws.magma.ca/pages/whoarewe/index_e.asp?section=1&page=108&language=e.

773 ■ Due East
Coyle Publishing, Inc.
362 Terry Fox Dr., Ste. 220
Kanata, ON, Canada K2K 2P5
Phone: (613)271-8903
Fax: (613)271-8905

Magazine featuring lifestyle topics for the whole family. **Freq:** Quarterly. **Key Personnel:** George W. Coyle, Publisher, gcoyle@coylepublishing.com. **Remarks:** Accepts advertising. **URL:** http://www.coylepublishing.com/dueeast/index.htm. **Circ:** 20,000

774 ■ Due West
Coyle Publishing, Inc.
362 Terry Fox Dr., Ste. 220
Kanata, ON, Canada K2K 2P5
Phone: (613)271-8903
Fax: (613)271-8905

Magazine featuring lifestyle topics for the whole family. **Freq:** Quarterly. **Key Personnel:** George W. Coyle, Publisher, gcoyle@coylepublishing.com. **Remarks:** Accepts advertising. **URL:** http://www.coylepublishing.com/duewest/index.htm. **Circ:** 32,000

775 ■ Fifty-Five Plus
Coyle Publishing, Inc.
362 Terry Fox Dr., Ste. 220
Kanata, ON, Canada K2K 2P5
Phone: (613)271-8903
Fax: (613)271-8905

Magazine featuring information about health, nutrition, fitness, travel, and finance for mature adult. **Founded:** 1988. **Freq:** 8/yr. **Key Personnel:** George W. Coyle, Publisher, gcoyle@coylepublishing.com. **Remarks:** Accepts advertising. **URL:** http://www.coylepublishing.com/fiftyfiveOttawa/index.htm. **Circ:** 100,000

776 ■ FSO
Coyle Publishing, Inc.
362 Terry Fox Dr., Ste. 220
Kanata, ON, Canada K2K 2P5
Phone: (613)271-8903
Fax: (613)271-8905

Magazine featuring information for the returning students of the City of Ottawa. **Freq:** Semiannual. **Key Personnel:** George W. Coyle, Publisher, phone (613)271-8903, gcoyle@coylepublishing.com. **Remarks:** Accepts advertising. **URL:** http://coylepublishing.com/fso/. **Circ:** 25,000

KITCHENER

777 ■ CJIQ-FM - 88.3
299 Doon Valley Dr.
Kitchener, ON, Canada N2G 4M4
Phone: (519)748-5220

Format: Classic Rock. **Owner:** Conestoga College, at above address. **Founded:** Jan. 8, 2001. **Key Personnel:** Mike Thurnell, Program Dir., mthurnell@conestogac.on.ca; Paul Scott, Coord., pdscott@conestogac.on.ca. **Wattage:** 4000. **Ad Rates:** Advertising accepted; rates available upon request. **URL:** http://www.cjiq.fm/.

LONDON

778 ■ International Journal of Manufacturing Research
Inderscience Publishers
c/o Dr. Lihui Wang, Ed.-in-Ch.
800 Collip Cir.
London, ON, Canada N6G 4X8
Publisher E-mail: editor@inderscience.com

Journal covering new developments in modern manufacturing research. **Founded:** 2006. **Freq:** 4/yr. **Key Personnel:** Dr. Lihui Wang, Editor-in-Chief, lihui.wang@nrc.gc.ca. **ISSN:** 1750-0591. **Subscription Rates:** EUR470 individuals includes surface mail, print only; EUR640 individuals print & online. **URL:** http://www.inderscience.com/browse/index.php?journalCODE=ijmr.

779 ■ CHJX-FM - 105.9
100 Fullarton St.
London, ON, Canada N6A 1K1
Phone: (519)679-9882
Fax: (519)679-2459

Format: Contemporary Christian. **URL:** http://www.gracefm.ca/index.cfm.

MARKHAM

780 ■ AES Contact
Palmeri Publishing Inc.
35-145 Royal Crest Ct.
Markham, ON, Canada L3R 9Z4
Phone: (905)489-1970
Fax: (905)489-1971
Free: (866)581-8949

Magazine featuring the study of dentistry. **Freq:** 3/yr. **Key Personnel:** Ettore Palmeri, President, ettore@palmeripublishing.com. **URL:** http://palmeripublishing.com/aes.html. **Circ:** 1,200

781 ■ Canadian Journal of Cosmetic Dentistry
Palmeri Publishing Inc.
35-145 Royal Crest Ct.
Markham, ON, Canada L3R 9Z4
Phone: (905)489-1970
Fax: (905)489-1971
Free: (866)581-8949

Journal focusing on cosmetic dentistry. **Freq:** Semiannual. **Key Personnel:** Ettore Palmeri, President, ettore@palmeripublishing.com. **URL:** http://palmeripublishing.com/cjcd.html. **Circ:** 15,000

782 ■ Journal of the Academy of General Dentistry
Palmeri Publishing Inc.
35-145 Royal Crest Ct.
Markham, ON, Canada L3R 9Z4
Phone: (905)489-1970
Fax: (905)489-1971
Free: (866)581-8949

Journal focusing on general dentistry. **Freq:** Quarterly. **Key Personnel:** Ettore Palmeri, President, ettore@palmeripublishing.com. **URL:** http://palmeripublishing.com/jagd.html. **Circ:** 15,000

783 ■ NJAGD Wisdom
Palmeri Publishing Inc.
35-145 Royal Crest Ct.
Markham, ON, Canada L3R 9Z4
Phone: (905)489-1970
Fax: (905)489-1971
Free: (866)581-8949

Magazine featuring dentistry. **Freq:** Quarterly. **Key Personnel:** Ettore Palmeri, President, ettore@palmeripublishing.com. **URL:** http://palmeripublishing.com/wisdom.html. **Circ:** 5,800

784 ■ Spectrum Denturism
Palmeri Publishing Inc.
35-145 Royal Crest Ct.
Markham, ON, Canada L3R 9Z4
Phone: (905)489-1970
Fax: (905)489-1971
Free: (866)581-8949

Magazine featuring dentistry. **Freq:** 4/yr. **Key Personnel:** Ettore Palmeri, President, ettore@palmeripublishing.com. **URL:** http://palmeripublishing.com/sd-denturism.html. **Circ:** 3,000

785 ■ Spectrum Dialogue
Palmeri Publishing Inc.
35-145 Royal Crest Ct.
Markham, ON, Canada L3R 9Z4
Phone: (905)489-1970
Fax: (905)489-1971
Free: (866)581-8949

Magazine featuring techno-clinical dentistry. **Freq:** 9/yr. **Key Personnel:** Ettore Palmeri, President, ettore@palmeripublishing.com. **URL:** http://palmeripublishing.com/sd.html. **Circ:** 24,000

786 ■ Spectrum Quebec
Palmeri Publishing Inc.
35-145 Royal Crest Ct.
Markham, ON, Canada L3R 9Z4
Phone: (905)489-1970
Fax: (905)489-1971
Free: (866)581-8949

Magazine featuring techno-clinical dentistry. **Freq:** Quarterly. **Key Personnel:** Ettore Palmeri, President, ettore@palmeripublishing.com. **URL:** http://palmeripublishing.com/sd-q.html. **Circ:** 5,600

787 ■ Team Work
Palmeri Publishing Inc.
35-145 Royal Crest Ct.
Markham, ON, Canada L3R 9Z4
Phone: (905)489-1970
Fax: (905)489-1971
Free: (866)581-8949

Magazine featuring dentistry. **Subtitle:** The Insiders' Guide to Dentistry. **Freq:** 6/yr. **Key Personnel:** Ettore Palmeri, President, ettore@palmeripublishing.com. **URL:** http://palmeripublishing.com/teamwork.html. **Circ:** 75,000

MISSISSAUGA

788 ■ The Chronicle of Cancer Therapy
Chronicle Information Resources Ltd.
2200 Dundas St. E, Ste. 200
Mississauga, ON, Canada L4X 2V3
Phone: (905)273-9116
Fax: (905)273-4322
Free: (866)63C-HRON
Publisher E-mail: health@chronicle.org

Newspaper providing news and information on practical therapeutics and clinical progress in cardiovascular medicine. **Freq:** 6/yr. **Trim Size:** 7 3/4 x 10 1/4. **Key Personnel:** Mitchell Shannon, Publisher, shannon@chronicle.org. **Remarks:** Accepts advertising. **URL:** http://ca-tx.chronicle.ca/. **Circ:** 2,774

789 ■ The Chronicle of Cardiovascular & Internal Medicine
Chronicle Information Resources Ltd.
2200 Dundas St. E, Ste. 200
Mississauga, ON, Canada L4X 2V3
Phone: (905)273-9116
Fax: (905)273-4322
Free: (866)63C-HRON
Publisher E-mail: health@chronicle.org

Newspaper providing news and information on practical therapeutics and clinical progress in cardiovascular medicine. **Freq:** 9/yr. **Key Personnel:** Mitchell Shannon, Publisher, shannon@chronicle.org. **Remarks:** Accepts advertising. **URL:** http://www.chronicle.ca/cardio.htm. **Circ:** 5,570

790 ■ The Chronicle of Healthcare Marketing
Chronicle Information Resources Ltd.
2200 Dundas St. E, Ste. 200
Mississauga, ON, Canada L4X 2V3
Phone: (905)273-9116
Fax: (905)273-4322
Free: (866)63C-HRON
Publisher E-mail: health@chronicle.org

Newspaper providing news and information on the world of healthcare business. **Freq:** 9/yr. **Key Personnel:** Mitchell Shannon, Publisher, shannon@chronicle.org. **Remarks:** Accepts advertising. **URL:** http://www.chronicle.ca/hcm.htm. **Circ:** 2,201

791 ■ The Chronicle of Skin & Allergy
Chronicle Information Resources Ltd.
2200 Dundas St. E, Ste. 200
Mississauga, ON, Canada L4X 2V3
Phone: (905)273-9116
Fax: (905)273-4322
Free: (866)63C-HRON
Publisher E-mail: health@chronicle.org

Newspaper providing news and information on practical therapeutics and clinical progress in dermatologic medicine. **Freq:** 9/yr. **Key Personnel:** Mitchell Shannon, Publisher, shannon@chronicle.org. **Remarks:** Accepts advertising. **URL:** http://www.chronicle.ca/skin.htm. **Circ:** 6,150

792 ■ The Chronicle of Urology & Sexual Medicine
Chronicle Information Resources Ltd.
2200 Dundas St. E, Ste. 200
Mississauga, ON, Canada L4X 2V3
Phone: (905)273-9116
Fax: (905)273-4322
Free: (866)63C-HRON
Publisher E-mail: health@chronicle.org

Newspaper providing news and information on practical progress in urology and sexual medicine. **Freq:** Bimonthly. **Key Personnel:** Mitchell Shannon, Publisher, shannon@chronicle.org. **Remarks:** Accepts advertising. **URL:** http://www.chronicle.ca/uro.htm. **Circ:** 5,184

793 ■ Physicians' Computing Chronicle
Chronicle Information Resources Ltd.
2200 Dundas St. E, Ste. 200
Mississauga, ON, Canada L4X 2V3
Phone: (905)273-9116
Fax: (905)273-4322
Free: (866)63C-HRON
Publisher E-mail: health@chronicle.org

Journal covering the medical knowledge management. **Freq:** 6/yr. **Key Personnel:** Mitchell Shannon, Publisher, shannon@chronicle.org. **Remarks:** Accepts advertising. **URL:** http://www.chronicle.ca/pcc.htm. **Circ:** 31,834

NIAGARA FALLS

794 ■ CFLZ-FM - 105.1
4668 St. Clair Ave.
Niagara Falls, ON, Canada L2E 6X7
Phone: (905)356-6710
Fax: (905)356-0644

Format: Adult Contemporary. **Key Personnel:** Andy Bilous, Sales Mgr., abilous@niagara.com; Mike Ryan, Program Dir., meryan@niagara.com; Chris Barnatt, Music Dir., phone (716)603-6300, cbarnatt@niagara.com. **Ad Rates:** Advertising accepted; rates available upon request. **URL:** http://www.river.fm/.

OTTAWA

795 ■ Journal of Wavelet Theory and Applications
Research India Publications
School of Mathematics & Statistics
Carleton University
Ottawa, ON, Canada K1S 5B6
Publisher E-mail: info@ripublication.com

Journal covering theories and applications of wavelets. **Freq:** 3/yr. **Key Personnel:** Prof. J.N. Pandey, Editor-in-Chief, jpandey@math.carleton.ca. **ISSN:** 0973-6336. **Subscription Rates:** US$180 libraries; US$180 institutions; US$140 individuals. **URL:** http://www.ripublication.com/jwta.htm.

796 ■ CIHT-FM - 89.9
6 Antares Dr., Phase 1, Unit 100
Ottawa, ON, Canada K2E 8A9
Phone: (613)723-8990
E-mail: advertise@hot899.com

Format: Contemporary Hit Radio (CHR). **Owner:** Newcap Broadcasting Ltd., 745 Windmill Rd., Dartmouth, NS, Canada B3B 1C2, (902)468-7557, Fax: (902)468-7558. **Founded:** Feb. 7, 2003. **Ad Rates:** Advertising accepted; rates available upon request. **URL:** http://www.hot899.com/default.asp?mn=1.2.

797 ■ CILV-FM - 88.5
6 Antares Dr., Phase 1, Unit 100
Ottawa, ON, Canada K2E 8A9
Phone: (613)688-8888
Fax: (613)723-7016

Format: Adult Contemporary. **Owner:** Newcap Broadcasting Ltd., 745 Windmill Rd., Dartmouth, NS, Canada B3B 1C2, (902)468-7557, Fax: (902)468-7558. **Key Personnel:** Scott Broderick, General Mgr., sbroderick@newcap.ca; Dan Youngs, Program Dir., dyoungs@newcap.ca; Lee Wagner, Operations Mgr., lwagner@newcap.ca. **URL:** http://www.live885.com/default.asp?mn=1.72.

798 ■ CJLL-FM - 97.9
30 Murray St., Ste. 100
Ottawa, ON, Canada K1N 5M4
Phone: (613)244-0979
Fax: (613)244-3858

Format: Ethnic; Talk. **URL:** http://www.chinradio.com/ottawa.php.

PARRY SOUND

799 ■ CHRZ-FM - 91.3
PO Box 482
Parry Sound, ON, Canada P2A 2X5
Phone: (705)746-4481
Fax: (705)746-4481

Format: Ethnic. **Ad Rates:** Advertising accepted; rates available upon request. **URL:** http://www.rez91.com/.

PEMBROKE

800 ■ CHRO-TV - 5
611 TV Tower Rd.
PO Box 1010
Pembroke, ON, Canada K8A 6Y6
Phone: (613)735-1036
Fax: (613)735-4374

Owner: CTVglobemedia Inc., 9 Channel Nine Ct., Scarborough, ON, Canada M1S 4B5, (416)332-5000, Fax: (416)332-5283. **Wattage:** 100,000 ERP. **URL:** http://www.achannel.ca/ottawa/.

801 ■ CIMY-FM - 104.9
84 Isabella St., 2nd Fl.
Pembroke, ON, Canada K8A 5S5
Phone: (613)735-1049
Fax: (613)732-4054

Format: Adult Contemporary. **Key Personnel:** Marc Poirier, General Mgr., marc@myfmradio.ca. **Ad Rates:** Advertising accepted; rates available upon request. **URL:** http://www.myfmradio.ca/1049/.

PETERBOROUGH

802 ■ CKKK-FM - 90.5
993 Talwood Dr., 2nd Fl.
Peterborough, ON, Canada K9J 7R8
Phone: (705)876-8600

Format: Religious; Classic Rock. **Operating Hours:** Continuous. **Key Personnel:** Rick Kirschner, General Mgr., rick@kaosradio.com; Darryl Parsons, Program Dir., darryl@kaosradio.com; Rick Amsbury, Promotions Dir., rickamsbury@kaosradio.com. **Ad Rates:** Advertising accepted; rates available upon request. **URL:** http://kaosradio.com/.

RENFREW

803 ■ CJHR-FM - 98.7
PO Box 945
Renfrew, ON, Canada K7V 4H4
Phone: (613)432-9873
E-mail: info@valleyheritageradio.ca

Format: Ethnic. **URL:** http://www.valleyheritageradio.ca/news.php.

SAINT CATHERINES

804 ■ CHSC-AM - 1220
36 Queenston St.
Saint Catherines, ON, Canada L2R 2Y9
Phone: (905)682-6692
Fax: (905)682-9434
Free: (888)334-3481

Format: Adult Contemporary; Oldies. **Key Personnel:** John Marshall, Contact; Chuck Lafleur, Contact. **Ad Rates:** Advertising accepted; rates available upon request. **URL:** http://www.1220chsc.ca/Home2.htm.

Circulation: ★ = ABC; △ = BPA; ♦ = CAC; • = CCAB; ❑ = VAC; ⊕ = PO Statement; ‡ = Publisher's Report; Boldface figures = sworn; Light figures = estimated.

STRATFORD

805 ■ CHGK-FM - 107.7
376 Romeo St. S
Stratford, ON, Canada N5A 4T9
Phone: (519)271-2450
Fax: (519)271-3102
Free: (866)649-1077
E-mail: mixmornings@1077mixfm.com

Format: Adult Contemporary. **Key Personnel:** Amber Cook, Contact, acook@1077mixfm.com; Elizabeth Cooper, Contact, elcooper@1077mixfm.com; Jamie Cootle, Contact, jcottle@1077mixfm.com. **Ad Rates:** Advertising accepted; rates available upon request. **URL:** http://www.1077mixfm.com/.

STRATHROY

806 ■ CJMI-FM - 105.7
125 Metcalfe St.
Strathroy, ON, Canada N7G 1M9
Phone: (519)246-6936
Fax: (519)245-6670

Format: Adult Contemporary. **Founded:** 2007. **Key Personnel:** Jeff Degraw, General Mgr., jeff.degraw@myfmradio.ca; Kevin McGowan, Program Dir., kevin@myfmradio.ca. **Ad Rates:** Advertising accepted; rates available upon request. **URL:** http://www.myfmradio.ca/1057/.

SUDBURY

807 ■ CICI-TV - 5
699 Frood Rd.
Sudbury, ON, Canada P3C 5A3
Phone: (705)674-8301
Fax: (705)674-2789

Owner: CTVglobemedia Inc., 9 Channel Nine Ct., Scarborough, ON, Canada M1S 4B5, (416)332-5000, Fax: (416)332-5283. **Ad Rates:** Advertising accepted; rates available upon request. **URL:** http://www.ctv.ca/servlet/ArticleNews/show/CTVShows/1065552962668_60956580.

TIMMINS

808 ■ CITO-TV - 3
681 Pine St. N
Timmins, ON, Canada P4N 7L6
Phone: (705)264-4211
Fax: (705)264-3266

Owner: CTVglobemedia Inc., 9 Channel Nine Ct., Scarborough, ON, Canada M1S 4B5, (416)332-5000, Fax: (416)332-5283. **Ad Rates:** Advertising accepted; rates available upon request. **URL:** http://www.ctv.ca/servlet/ArticleNews/show/CTVShows/1065552962668_60956580.

TORONTO

809 ■ Building Strategies
MediaEDGE
5255 Yonge St., Ste. 1000
Toronto, ON, Canada M2N 6P4
Phone: (416)512-8186
Fax: (416)512-8344
Free: (866)216-0860
Publisher E-mail: info@mediaedge.ca

Magazine featuring commercial real estate developments. **Freq:** 4/yr. **Trim Size:** 8.75 x 12.25. **Key Personnel:** Chuck Nervick, Publisher, phone (416)512-8186, chuckn@buildingstrategies.ca. **Remarks:** Accepts advertising. **URL:** http://www.mediaedge.ca/bs.shtml. **Circ:** 7,000

810 ■ Canadian Apartment Magazine
MediaEDGE
5255 Yonge St., Ste. 1000
Toronto, ON, Canada M2N 6P4
Phone: (416)512-8186
Fax: (416)512-8344
Free: (866)216-0860
Publisher E-mail: info@mediaedge.ca

Magazine featuring information for property owners and managers of multi-residential market in Canada. **Freq:** 6/yr. **Trim Size:** 8.125 x 10.875. **Key Personnel:** Marc Cote, Publisher, phone (416)512-8186, marcc@mediaedge.ca. **Subscription Rates:** C$44.94 individuals; C$80.79 two years. **Remarks:** Accepts advertising. **URL:** http://www.mediaedge.ca/cam.shtml. **Circ:** 7,000

811 ■ Canadian Gaming Business
MediaEDGE
5255 Yonge St., Ste. 1000
Toronto, ON, Canada M2N 6P4
Phone: (416)512-8186
Fax: (416)512-8344
Free: (866)216-0860
Publisher E-mail: info@mediaedge.ca

Magazine featuring the gaming industry in Canada. **Subtitle:** Canada's Premier Gaming Industry Magazine. **Trim Size:** 8.125 x 10.875. **Key Personnel:** Chuck Nervick, Publisher, phone (416)512-8186, chuckn@buildingstrategies.ca. **Remarks:** Accepts advertising. **URL:** http://www.canadiangamingbusiness.com/. **Circ:** 4,000

812 ■ Canadian Healthcare Facilities
MediaEDGE
5255 Yonge St., Ste, 1000
Toronto, ON, Canada M2N 6P4
Phone: (416)512-8186
Fax: (416)512-8344
Free: (866)216-0860
Publisher E-mail: info@mediaedge.ca

Journal focusing on Canadian healthcare facility. **Subtitle:** Journal of Canadian Healthcare Engineering Society. **Freq:** 4/yr. **Trim Size:** 8.125 x 10.875. **Key Personnel:** Andrea Civichino, Editor, phone (416)512-8186, andreac@mediaedge.ca. **Remarks:** Accepts advertising. **URL:** http://www.mediaedge.ca/ches.shtml. **Circ:** 2,500

813 ■ Canadian Property Management
MediaEDGE
5255 Yonge St., Ste. 1000
Toronto, ON, Canada M2N 6P4
Phone: (416)512-8186
Fax: (416)512-8344
Free: (866)216-0860
Publisher E-mail: info@mediaedge.ca

Magazine featuring property management industry. **Freq:** 8/yr. **Trim Size:** 8.125 x 10.875. **Key Personnel:** Tony Robinson, Publisher, phone (416)512-8186, tonyr@mediaedge.ca. **Subscription Rates:** C$53.45 individuals; C$96.30 two years. **Remarks:** Accepts advertising. **URL:** http://www.mediaedge.ca/cpm.shtml; http://www.canadianpropertymanagement.ca/. **Circ:** 12,500

814 ■ CONDOBUSINESS
MediaEDGE
5255 Yonge St., Ste. 1000
Toronto, ON, Canada M2N 6P4
Phone: (416)512-8186
Fax: (416)512-8344
Free: (866)216-0860
Publisher E-mail: info@mediaedge.ca

Magazine featuring information for condominium owner. **Freq:** 8/yr. **Key Personnel:** Steve McLenden, Publisher, phone (416)512-8186, stevem@mediaedge.ca. **Subscription Rates:** C$136.65 individuals; C$246.05 two years. **Remarks:** Accepts advertising. **URL:** http://www.condobusiness.ca/. **Circ:** (Not Reported)

815 ■ Construction Business
MediaEDGE
5255 Yonge St., Ste. 1000
Toronto, ON, Canada M2N 6P4
Phone: (416)512-8186
Fax: (416)512-8344
Free: (866)216-0860
Publisher E-mail: info@mediaedge.ca

Magazine featuring construction business. **Freq:** 6/yr. **Trim Size:** 8.75 x 12.25. **Key Personnel:** Dan Gnocato, Publisher, phone (604)739-2115, dang@mediaedge.ca. **Subscription Rates:** C$25.63 individuals; C$51.26 two years. **Remarks:** Accepts advertising. **URL:** http://www.mediaedge.ca/constructb.shtml. **Circ:** 3,100

816 ■ Corporate Meetings & Events
MediaEDGE
5255 Yonge St., Ste. 1000
Toronto, ON, Canada M2N 6P4
Phone: (416)512-8186
Fax: (416)512-8344
Free: (866)216-0860
Publisher E-mail: info@mediaedge.ca

Magazine featuring information for the organizer of corporate events. **Freq:** 6/yr. **Trim Size:** 8 3/8 x 10 7/8. **Key Personnel:** Chuck Nervick, Publisher, phone (416)512-8186, chuckn@buildingstrategies.ca. **Remarks:** Accepts advertising. **URL:** http://www.mediaedge.ca/cme.shtml. **Circ:** 10,000

817 ■ Correo Canadiense
Multicom Media Services Ltd.
101 Wingold Ave.
Toronto, ON, Canada M6B 1P8
Phone: (416)785-4300
Fax: (416)785-4310
Free: (877)503-5077
Publisher E-mail: info@multicommedia.ca

Newspaper covering special events, campaigns, and contests in Toronto area. **Founded:** 2001. **Freq:** Weekly (Fri.). **Key Personnel:** Lori Abittan, Pres./Publisher. **URL:** http://www.multicommedia.ca/index.htm?id=3. **Circ:** 20,000

818 ■ Energy Procurement & Conservation
MediaEDGE
5255 Yonge St., Ste. 1000
Toronto, ON, Canada M2N 6P4
Phone: (416)512-8186
Fax: (416)512-8344
Free: (866)216-0860
Publisher E-mail: info@mediaedge.ca

Magazine featuring today's market conditions related to energy. **Freq:** 4/yr. **Trim Size:** 8.125 x 10.875. **Key Personnel:** Chuck Nervick, VP, phone (416)512-8186, chuckn@buildingstrategies.ca. **Remarks:** Accepts advertising. **URL:** http://www.mediaedge.ca/epc.shtml. **Circ:** 6,000

819 ■ Engineering Business
MediaEDGE
5255 Yonge St., Ste. 1000
Toronto, ON, Canada M2N 6P4
Phone: (416)512-8186
Fax: (416)512-8344
Free: (866)216-0860
Publisher E-mail: info@mediaedge.ca

Magazine featuring Canada's consulting engineering community. **Subtitle:** Canada's Consulting Engineering Magazine. **Freq:** 6/yr. **Trim Size:** 8.75 x 12.25. **Key Personnel:** Sean Foley, Publisher, phone (416)512-8186, sean@engineeringbusiness.ca. **Remarks:** Accepts advertising. **URL:** http://www.mediaedge.ca/eb.shtml. **Circ:** 10,500

820 ■ Food Safety & Quality
MediaEDGE
5255 Yonge St., Ste. 1000
Toronto, ON, Canada M2N 6P4
Phone: (416)512-8186
Fax: (416)512-8344
Free: (866)216-0860
Publisher E-mail: info@mediaedge.ca

Magazine featuring facts on food safety issues in Canada. **Freq:** 4/yr. **Trim Size:** 7 7/8 x 10 7/8. **Key Personnel:** Andrew Feldman, Assoc. Publisher, phone (416)512-8186, andyf@mediaedge.ca. **Subscription Rates:** C$40 individuals; C$90 two years. **Remarks:** Accepts advertising. **URL:** http://www.mediaedge.ca/cfsm.shtml. **Circ:** 7,000

821 ■ Gender and Language
Equinox Publishing Ltd.
Sidney Smith Hall
Department of Anthropology
100 St. George St.
Toronto, ON, Canada M5S 3G3
Publication E-mail: genderandlanguage@utoronto.ca

Journal covering gender and language. **Freq:** 2/yr (January and June). **Key Personnel:** Bonnie McElhinny, Editor, bonnie.mcelhinny@utoronto.ca; Sara

Mills, Editor, s.l.mills@shu.ac.uk. **ISSN:** 1747-6321. **Subscription Rates:** 90 institutions; US$165 institutions; 40 individuals; US$65 individuals; EUR55 individuals; 25 students; US$45 students; EUR38 students. **Remarks:** Accepts advertising. **URL:** http://www.equinoxjournals.com/ojs/index.php/GL. **Circ:** (Not Reported)

822 ■ The Grind
Paton Publishing
2802 Lakeshore Blvd. W
Toronto, ON, Canada M8V 1H5
Phone: (416)503-4576
Fax: (416)503-8474
Publisher E-mail: info@patonpublishing.com

Magazine featuring life, work, and everything in between for teens. **Freq:** Semiannual. **Key Personnel:** Anne Lovegrove, Contact, phone (416)503-4576. **URL:** http://www.patonpublishing.com/our_mag/index.html. **Circ:** Combined 150,000

823 ■ Healthcare Facilities Management
MediaEDGE
5255 Yonge St., Ste. 1000
Toronto, ON, Canada M2N 6P4
Phone: (416)512-8186
Fax: (416)512-8344
Free: (866)216-0860
Publisher E-mail: info@mediaedge.ca

Magazine featuring Canadian long-term health care sector. **Freq:** 4/yr. **Trim Size:** 8.125 x 10.875. **Key Personnel:** Andrea Civichino, Editor, phone (416)512-8186, andreac@mediaedge.ca. **Remarks:** Accepts advertising. **URL:** http://www.mediaedge.ca/hfm.shtml. **Circ:** 3,500

824 ■ Healthcare Policy
Longwoods Publishing Corp.
260 Adelaide St. E, No. 8
Toronto, ON, Canada M5A 1N1
Phone: (416)864-9667
Fax: (416)368-4443

Journal covering healthcare policy research. **Freq:** Quarterly. **Key Personnel:** Dr. Brian Hutchison, Editor-in-Chief. **ISSN:** 1715-6572. **Subscription Rates:** C$105 individuals online only; C$155 individuals online and print; C$460 institutions online only; C$562 institutions print and online. **Remarks:** Accepts advertising. **URL:** http://www.longwoods.com/home.php?cat=247. **Circ:** (Not Reported)

825 ■ Healthcare Quarterly
Longwoods Publishing Corp.
260 Adelaide St. E, No. 8
Toronto, ON, Canada M5A 1N1
Phone: (416)864-9667
Fax: (416)368-4443

Journal covering best practices, policy, and innovations in the administration of healthcare. **Freq:** Quarterly. **Key Personnel:** Dr. Peggy Leatt, Editor-in-Chief. **ISSN:** 1710-2774. **Subscription Rates:** C$90 individuals online only; C$107 individuals online and print; C$310 institutions online only; C$430 institutions print and online. **Remarks:** Accepts advertising. **URL:** http://www.longwoods.com/home.php?cat=249. **Circ:** (Not Reported)

826 ■ HealthcarePapers
Longwoods Publishing Corp.
260 Adelaide St. E, No. 8
Toronto, ON, Canada M5A 1N1
Phone: (416)864-9667
Fax: (416)368-4443

Journal covering reviews of new models in healthcare. **Freq:** Quarterly. **Key Personnel:** Dr. Peggy Leatt, Editor-in-Chief. **ISSN:** 1488-917X. **Subscription Rates:** C$102 individuals online only; C$143 individuals online and print; C$430 institutions online only; C$533 institutions print and online. **Remarks:** Accepts advertising. **URL:** http://www.longwoods.com/home.php?cat=250. **Circ:** (Not Reported)

827 ■ Home and Community Care Digest
Longwoods Publishing Corp.
260 Adelaide St. E, No. 8
Toronto, ON, Canada M5A 1N1
Phone: (416)864-9667
Fax: (416)368-4443

Journal covering research pertaining to the financing, delivery, and organization of home and community-based health care activities. **Freq:** Quarterly. **Key Personnel:** Dr. Peter C. Coyote, Contact, peter.coyte@utoronto.ca. **Subscription Rates:** C$300 individuals online only; C$600 institutions online only. **URL:** http://www.longwoods.com/home.php?cat=150.

828 ■ ImageMakers
MediaEDGE
5255 Yonge St., Ste. 1000
Toronto, ON, Canada M2N 6P4
Phone: (416)512-8186
Fax: (416)512-8344
Free: (866)216-0860
Publisher E-mail: info@mediaedge.ca

Magazine featuring information for commercial image making professionals. **Freq:** 4/yr. **Trim Size:** 8 3/8 x 10 7/8. **Key Personnel:** Anne Haapanen, Editor, phone (905)699-2175, anneh@mediaedge.ca. **Remarks:** Accepts advertising. **URL:** http://www.mediaedge.ca/im.shtml. **Circ:** 8,000

829 ■ Inside Moto-X
Inside Track Publications
5 Lower Sherbourne St., Ste. 203
Toronto, ON, Canada M5A 2P3
Phone: (416)962-7223
Fax: (416)962-7208
Publication E-mail: imx@insidemotocross.ca
Publisher E-mail: ads@insidetracknews.com

Magazine featuring Canadian motocross amateur and professional racing. **Freq:** 6/yr. **Key Personnel:** John Hopkins, Editor. **Subscription Rates:** US$2.50 single issue; US$14.99 individuals. **Remarks:** Accepts advertising. **URL:** http://www.insidemotocross.ca/. **Circ:** 20,000

830 ■ Inside Motorcycles
Inside Track Publications
5 Lower Sherbourne St., Ste. 203
Toronto, ON, Canada M5A 2P3
Phone: (416)962-7223
Fax: (416)962-7208
Publisher E-mail: ads@insidetracknews.com

Magazine for motorcycle riders and racing enthusiasts. **Founded:** 1998. **Freq:** 10/yr. **Key Personnel:** John Hopkins, Editor, editor@insidemotorcycles.com. **Subscription Rates:** US$24.99 individuals; US$2.50 single issue. **Remarks:** Accepts advertising. **URL:** http://www.insidemotorcycles.com/. **Circ:** 20,000

831 ■ Law & Governance
Longwoods Publishing Corp.
260 Adelaide St. E, No. 8
Toronto, ON, Canada M5A 1N1
Phone: (416)864-9667
Fax: (416)368-4443

Journal covering law and governance. **Freq:** 10/yr. **Key Personnel:** Kevin Smith, Editorial Advisory Board. **ISSN:** 1710-3363. **Subscription Rates:** C$260 individuals online only; C$520 institutions online only. **Remarks:** Accepts advertising. **URL:** http://www.longwoods.com/home.php?cat=120. **Circ:** (Not Reported)

832 ■ Nove Ilhas/O Correio Canadiano
Multicom Media Services Ltd.
101 Wingold Ave.
Toronto, ON, Canada M6B 1P8
Phone: (416)785-4300
Fax: (416)785-4310
Free: (877)503-5077
Publisher E-mail: info@multicommedia.ca

Newspaper covering current political, sports, and community events in Toronto. **Freq:** Weekly (Tues.). **Key Personnel:** Joe March, Contact, jmarch@multicommedia.ca. **Subscription Rates:** C$110 individuals. **URL:** http://www.multicommedia.ca/index.htm?id=5.

833 ■ Nursing Leadership
Longwoods Publishing Corp.
260 Adelaide St. E, No. 8
Toronto, ON, Canada M5A 1N1
Phone: (416)864-9667
Fax: (416)368-4443

Journal covering nursing management, practice, education, and research. **Freq:** Quarterly. **Key Personnel:** Dorothy Pringle, PhD, Editor-in-Chief; Dianne Foster Kent, Managing Editor, phone (416)864-9667, dkent@longwoods.com. **ISSN:** 1910-622X. **Subscription Rates:** C$82 individuals online only; C$98 individuals online and print; C$307 institutions online only; C$410 institutions print and online. **Remarks:** Accepts advertising. **URL:** http://www.longwoods.com/home.php?cat=252. **Circ:** (Not Reported)

834 ■ Plan Bleu
Spafax
1179 King St., Ste. 101
Toronto, ON, Canada M6K 3C5
Phone: (416)350-2425
Fax: (416)350-2440
Publisher E-mail: kkopvillem@spafax.net

Magazine featuring management and in-strategic marketing partnerships. **Key Personnel:** Niall McBain, CEO. **URL:** http://www.spafax.com/.

835 ■ POP!
Paton Publishing
2802 Lakeshore Blvd. W
Toronto, ON, Canada M8V 1H5
Phone: (416)503-4576
Fax: (416)503-8474
Publisher E-mail: info@patonpublishing.com

Magazine featuring science, history, literature, social studies, and world issues in an entertaining way for kids and youths. **Freq:** Quarterly. **Key Personnel:** Anne Lovegrove, Contact, phone (416)503-4576. **URL:** http://www.patonpublishing.com/our_mag/index_pop.html. **Circ:** Combined 325,000

836 ■ POP! JR.
Paton Publishing
2802 Lakeshore Blvd. W
Toronto, ON, Canada M8V 1H5
Phone: (416)503-4576
Fax: (416)503-8474
Publisher E-mail: info@patonpublishing.com

Magazine featuring interactive puzzles and games and read-out-loud stories for the young minds. **Freq:** Semiannual. **Key Personnel:** Anne Lovegrove, Contact, phone (416)503-4576. **URL:** http://www.patonpublishing.com/our_mag/index_popjr.html. **Circ:** 125,000

837 ■ PowerPlay
Paton Publishing
2802 Lakeshore Blvd. W
Toronto, ON, Canada M8V 1H5
Phone: (416)503-4576
Fax: (416)503-8474
Publisher E-mail: info@patonpublishing.com

Magazine featuring information for young hockey players. **Freq:** 4/yr. **Key Personnel:** Anne Lovegrove, Contact, phone (416)503-4576. **URL:** http://www.patonpublishing.com/our_mag/index_pp.html. **Circ:** Combined 125,000

838 ■ Property Management Report
MediaEDGE
5255 Yonge St., Ste. 1000
Toronto, ON, Canada M2N 6P4
Phone: (416)512-8186
Fax: (416)512-8344
Free: (866)216-0860
Publisher E-mail: info@mediaedge.ca

Magazine featuring information about development and commercial real estate in the Greater Toronto

Circulation: ★ = ABC; △ = BPA; ♦ = CAC; • = CCAB; ❑ = VAC; ⊕ = PO Statement; ‡ = Publisher's Report; Boldface figures = sworn; Light figures = estimated.

Area, Hamilton, and Niagara Region. **Freq:** 8/yr. **Trim Size:** 8.125 x 10.875. **Key Personnel:** Tony Robinson, Publisher, phone (416)512-8186, tonyr@mediaedge.ca. **Remarks:** Accepts advertising. **URL:** http://www.mediaedge.ca/pmr.shtml. **Circ:** 12,500

839 ■ Pure Canada
Spafax
1179 King St., Ste. 101
Toronto, ON, Canada M6K 3C5
Phone: (416)350-2425
Fax: (416)350-2440
Publisher E-mail: kkopvillem@spafax.net

Travel magazine featuring Canadian destinations and tourism offerings. **Key Personnel:** Niall McBain, CEO. **URL:** http://www.spafax.com/.

840 ■ SBC Kiteboard
SBC Media Group
2255B Queen St. E, Ste. 3266
Toronto, ON, Canada M4E 1G3
Phone: (416)406-2400
Fax: (416)406-0656
Publisher E-mail: subscriptions@sbcmedia.com

Magazine featuring kiteboarding. **Freq:** 4/yr. **Subscription Rates:** US$17.98 individuals; C$17.98 Canada; US$50 other countries; US$29.98 two years; C$29.98 Canada two years; US$70 other countries two years. **Remarks:** Accepts advertising. **URL:** http://www.sbcmedia.com/sites/kiteboard/i_kiteboard.html; http://www.sbckiteboard.com/. **Circ:** (Not Reported)

841 ■ SBC Skateboard
SBC Media Group
2255B Queen St. E, Ste. 3266
Toronto, ON, Canada M4E 1G3
Phone: (416)406-2400
Fax: (416)406-0656
Publisher E-mail: subscriptions@sbcmedia.com

Magazine featuring skateboarding. **Founded:** 1998. **Freq:** 5/yr. **Key Personnel:** Mike Moore, Contact, phone (416)406-2400, mike@sbcmedia.com. **Subscription Rates:** US$22.49 individuals; C$22.49 Canada; US$50 other countries; US$35.98 two years; C$35.98 Canada two years; US$70 other countries two years. **Remarks:** Accepts advertising. **URL:** http://www.sbcmedia.com/sites/skateboard/i_skateboard.html; http://www.sbcskateboard.com/. **Circ:** (Not Reported)

842 ■ SBC Skier
SBC Media Group
2255B Queen St. E, Ste. 3266
Toronto, ON, Canada M4E 1G3
Phone: (416)406-2400
Fax: (416)406-0656
Publisher E-mail: subscriptions@sbcmedia.com

Magazine featuring skiing. **Freq:** 4/yr. **Key Personnel:** Mike Moore, Contact, phone (416)406-2400, mike@sbcmedia.com. **Subscription Rates:** US$17.98 individuals; C$17.98 Canada; US$50 other countries; US$28.77 two years; C$28.77 Canada two years; US$70 other countries two years. **Remarks:** Accepts advertising. **URL:** http://www.sbcmedia.com/sites/skier/i_skier.html; http://www.sbcskier.com/advertising.php. **Circ:** (Not Reported)

843 ■ SBC Wakeboard
SBC Media Group
2255B Queen St. E, Ste. 3266
Toronto, ON, Canada M4E 1G3
Phone: (416)406-2400
Fax: (416)406-0656
Publisher E-mail: subscriptions@sbcmedia.com

Magazine featuring wakeboard hotspot in Canada. **Founded:** 1999. **Freq:** Semiannual. **Subscription Rates:** US$17.98 individuals; C$17.98 Canada; US$50 other countries; US$28.77 two years; C$28.77 Canada two years; US$70 other countries two years. **Remarks:** Accepts advertising. **URL:** http://www.sbcmedia.com/sites/wakeboard/i_wakeboard.html; http://www.sbcwakeboard.com/. **Circ:** (Not Reported)

844 ■ Snowboard Canada
SBC Media Group
2255B Queen St. E, Ste. 3266
Toronto, ON, Canada M4E 1G3
Phone: (416)406-2400
Fax: (416)406-0656
Publisher E-mail: subscriptions@sbcmedia.com

Magazine featuring snowboarding. **Founded:** 1992. **Freq:** 4/yr. **Key Personnel:** Mike Moore, Contact, phone (416)406-2400, mike@sbcmedia.com. **Subscription Rates:** US$17.98 individuals; C$17.98 Canada; US$50 other countries; US$28.77 two years; C$28.77 Canada two years; US$70 other countries two years. **Remarks:** Accepts advertising. **URL:** http://www.sbcmedia.com/sites/snowboard/i_snowboard.html; http://www.snowboardcanada.com/. **Circ:** (Not Reported)

845 ■ Start Here
Spafax
1179 King St., Ste. 101
Toronto, ON, Canada M6K 3C5
Phone: (416)350-2425
Fax: (416)350-2440
Publisher E-mail: kkopvillem@spafax.net

Magazine featuring information about business and technology. **Freq:** Semiannual. **Key Personnel:** Niall McBain, CEO. **URL:** http://www.spafax.com/.

846 ■ Tandem
Multicom Media Services Ltd.
101 Wingold Ave.
Toronto, ON, Canada M6B 1P8
Phone: (416)785-4300
Fax: (416)785-4310
Free: (877)503-5077
Publisher E-mail: info@multicommedia.ca

Newspaper covering lifestyle and business development issues. **Founded:** 1995. **Freq:** Weekly (Fri.). **Key Personnel:** Paola Bernardini, Interim Ed.-in-Ch. **URL:** http://www.multicommedia.ca/index.htm?id=2. **Circ:** 62,519

847 ■ Vaughan Today
Multicom Media Services Ltd.
101 Wingold Ave.
Toronto, ON, Canada M6B 1P8
Phone: (416)785-4300
Fax: (416)785-4310
Free: (877)503-5077
Publisher E-mail: info@multicommedia.ca

Newspaper covering the city of Vaughan. **Freq:** Weekly (Fri.). **Key Personnel:** Joe March, Contact, jmarch@multicommedia.ca. **URL:** http://www.multicommedia.ca/index.htm?id=26. **Circ:** 72,157

848 ■ World Health & Population
Longwoods Publishing Corp.
260 Adelaide St. E, No. 8
Toronto, ON, Canada M5A 1N1
Phone: (416)864-9667
Fax: (416)368-4443

Journal covering best practices, policy, and innovations in healthcare. **Freq:** Quarterly. **Key Personnel:** John E. Paul, PhD, Editor-in-Chief, phone (919)966-7373, paulj@email.unc.edu; Prof. Sagar C. Jain, PhD, Founding Ed.-in-Ch.; Amir A. Khaliq, PhD, Assoc. Ed. **ISSN:** 1718-3340. **Subscription Rates:** C$40 individuals online only; C$200 individuals online and print; C$240 institutions online; C$340 institutions online and print. **Remarks:** Accepts advertising. **URL:** http://www.longwoods.com/home.php?cat=381. **Circ:** (Not Reported)

849 ■ YTV Whoa!
Paton Publishing
2802 Lakeshore Blvd. W
Toronto, ON, Canada M8V 1H5
Phone: (416)503-4576
Fax: (416)503-8474
Publisher E-mail: info@patonpublishing.com

Magazine featuring pop culture. **Freq:** Quarterly. **Key Personnel:** Anne Lovegrove, Contact, phone (416)503-4576. **URL:** http://www.patonpublishing.com/our_mag/index_whoa.html. **Circ:** Combined 325,000

850 ■ CJGV-FM - 99.1
32 Atlantic Ave.
Toronto, ON, Canada M6K 1X8
Fax: (416)530-6885
Free: (866)537-2397

Format: Adult Album Alternative. **Owner:** Corus Entertainment Inc., at above address. **URL:** http://www.corusent.com/radio/manitoba/index.asp?lineofbusiness=radio&compa nyid=144.

851 ■ CKXT-TV - 52
25 Ontario St.
Toronto, ON, Canada M5A 4L6
Phone: (416)601-0010
Fax: (416)601-0004
E-mail: feedback@suntv.canoe.ca

Founded: Sept. 19, 2003. **Key Personnel:** Duane Parks, Retail Sales Mgr., phone (416)933-5693; Christina Fagan, Sen. Mktg. Specialist, phone (416)933-5674. **Ad Rates:** Advertising accepted; rates available upon request. **URL:** http://suntv.canoe.ca/about_suntv.html.

VERMILION BAY

852 ■ CKQV-FM - 103.3
78 Spruce St.
PO Box 459
Vermilion Bay, ON, Canada P0V 2V0
Phone: (807)227-9988
Fax: (807)227-9985
Free: (866)338-9969
E-mail: info@q104fm.ca

Format: Adult Contemporary. **Key Personnel:** Rick Doucet, General Mgr., rick@q104fm.ca; Ken O'Neil, Program Dir., ken@q104fm.ca; Crash Adams, Music Dir., crash@q104fm.ca. **URL:** http://www.q104fm.ca/.

WASAGA BEACH

853 ■ CHGB-FM - 97.7
1383 Mosley St.
PO Box 476
Wasaga Beach, ON, Canada L9Z 1A5
Phone: (705)422-0970
Fax: (705)422-0468
E-mail: info@977thebeach.ca

Format: Adult Contemporary. **Key Personnel:** Deb Shaw, Station Mgr.; Rick Ringer, Oper. Mgr.; John O'Mara, Creative/Production Dir. **Ad Rates:** Advertising accepted; rates available upon request. **URL:** http://www.977thebeach.ca/.

WOODSTOCK

854 ■ CIHR-FM - 104.7
223 Norwich Ave.
Woodstock, ON, Canada N4S 3V8
Phone: (519)537-8400
Fax: (519)537-8600

Format: Adult Contemporary. **Key Personnel:** Dan Henry, Program Dir., dan@1047.ca; Jody Hug, Music Dir., jody@1047.ca; Adam Nyp, News Dir., adam@1047.ca. **URL:** http://www.1047.ca/.

855 ■ CJFH-FM - 94.3
535 Mill St.
TA Travel Ctr., RR No. 1
Woodstock, ON, Canada N4S 7V6
Phone: (519)539-2304
Fax: (519)539-2011
E-mail: info@hopefm.ca

Format: Religious. **Key Personnel:** Gary Hill, Station Mgr., gary@hopefm.ca. **URL:** http://www.hopefm.ca/index2.htm.

PRINCE EDWARD ISLAND

CHARLOTTETOWN

856 ■ CBCT-TV - 13
430 University Ave.
PO Box 2230
Charlottetown, PE, Canada C1A 8B9
Phone: (902)629-6400

Owner: Canadian Broadcasting Corporation, PO

Box 500, Sta. A, Toronto, ON, Canada M5W 1E6, Free: (866)306-4636. **Key Personnel:** Andrew Cochran, Regional Dir.; Henk van Leeuwen, Managing Editor; Donna Allen, Sen. Producer. **URL:** http://www.cbc.ca/pei/.

857 ■ CHTN-FM - 100.3
90 University Ave., Ste. 320
Atlantic Technology Ctr.
Charlottetown, PE, Canada C1A 4K9

Format: Adult Contemporary. **Owner:** Newcap Radio, 745 Windmill Rd., Dartmouth, NS, Canada B3B 1C2, (902)468-7557, Fax: (902)468-7558. **Key Personnel:** Jennifer Evans, Gen. Mgr./Sales Mgr.; Gerard Murphy, Program Dir.; Scott Chapman, News Dir. **Ad Rates:** Advertising accepted; rates available upon request. **URL:** http://www.ocean1003.com/.

QUEBEC

ALMA

858 ■ CKYK-FM - 95.7
460 Sacre-Coeur Quest
Bur. 200
Alma, QC, Canada G8B 1L9
Phone: (418)662-6888
Fax: (418)662-6070

Format: Adult Contemporary. **Owner:** Radio Nord Communications, 1 Pl. Ville Marie, Office 1523, Montreal, QC, Canada H3B 2B5, (514)866-8686, Fax: (514)866-8056. **URL:** http://www.kykfm.com/.

CARLETON-SUR-MER

859 ■ CHAU-TV - 5
349, blvd. Perron
Carleton-sur-Mer, QC, Canada G0C 1J0
Phone: (418)364-3344
Fax: (418)364-7168
E-mail: info@chautva.com

URL: http://www.chautva.com/.

CHARLESBOURG

860 ■ CIMI-FM - 103.7
4500, Boul. Henri-Bourassa bur.103
Charlesbourg, QC, Canada G1H 3A5
Phone: (418)670-9001
Fax: (418)623-2538
E-mail: radio@cimifm.com

Format: Talk; Alternative/New Music/Progressive. **Key Personnel:** M. Remi Savard, Sales Mgr., vente@cimifm.com; Stephanie Leveille, Promotions Dir., steph_leveille@cimifm.com. **URL:** http://www.cimifm.com/.

GATINEAU

861 ■ CFTX-FM - 96.5
171-A, rue Jean-Proulx
Gatineau, QC, Canada J8Z 1W5
Phone: (819)770-9650

Format: Contemporary Hit Radio (CHR). **Key Personnel:** Robert H. Parent, VP/Dir. Gen., rparent@rncmedia.ca; Benoit Vanasse, Music Dir., bvanasse@rncmedia.ca. **URL:** http://www.tagradio.fm/.

862 ■ CJRC-FM - 104.7
150, rue Edmonton
Gatineau, QC, Canada J8Y 3S6
Phone: (819)561-8801

Format: Talk. **Owner:** Corus Entertainment, Inc., 32 Atlantic Ave., Toronto, ON, Canada M6K 1X8, Fax: (416)530-6885, Free: (866)537-2397. **Key Personnel:** Sylvie Charette, Gen. Dir., scharette@1047fm.ca; Eric Poirier, Sales Mgr., epoirier@1047fm.ca; Gaston Tousignant, Tech. Dir., gtousignant@1047fm.ca. **URL:** http://www.cjrc1150.com.

KAWAWACHIKAMACH

863 ■ CJCK-FM - 89.9
PO Box 5111
Kawawachikamach, QC, Canada G0G 2Z0
Phone: (418)585-2686
Fax: (418)585-3130
E-mail: kawawa@naskapi.ca

Format: Country; Classic Rock. **Owner:** Naskapi Development Corporation, at above address. **Founded:** 1980. **URL:** http://www.naskapi.ca/en/our_community/ckjc.htm.

MONTREAL

864 ■ CBMT - 6
PO Box 6000
Montreal, QC, Canada H3C 3A8
Phone: (514)597-6000

Owner: Canadian Broadcasting Corporation, PO Box 500, Sta. A, Toronto, ON, Canada M5W 1E6, Free: (866)306-4636. **Key Personnel:** Catherine Megelas, Communications Off.; Patricia Pleszczynska, Director; Shelagh Kinch, Managing Editor, phone (514)597-5608, fax (514)597-6354. **Ad Rates:** Advertising accepted; rates available upon request. **URL:** http://www.cbc.ca/montreal/.

865 ■ CHMP-FM - 98.5
800 Rue de la Gauchetiere Ouest
Bureau 1100
Montreal, QC, Canada H5A 1M1
Phone: (514)787-7799

Format: Talk. **Owner:** Corus Entertainment, Inc., 32 Atlantic Ave., Toronto, ON, Canada M6K 1X8, Fax: (416)530-6885, Free: (866)537-2397. **Key Personnel:** Robert Poissant, Sales Mgr., phone (514)787-7840, robert@985fm.ca; Catherine Jonckeau, Assoc. Sales, phone (514)787-7841, catherine.jonckeau@985fm.ca; Melissa Glaude, Dir. of Accounts, phone (514)787-7846, melissa@985fm.ca. **URL:** http://www.fm985.ca/.

866 ■ CINF-AM - 690
800 Rue de la Gauchetiere Ouest
bureau 1100
Montreal, QC, Canada H5A 1M1
Phone: (514)787-0690

Format: News. **Owner:** Corus Entertainment, Inc., 32 Atlantic Ave., Toronto, ON, Canada M6K 1X8, Fax: (416)530-6885, Free: (866)537-2397. **Operating Hours:** Continuous. **Wattage:** 50,000. **URL:** http://www.info690.com/.

867 ■ CINW-AM - 940
800 Rue de la Gauchetiere Ouest
Bureau 1100
Montreal, QC, Canada H5A 1M1
Phone: (514)849-0940

Format: News. **Owner:** Corus Entertainment, Inc., 32 Atlantic Ave., Toronto, ON, Canada M6K 1X8, Fax: (416)530-6885, Free: (866)537-2397. **Operating Hours:** Continuous. **Key Personnel:** Melissa Glaude, Dir. of Accounts, melissa@985fm.ca. **Wattage:** 50,000. **URL:** http://www.940montreal.com/.

QUEBEC

868 ■ CJEC-FM - 91.9
1305 chemin Ste.-Foy, 4ieme etage
Quebec, QC, Canada G1S 4Y5
Phone: (418)688-0919
Fax: (418)527-0919

Format: Adult Contemporary. **Owner:** Cogeco Inc., 5 Pl. Ville-Marie, Ste. 915, Montreal, QC, Canada H3B 2G2, (514)874-2600. **URL:** http://www.rythmefm.com/quebec/.

869 ■ CJSQ-FM - 92.7
2525 Blvd. Laurier
Quebec, QC, Canada G1V 2L2
Phone: (418)650-9270

Format: Classical. **Operating Hours:** Continuous. **URL:** http://www.cjpx.ca/indexQc.php.

870 ■ CKJF-FM - 90.3
3208 des sumacs
Quebec, QC, Canada G1G 1X4
Phone: (418)623-7676
Fax: (418)623-8499
E-mail: info@radiotouristique.ca

Format: Information. **Operating Hours:** Continuous. **URL:** http://www.radiotouristique.ca/cgi-cs/cs.waframe.index?lang=2.

RIVIERE-DU-LOUP

871 ■ CKRT-TV - 7
15, de le Chute
Riviere-du-Loup, QC, Canada G5R 5B7
Phone: (418)867-8080
Fax: (418)867-4710
E-mail: info@ckrt.ca

Key Personnel: Marc Simard, President; Stephane Gregoire, VP, Finance; Germain Gelinas, Tech. Dir. **URL:** http://www.ckrt.ca/.

SAINT-JEAN-SUR-RICHELIEU

872 ■ CFZZ-FM - 104.1
104 rue Richelieu
Saint-Jean-sur-Richelieu, QC, Canada J3B 6X3
Phone: (450)346-0104
Fax: (450)348-2274

Format: Oldies. **Owner:** Astral Media, 2100 rue, Ste.-Catherine Quest, Bureau 1000, Montreal, QC, Canada H3H 2T3, (514)939-5000, Fax: (514)939-1515. **URL:** http://www.boomfm.com/bienvenue.asp.

SAINTE-FOY

873 ■ CFCM-TV - 4
1000, ave. Myrand
Sainte-Foy, QC, Canada G1V 2W3
Phone: (418)688-9330
E-mail: administration@tele-4.tva.ca

Owner: TVA Group Inc., 1600, de Maisonneuve Blvd. E, Montreal, QC, Canada H2L 4P2, (514)526-9251. **URL:** http://tva.canoe.com/stations/cfcm/.

SASKATCHEWAN

MOOSE JAW

874 ■ CILG-FM - 100.7
1704 Main St. N
Moose Jaw, SK, Canada S6J 1L4
Phone: (306)692-1007
Fax: (306)692-8880
Free: (800)746-5759

Format: Country. **Key Personnel:** Dustin Dion, Contact, ddion@goldenwestradio.com. **URL:** http://www.country100fm.com/.

REGINA

875 ■ CBKR-FM - 102.5
2440 Broad St.
Regina, SK, Canada S4P 4A1
Phone: (306)347-9540

Format: News; Information. **Owner:** Canadian Broadcasting Corporation, PO Box 500, Sta. A, Toronto, ON, Canada M5W 1E6, Free: (866)306-4636. **URL:** http://www.cbc.ca/sask/.

876 ■ CBKT - 9
2440 Broad St.
Regina, SK, Canada S4P 4A1
Phone: (306)347-9540
Fax: (306)347-9635

Owner: Canadian Broadcasting Corporation, PO Box 500, Sta. A, Toronto, ON, Canada M5W 1E6, Free: (866)306-4636. **Key Personnel:** Lenora Sturge, Prog. & Mktg. Coord., phone (306)347-9714. **Ad Rates:** Advertising accepted; rates available upon request. **URL:** http://www.cbc.ca/sask/.

Circulation: ★ = ABC; △ = BPA; ♦ = CAC; • = CCAB; ☐ = VAC; ⊕ = PO Statement; ‡ = Publisher's Report; Boldface figures = sworn; Light figures = estimated.

SASKATOON

877 ■ CJMK-FM - 98.3
366 3rd Ave. S
Saskatoon, SK, Canada S7K 1M5
Phone: (306)244-1975
Fax: (306)665-5501
E-mail: magic@magic983.fm
Format: Adult Contemporary. **Key Personnel:** Vic Dubois, General Mgr., vicdubois@saskatoonmediagroup.com; Ken McFarlane, General Sales Mgr., kmcfarlane@saskatoonmediagroup.com; Myles Myrol, Retail Sales Mgr., myles@saskatoonmediagroup.com. **URL:** http://www.magic983.fm/.

AUSTRALIA

ADELAIDE

878 ■ SBS Television - 28
Locked Bag 028
Crows Nest, New South Wales 1585, Australia
Ph: 61 2 94302828
Fax: 61 2 94303047

Owner: Special Broadcasting Service, at above address. **URL:** http://www20.sbs.com.au/transmissions/index.php?pid=1&sid=1.

AIRLIE BEACH

879 ■ SBS Television - 34
Locked Bag 028
Crows Nest, New South Wales 1585, Australia
Ph: 61 2 94302828
Fax: 61 2 94303047

Owner: Special Broadcasting Service, at above address. **Wattage:** 600 ERP. **URL:** http://www20.sbs.com.au/transmissions/index.php?pid=1&sid=1.

ALBURY

880 ■ SBS Television - 53
Locked Bag 028
Crows Nest, New South Wales 1585, Australia
Ph: 61 2 94302828
Fax: 61 2 94303047

Owner: Special Broadcasting Service, at above address. **Wattage:** 600 ERP. **URL:** http://www20.sbs.com.au/transmissions/index.php?pid=1&sid=1.

ALEXANDRA

881 ■ SBS Television - 68
Locked Bag 028
Crows Nest, New South Wales 1585, Australia
Ph: 61 2 94302828
Fax: 61 2 94303047

Owner: Special Broadcasting Service, at above address. **Wattage:** 20,000 ERP. **URL:** http://www20.sbs.com.au/transmissions/index.php?pid=1&sid=1.

ARMIDALE

882 ■ SBS Television - 28
Locked Bag 028
Crows Nest, New South Wales 1585, Australia
Ph: 61 2 94302828
Fax: 61 2 94303047

Owner: Special Broadcasting Service, at above address. **Wattage:** 120,000 ERP. **URL:** http://www20.sbs.com.au/transmissions/index.php?pid=1&sid=1.

883 ■ SBS Television - 30
Locked Bag 028
Crows Nest, New South Wales 1585, Australia
Ph: 61 2 94302828
Fax: 61 2 94303047

Owner: Special Broadcasting Service, at above address. **Wattage:** 120,000 ERP. **URL:** http://www20.sbs.com.au/transmissions/index.php?pid=1&sid=1.

ASHFORD

884 ■ SBS Television - 54
Locked Bag 028
Crows Nest, New South Wales 1585, Australia
Ph: 61 2 94302828
Fax: 61 2 94303047

Owner: Special Broadcasting Service, at above address. **Wattage:** 200 ERP. **URL:** http://www20.sbs.com.au/transmissions/index.php?pid=1&sid=1.

AYR

885 ■ SBS Television - 57
Locked Bag 028
Crows Nest, New South Wales 1585, Australia
Ph: 61 2 94302828
Fax: 61 2 94303047

Owner: Special Broadcasting Service, at above address. **Wattage:** 20,000 ERP. **URL:** http://www20.sbs.com.au/transmissions/index.php?pid=1&sid=1.

BABINDA

886 ■ SBS Television - 45
Locked Bag 028
Crows Nest, New South Wales 1585, Australia
Ph: 61 2 94302828
Fax: 61 2 94303047

Owner: Special Broadcasting Service, at above address. **Wattage:** 2000 ERP. **URL:** http://www20.sbs.com.au/transmissions/index.php?pid=1&sid=1.

BAIRNSDALE

887 ■ SBS Television - 54
Locked Bag 028
Crows Nest, New South Wales 1585, Australia
Ph: 61 2 94302828
Fax: 61 2 94303047

Owner: Special Broadcasting Service, at above address. **Wattage:** 40,000 ERP. **URL:** http://www20.sbs.com.au/transmissions/index.php?pid=1&sid=1.

BALLARAT

888 ■ SBS Television - 30
Locked Bag 028
Crows Nest, New South Wales 1585, Australia
Ph: 61 2 94302828
Fax: 61 2 94303047

Owner: Special Broadcasting Service, at above address. **Wattage:** 2,000,000 ERP. **URL:** http://www20.sbs.com.au/transmissions/index.php?pid=1&sid=1.

BATEMANS BAY

889 ■ SBS Television - 55
Locked Bag 028
Crows Nest, New South Wales 1585, Australia
Ph: 61 2 94302828
Fax: 61 2 94303047

Owner: Special Broadcasting Service, at above address. **Wattage:** 10,000 ERP. **URL:** http://www20.sbs.com.au/transmissions/index.php?pid=1&sid=1.

BATHURST

890 ■ SBS Television - 46
Locked Bag 028
Crows Nest, New South Wales 1585, Australia
Ph: 61 2 94302828
Fax: 61 2 94303047

Owner: Special Broadcasting Service, at above address. **Wattage:** 2000 ERP. **URL:** http://www20.sbs.com.au/transmissions/index.php?pid=1&sid=1.

BEGA

891 ■ SBS Television - 43
Locked Bag 028
Crows Nest, New South Wales 1585, Australia
Ph: 61 2 94302828
Fax: 61 2 94303047

Owner: Special Broadcasting Service, at above address. **Wattage:** 10,000 ERP. **URL:** http://www20.sbs.com.au/transmissions/index.php?pid=1&sid=1.

BELL

892 ■ SBS Television - 53
Locked Bag 028
Crows Nest, New South Wales 1585, Australia
Ph: 61 2 94302828
Fax: 61 2 94303047

Owner: Special Broadcasting Service, at above address. **Wattage:** 10 ERP. **URL:** http://www20.sbs.com.au/transmissions/index.php?pid=1&sid=1.

BENDIGO

893 ■ SBS Television - 29
Locked Bag 028
Crows Nest, New South Wales 1585, Australia
Ph: 61 2 94302828
Fax: 61 2 94303047

Owner: Special Broadcasting Service, at above address. **Wattage:** 2,000,000 ERP. **URL:** http://www20.sbs.com.au/transmissions/index.php?pid=1&sid=1.

Circulation: ★ = ABC; △ = BPA; ♦ = CAC; • = CCAB; ❑ = VAC; ⊕ = PO Statement; ‡ = Publisher's Report; Boldface figures = sworn; Light figures = estimated.

BLACKWATER

894 ■ SBS Television - 43
Locked Bag 028
Crows Nest, New South Wales 1585, Australia
Ph: 61 2 94302828
Fax: 61 2 94303047
Owner: Special Broadcasting Service, at above address. **Wattage:** 20,000 ERP. **URL:** http://www20.sbs.com.au/transmissions/index.php?pid=1&sid=1.

BONNIE DOON

895 ■ SBS Television - 55
Locked Bag 028
Crows Nest, New South Wales 1585, Australia
Ph: 61 2 94302828
Fax: 61 2 94303047
Owner: Special Broadcasting Service, at above address. **Wattage:** 150 ERP. **URL:** http://www20.sbs.com.au/transmissions/index.php?pid=1&sid=1.

BOONAH

896 ■ SBS Television - 54
Locked Bag 028
Crows Nest, New South Wales 1585, Australia
Ph: 61 2 94302828
Fax: 61 2 94303047
Owner: Special Broadcasting Service, at above address. **Wattage:** 80 ERP. **URL:** http://www20.sbs.com.au/transmissions/index.php?pid=1&sid=1.

BOUDDI

897 ■ SBS Television - 64
Locked Bag 028
Crows Nest, New South Wales 1585, Australia
Ph: 61 2 94302828
Fax: 61 2 94303047
Owner: Special Broadcasting Service, at above address. **Wattage:** 5100 ERP. **URL:** http://www20.sbs.com.au/transmissions/index.php?pid=1&sid=1.

BOWEN

898 ■ SBS Television - 48
Locked Bag 028
Crows Nest, New South Wales 1585, Australia
Ph: 61 2 94302828
Fax: 61 2 94303047
Owner: Special Broadcasting Service, at above address. **Wattage:** 10,000 ERP. **URL:** http://www20.sbs.com.au/transmissions/index.php?pid=1&sid=1.

BOWRAL

899 ■ SBS Television - 30
Locked Bag 028
Crows Nest, New South Wales 1585, Australia
Ph: 61 2 94302828
Fax: 61 2 94303047
Owner: Special Broadcasting Service, at above address. **Wattage:** 4000 ERP. **URL:** http://www20.sbs.com.au/transmissions/index.php?pid=1&sid=1.

BRAIDWOOD

900 ■ SBS Television - 54
Locked Bag 028
Crows Nest, New South Wales 1585, Australia
Ph: 61 2 94302828
Fax: 61 2 94303047
Owner: Special Broadcasting Service, at above address. **Wattage:** 480 ERP. **URL:** http://www20.sbs.com.au/transmissions/index.php?pid=1&sid=1.

BRIGHT

901 ■ SBS Television - 29
Locked Bag 028
Crows Nest, New South Wales 1585, Australia
Ph: 61 2 94302828
Fax: 61 2 94303047
Owner: Special Broadcasting Service, at above address. **Wattage:** 200 ERP. **URL:** http://www20.sbs.com.au/transmissions/index.php?pid=1&sid=1.

BROKEN HILL

902 ■ SBS Television - 44
Locked Bag 028
Crows Nest, New South Wales 1585, Australia
Ph: 61 2 94302828
Fax: 61 2 94303047
Owner: Special Broadcasting Service, at above address. **Wattage:** 40,000 ERP. **URL:** http://www20.sbs.com.au/transmissions/index.php?pid=1&sid=1.

BRUTHEN

903 ■ SBS Television - 50
Locked Bag 028
Crows Nest, New South Wales 1585, Australia
Ph: 61 2 94302828
Fax: 61 2 94303047
Owner: Special Broadcasting Service, at above address. **Wattage:** 200 ERP. **URL:** http://www20.sbs.com.au/transmissions/index.php?pid=1&sid=1.

BUNBURY

904 ■ SBS Television - 33
Locked Bag 028
Crows Nest, New South Wales 1585, Australia
Ph: 61 2 94302828
Fax: 61 2 94303047
Owner: Special Broadcasting Service, at above address. **Wattage:** 600,000 ERP. **URL:** http://www20.sbs.com.au/transmissions/index.php?pid=1&sid=1.

CANBERRA

905 ■ SBS Television - 28
Locked Bag 028
Crows Nest, New South Wales 1585, Australia
Ph: 61 2 94302828
Fax: 61 2 94303047
Owner: Special Broadcasting Service, at above address. **Wattage:** 600,000 ERP. **URL:** http://www20.sbs.com.au/transmissions/index.php?pid=1&sid=1.

CARNARVON

906 ■ SBS Television - 12
Locked Bag 028
Crows Nest, New South Wales 1585, Australia
Ph: 61 2 94302828
Fax: 61 2 94303047
Owner: Special Broadcasting Service, at above address. **Wattage:** 5000 ERP. **URL:** http://www20.sbs.com.au/transmissions/index.php?pid=1&sid=1.

CHURCHILL

907 ■ SBS Television - 52
Locked Bag 028
Crows Nest, New South Wales 1585, Australia
Ph: 61 2 94302828
Fax: 61 2 94303047
Owner: Special Broadcasting Service, at above address. **Wattage:** 500 ERP. **URL:** http://www20.sbs.com.au/transmissions/index.php?pid=1&sid=1.

CLARE

908 ■ SBS Television - 57
Locked Bag 028
Crows Nest, New South Wales 1585, Australia
Ph: 61 2 94302828
Fax: 61 2 94303047
Owner: Special Broadcasting Service, at above address. **Wattage:** 100 ERP. **URL:** http://www20.sbs.com.au/transmissions/index.php?pid=1&sid=1.

COBAR

909 ■ SBS Television - 12
Locked Bag 028
Crows Nest, New South Wales 1585, Australia
Ph: 61 2 94302828
Fax: 61 2 94303047
Owner: Special Broadcasting Service, at above address. **Wattage:** 80 ERP. **URL:** http://www20.sbs.com.au/transmissions/index.php?pid=1&sid=1.

COBDEN

910 ■ SBS Television - 67
Locked Bag 028
Crows Nest, New South Wales 1585, Australia
Ph: 61 2 94302828
Fax: 61 2 94303047
Owner: Special Broadcasting Service, at above address. **Wattage:** 10,000 ERP. **URL:** http://www20.sbs.com.au/transmissions/index.php?pid=1&sid=1.

COFFS HARBOUR

911 ■ SBS Television - 69
Locked Bag 028
Crows Nest, New South Wales 1585, Australia
Ph: 61 2 94302828
Fax: 61 2 94303047
Owner: Special Broadcasting Service, at above address. **Wattage:** 300 ERP. **URL:** http://www20.sbs.com.au/transmissions/index.php?pid=1&sid=1.

COLAC

912 ■ SBS Television - 55
Locked Bag 028
Crows Nest, New South Wales 1585, Australia
Ph: 61 2 94302828
Fax: 61 2 94303047
Owner: Special Broadcasting Service, at above address. **Wattage:** 10,000 ERP. **URL:** http://www20.sbs.com.au/transmissions/index.php?pid=1&sid=1.

CONDOBOLIN

913 ■ SBS Television - 56
Locked Bag 028
Crows Nest, New South Wales 1585, Australia
Ph: 61 2 94302828
Fax: 61 2 94303047
Owner: Special Broadcasting Service, at above address. **Wattage:** 200 ERP. **URL:** http://www20.sbs.com.au/transmissions/index.php?pid=1&sid=1.

COOLAH

914 ■ SBS Television - 53
Locked Bag 028
Crows Nest, New South Wales 1585, Australia
Ph: 61 2 94302828
Fax: 61 2 94303047
Owner: Special Broadcasting Service, at above address. **Wattage:** 240 ERP. **URL:** http://www20.sbs.com.au/transmissions/index.php?pid=1&sid=1.

COOMA

915 ■ SBS Television - 53
Locked Bag 028
Crows Nest, New South Wales 1585, Australia
Ph: 61 2 94302828
Fax: 61 2 94303047
Owner: Special Broadcasting Service, at above address. **Wattage:** 200 ERP. **URL:** http://www20.sbs.com.au/transmissions/index.php?pid=1&sid=1.

COWELL

916 ■ SBS Television - 58
Locked Bag 028
Crows Nest, New South Wales 1585, Australia
Ph: 61 2 94302828
Fax: 61 2 94303047
Owner: Special Broadcasting Service, at above address. **Wattage:** 15,000 ERP. **URL:** http://www20.sbs.com.au/transmissions/index.php?pid=1&sid=1.

COWRA

917 ■ SBS Television - 45
Locked Bag 028
Crows Nest, New South Wales 1585, Australia
Ph: 61 2 94302828
Fax: 61 2 94303047
Owner: Special Broadcasting Service, at above

address. **Wattage:** 60 ERP. **URL:** http://www20.sbs.com.au/transmissions/index.php?pid=1&sid=1.

CROWS NEST

SBS Television - See Adelaide
SBS Television - See Airlie Beach
SBS Television - See Albury
SBS Television - See Alexandra
SBS Television - See Armidale
SBS Television - See Armidale
SBS Television - See Ashford
SBS Television - See Ayr
SBS Television - See Babinda
SBS Television - See Bairnsdale
SBS Television - See Ballarat
SBS Television - See Batemans Bay
SBS Television - See Bathurst
SBS Television - See Bega
SBS Television - See Bell
SBS Television - See Bendigo
SBS Television - See Blackwater
SBS Television - See Bonnie Doon
SBS Television - See Boonah
SBS Television - See Bouddi
SBS Television - See Bowen
SBS Television - See Bowral
SBS Television - See Braidwood
SBS Television - See Bright
SBS Television - See Broken Hill
SBS Television - See Bruthen
SBS Television - See Bunbury
SBS Television - See Canberra
SBS Television - See Carnarvon
SBS Television - See Churchill
SBS Television - See Clare
SBS Television - See Cobar
SBS Television - See Cobden
SBS Television - See Coffs Harbour
SBS Television - See Colac
SBS Television - See Condobolin
SBS Television - See Coolah
SBS Television - See Cooma
SBS Television - See Cowell
SBS Television - See Cowra
SBS Television - See Eildon
SBS Television - See Fraser
SBS Television - See Naracoorte
SBS Television - See Port Lincoln
SBS Television - See Tennant Creek
SBS Television - See Toodyay
SBS Television - See Tuggeranong

918 ■ SBS Television - Acton Road - 28
Locked Bag 028
Crows Nest, New South Wales 1585, Australia
Ph: 61 2 94302828
Fax: 61 2 94303047

Owner: Special Broadcasting Service, at above address. **Wattage:** 160 ERP. **URL:** http://www20.sbs.com.au/transmissions/index.php?pid=1&sid=1.

919 ■ SBS Television - Caralue Bluff - 62
Locked Bag 028
Crows Nest, New South Wales 1585, Australia
Ph: 61 2 94302828
Fax: 61 2 94303047

Owner: Special Broadcasting Service, at above address. **Wattage:** 20,000 ERP. **URL:** http://www20.sbs.com.au/transmissions/index.php?pid=1&sid=1.

920 ■ SBS Television - Central Tablelands - 30
Locked Bag 028
Crows Nest, New South Wales 1585, Australia
Ph: 61 2 94302828
Fax: 61 2 94303047

Owner: Special Broadcasting Service, at above address. **Wattage:** 2,000,000 ERP. **URL:** http://www20.sbs.com.au/transmissions/index.php?pid=1&sid=1.

921 ■ SBS Television - Central Western Slopes - 29
Locked Bag 028
Crows Nest, New South Wales 1585, Australia
Ph: 61 2 94302828
Fax: 61 2 94303047

Owner: Special Broadcasting Service, at above address. **Wattage:** 1,000,000 ERP. **URL:** http://www20.sbs.com.au/transmissions/index.php?pid=1&sid=1.

922 ■ SBS Television - Weston Creek/Woden - 58
Locked Bag 028
Crows Nest, New South Wales 1585, Australia
Ph: 61 2 94302828
Fax: 61 2 94303047

Owner: Special Broadcasting Service, at above address. **Wattage:** 60 ERP. **URL:** http://www20.sbs.com.au/transmissions/index.php?pid=1&sid=1.

EAST SYDNEY

923 ■ Cherrie
Evolution Publishing
140 William St., Level 3
East Sydney, New South Wales 2010, Australia
Ph: 61 2 93608934
Fax: 61 2 93609497

Magazine featuring what's hot in music, entertainment, arts, pop culture and fashion, as well as the political and social issues taking place across the world. **Freq:** Monthly. **Key Personnel:** Katrina Fox, Editor, phone 61 2 93608934. **Subscription Rates:** $A 60 individuals. **Remarks:** Accepts advertising. **URL:** http://cherrie.e-p.net.au. . **Ad Rates:** BW: $A 2,160. **Circ:** Combined 19,000

924 ■ Fellow Traveller
Evolution Publishing
140 William St., Level 3
East Sydney, New South Wales 2010, Australia
Ph: 61 2 93608934
Fax: 61 2 93609497

Magazine featuring guide of places to stay, play, dine and shop for gay and lesbian travellers. **Trim Size:** 80 x 188 mm. **Subscription Rates:** Free. **Remarks:** Accepts advertising. **URL:** http://fellowtraveller.com.au/. **Circ:** Combined 30,000

925 ■ SX News
Evolution Publishing
140 William St., Level 3
East Sydney, New South Wales 2010, Australia
Ph: 61 2 93608934
Fax: 61 2 93609497

Magazine featuring news and features, fashion, style and entertainment, music and film reviews, interviews and competitions concerning the gays and lesbian in Australia. **Freq:** Weekly. **Subscription Rates:** $A 80 individuals. **Remarks:** Accepts advertising. **URL:** http://sxnews.e-p.net.au/. . **Ad Rates:** 4C: $A 1,730. **Circ:** Combined 25,000

EILDON

926 ■ SBS Television - 30
Locked Bag 028
Crows Nest, New South Wales 1585, Australia
Ph: 61 2 94302828
Fax: 61 2 94303047

Owner: Special Broadcasting Service, at above address. **Wattage:** 400 ERP. **URL:** http://www20.sbs.com.au/transmissions/index.php?pid=1&sid=1.

FORTITUDE VALLEY

927 ■ Queensland Pride
Evolution Publishing
83 Alfred St., Ste. 2
Fortitude Valley, Queensland 4006, Australia
Ph: 61 7 32160860

Magazine featuring news, entertainment, lifestyle, and community information for gays and lesbian. **Founded:** Dec. 1990. **Freq:** Monthly. **Subscription Rates:** Free. **Remarks:** Accepts advertising. **URL:** http://qlp.e-p.net.au/. . **Ad Rates:** 4C: $A 1,384. **Circ:** (Not Reported)

FRASER

928 ■ SBS Television - 53
Locked Bag 028
Crows Nest, New South Wales 1585, Australia
Ph: 61 2 94302828
Fax: 61 2 94303047

Owner: Special Broadcasting Service, at above address. **Wattage:** 10 ERP. **URL:** http://www20.sbs.com.au/transmissions/index.php?pid=1&sid=1.

HAMILTON

929 ■ Beanscene
Insight Publishing Pty. Ltd.
Level 1, 468 Kingsford Smith Dr.
PO Box 886
Hamilton, Queensland 4007, Australia
Ph: 61 7 36301388
Fax: 61 7 36301344
Publisher E-mail: admin@insightpublishing.com.au

Journal featuring coffee culture. **Freq:** Monthly. **Remarks:** Accepts advertising. **URL:** http://www.insightpublishing.com.au/. **Circ:** (Not Reported)

930 ■ Insight Into Healing
Insight Publishing Pty. Ltd.
Level 1, 468 Kingsford Smith Dr.
PO Box 886
Hamilton, Queensland 4007, Australia
Ph: 61 7 36301388
Fax: 61 7 36301344
Publisher E-mail: admin@insightpublishing.com.au

Magazine featuring topics about alternative living and thinking such as Druid, Magic, Pagan, Astrology, Feng Shui, Numerology, Organic, Psychic and more. **Freq:** Annual. **Remarks:** Accepts advertising. **URL:** http://www.insightpublishing.com.au/. **Circ:** (Not Reported)

MELBOURNE

931 ■ Melbourne Community Voice
Evolution Publishing
365 Lt. Collins St., Level 7, Ste. 8
Melbourne, Victoria 3000, Australia
Ph: 61 3 96022333

Magazine featuring news and features, social pages, fashion, beauty, sport, travel, automotive real estate, interviews, and reviews for gays and lesbians. **Freq:** Weekly. **Subscription Rates:** $A 80 individuals. **Remarks:** Accepts advertising. **URL:** http://mcv.e-p.net.au/. **Also known as:** MCV. . **Ad Rates:** 4C: $A 1,730. **Circ:** Combined 20,000

NARACOORTE

932 ■ SBS Television - 54
Locked Bag 028
Crows Nest, New South Wales 1585, Australia
Ph: 61 2 94302828
Fax: 61 2 94303047

Owner: Special Broadcasting Service, at above address. **Wattage:** 300 ERP. **URL:** http://www20.sbs.com.au/transmissions/index.php?pid=1&sid=1.

Circulation: ★ = ABC; △ = BPA; ♦ = CAC; • = CCAB; ❑ = VAC; ⊕ = PO Statement; ‡ = Publisher's Report; Boldface figures = sworn; Light figures = estimated.

PERTH

933 ■ Australian Longwall Magazine
Aspermont Limited
613-619 Wellington St.
Perth, Western Australia 6000, Australia
Ph: 61 8 62639100
Fax: 61 8 62639148
Publication E-mail: editorial@longwalls.com
Publisher E-mail: contact@aspermont.com

Magazine featuring information on the longwall sector in Australia. **Freq:** 3/yr. **Trim Size:** 210 x 297 mm. **Key Personnel:** Angie Bahr, Editor. **Subscription Rates:** $A 46.20 individuals; $A 69 U.S. **Remarks:** Accepts advertising. **URL:** http://www.longwallmagazine.net/. **Circ:** 4,000

934 ■ Australia's Mining Monthly
Aspermont Limited
613-619 Wellington St.
Perth, Western Australia 6000, Australia
Ph: 61 8 62639100
Fax: 61 8 62639148
Publication E-mail: editorial@miningmonthly.com
Publisher E-mail: contact@aspermont.com

Magazine featuring mining industry in Australia. **Freq:** Monthly. **Trim Size:** 210 x 297 mm. **Key Personnel:** Noel Dyson, Editor. **Subscription Rates:** $A 132 individuals; $A 270 U.S. **Remarks:** Accepts advertising. **URL:** http://www.miningmonthly.com/. **Circ:** ★8,769

935 ■ Cranes and Lifting Australia
Aspermont Limited
613-619 Wellington St.
Perth, Western Australia 6000, Australia
Ph: 61 8 62639100
Fax: 61 8 62639148
Publication E-mail: editorial@cranesandlifting.net
Publisher E-mail: contact@aspermont.com

Magazine covering all classes of mobile and fixed cranes, construction and workshop cranes, truck-loading cranes. **Freq:** Quarterly. **Trim Size:** 210 x 297 mm. **Key Personnel:** Chris Le Messurier, Advertising Sales Mgr., advertising@cranesandlifting.net. **Remarks:** Accepts advertising. **URL:** http://www.cranesandlifting.net/. **Circ:** (Not Reported)

936 ■ RESOURCESTOCKS
Aspermont Limited
613-619 Wellington St.
Perth, Western Australia 6000, Australia
Ph: 61 8 62639100
Fax: 61 8 62639148
Publication E-mail: editorial.resourcestocks@aspermont.com
Publisher E-mail: contact@aspermont.com

Journal featuring resource investment in Australia. **Freq:** Monthly. **Trim Size:** 210 x 297 mm. **Key Personnel:** Ron Berryman, Editor. **Subscription Rates:** $A 132 individuals; $A 168 U.S. **Remarks:** Accepts advertising. **URL:** http://www.resourcestocks.com.au/. **Circ:** 11,000

PORT LINCOLN

937 ■ SBS Television - 54
Locked Bag 028
Crows Nest, New South Wales 1585, Australia
Ph: 61 2 94302828
Fax: 61 2 94303047

Owner: Special Broadcasting Service, at above address. **Wattage:** 400 ERP. **URL:** http://www20.sbs.com.au/transmissions/index.php?pid=1&sid=1.

STRAWBERRY HILLS

938 ■ ADF Health
Australasian Medical Publishing Company Ltd.
Locked Bag 3030
Strawberry Hills, New South Wales 2012, Australia
Ph: 61 2 9562 6666
Fax: 61 2 9562 6699
Publisher E-mail: medjaust@ampco.com.au

Journal featuring opinions and views about military medicine and the health. **Freq:** Semiannual. **Key Personnel:** Commander M.C. O'Connor, Editor. **ISSN:** 1443-1033. **Remarks:** Accepts advertising. **URL:** http://www.defence.gov.au/health/infocentre/journals/i-ADFHJ.htm. **Circ:** (Not Reported)

939 ■ Australian Health Review
Australasian Medical Publishing Company Ltd.
Locked Bag 3030
Strawberry Hills, New South Wales 2012, Australia
Ph: 61 2 9562 6666
Fax: 61 2 9562 6699
Publisher E-mail: medjaust@ampco.com.au

Journal featuring information about the health care industry. **Freq:** Quarterly. **Key Personnel:** Prof. Sandra Leggat, Editor. **Subscription Rates:** $A 167 single issue; $A 221 single issue other countries; $A 187 single issue library; $A 251 single issue library (other countries); $A 533 individuals; $A 707 other countries; $A 598 libraries; $A 803 libraries other countries. **Remarks:** Accepts advertising. **URL:** http://www.aushealthreview.com.au/publications/articles/. **Circ:** (Not Reported)

940 ■ Critical Care and Resuscitation
Australasian Medical Publishing Company Ltd.
Locked Bag 3030
Strawberry Hills, New South Wales 2012, Australia
Ph: 61 2 9562 6666
Fax: 61 2 9562 6699
Publisher E-mail: medjaust@ampco.com.au

Journal featuring information about critical care and intensive care medicine. **Freq:** Quarterly. **Trim Size:** A4. **Key Personnel:** Prof. Rinaldo Bellomo, Editor, phone 61 3 94965992, fax 61 3 94963932, rinaldo.bellomo@austin.org.au. **ISSN:** 1441-2772. **Subscription Rates:** $A 110 individuals; $A 110 other countries; $A 165 institutions; $A 200 institutions, other countries. **Remarks:** Accepts advertising. **URL:** http://www.anzca.edu.au/jficm/resources/ccr/. **Circ:** (Not Reported)

SYDNEY

941 ■ International Journal of Abrasive Technology
Inderscience Publishers
c/o Prof. Jun Wang, Ed.-in-Ch.
The University of New South Wales
School of Mechanical & Manufacturing Engineering
Sydney, New South Wales 2052, Australia
Publisher E-mail: editor@inderscience.com

Journal covering areas of abrasive technology. **Founded:** 2007. **Freq:** 4/yr. **Key Personnel:** Prof. Jun Wang, Editor-in-Chief, jun.wang@unsw.edu.au. **ISSN:** 1752-2641. **Subscription Rates:** EUR470 individuals includes surface mail, print only; EUR640 individuals print & online. **URL:** http://www.inderscience.com/browse/index.php?journalCODE=ijat.

942 ■ miceAsia.net
Business & Tourism Publishing
189 Kent St., Ste. 3
Sydney, New South Wales 2000, Australia
Ph: 61 2 82644444
Fax: 61 2 82644401

Magazine for business events industry through South East Asia. **Founded:** 2006. **Freq:** Annual. **Key Personnel:** Helen Batt-Rawden, Publisher/Mng. Dir., helen@btp.net.au; Brad Foster, Managing Editor, brad@btp.net.au. **Remarks:** Accepts advertising. **URL:** http://www.btp.net.au/miceasianet/miceasia-home.aspx. **Circ:** (Not Reported)

943 ■ mice.net
Business & Tourism Publishing
189 Kent St., Ste. 3
Sydney, New South Wales 2000, Australia
Ph: 61 2 82644444
Fax: 61 2 82644401

Magazine for the Australian business events community. **Founded:** 2000. **Freq:** Bimonthly. **Key Personnel:** Helen Batt-Rawden, Publisher/Mng. Dir., helen@btp.net.au; Brad Foster, Managing Editor, brad@btp.net.au. **Remarks:** Accepts advertising. **URL:** http://www.btp.net.au/micenet/micenet-home.aspx. **Circ:** ★13,210

944 ■ miceNZ.net
Business & Tourism Publishing
189 Kent St., Ste. 3
Sydney, New South Wales 2000, Australia
Ph: 61 2 82644444
Fax: 61 2 82644401

Magazine for the New Zealand business events community. **Founded:** 2005. **Freq:** Quarterly. **Key Personnel:** Helen Batt-Rawden, Publisher/Mng. Dir., helen@btp.net.au; Brad Foster, Managing Editor, brad@btp.net.au. **Remarks:** Accepts advertising. **URL:** http://www.btp.net.au/micenznet/micenznet-home.aspx. **Circ:** 6,944

TENNANT CREEK

945 ■ SBS Television - 10
Locked Bag 028
Crows Nest, New South Wales 1585, Australia
Ph: 61 2 94302828
Fax: 61 2 94303047

Owner: Special Broadcasting Service, at above address. **Wattage:** 2000 ERP. **URL:** http://www20.sbs.com.au/transmissions/index.php?pid=1&sid=1.

TOODYAY

946 ■ SBS Television - 34
Locked Bag 028
Crows Nest, New South Wales 1585, Australia
Ph: 61 2 94302828
Fax: 61 2 94303047

Owner: Special Broadcasting Service, at above address. **Wattage:** 50 ERP. **URL:** http://www20.sbs.com.au/transmissions/index.php?pid=1&sid=1.

TOOWOOMBA

947 ■ International Journal of Information Quality
Inderscience Publishers
c/o Dr. Latif Al-Hakim, Ed.-in-Ch.
University of Southern Queensland
School of Management & Marketing, Faculty of Business
Toowoomba, Queensland 4350, Australia
Publisher E-mail: editor@inderscience.com

Journal covering implementation of information quality systems. **Founded:** 2007. **Freq:** 4/yr. **Key Personnel:** Dr. Latif Al-Hakim, Editor-in-Chief, hakim@usq.edu.au. **ISSN:** 1751-0457. **Subscription Rates:** EUR470 individuals includes surface mail, print only; EUR640 individuals print and online. **URL:** http://www.inderscience.com/browse/index.php?journalCODE=ijiq.

TUGGERANONG

948 ■ SBS Television - 54
Locked Bag 028
Crows Nest, New South Wales 1585, Australia
Ph: 61 2 94302828
Fax: 61 2 94303047

Owner: Special Broadcasting Service, at above address. **Wattage:** 640 ERP. **URL:** http://www20.sbs.com.au/transmissions/index.php?pid=1&sid=1.

AUSTRIA

GRAZ

949 ■ European Journal of International Management
Inderscience Publishers
c/o Vlad Vaiman, Exec. Ed.
FH JOANNEUM University of Applied Sciences
Department of International Management
Eggenberger Allee 11
A-8020 Graz, Austria
Publication E-mail: ejim@inderscience.com
Publisher E-mail: editor@inderscience.com

Journal covering issues in international management theory & practice. **Founded:** 2007. **Freq:** 4/yr. **Key Personnel:** Vlad Vaiman, Exec. Ed., editors@ejim-global.org. **ISSN:** 1751-6757. **Subscription Rates:** EUR470 individuals includes surface mail, print only; EUR640 individuals print & online. **URL:** http://www.

inderscience.com/browse/index.php?journalCODE=ejim.

BRAZIL

RIO DE JANEIRO

950 ■ International Journal of High Performance System Architecture
Inderscience Publishers
c/o Dr. Nadia Nedjah, Ed.-in-Ch.
State University of Rio de Janeiro
Rua Sao Francisco Xavier, 524
5022-D Maracana
20550-900 Rio de Janeiro, Rio de Janeiro, Brazil
Publisher E-mail: editor@inderscience.com

Journal covering designs & implementation of high performance architecture. **Founded:** 2007. **Freq:** 4/yr. **Key Personnel:** Dr. Nadia Nedjah, Editor-in-Chief, nadia@eng.uerj.br. **ISSN:** 1751-6528. **Subscription Rates:** EUR470 individuals includes surface mail, print only; EUR640 individuals print and online. **URL:** http://www.inderscience.com/browse/index.php?journalCODE=ijhpsa.

951 ■ International Journal of Innovative Computing and Applications
Inderscience Publishers
c/o Dr. Nadia Nedjah, Ed.-in-Ch.
State University of Rio de Janeiro
Rua Sao Francisco Xavier, 524
5022-D Maracana
20550-900 Rio de Janeiro, Rio de Janeiro, Brazil
Publisher E-mail: editor@inderscience.com

Journal covering all new computing paradigms. **Founded:** 2007. **Freq:** 4/yr. **Key Personnel:** Dr. Nadia Nedjah, Editor-in-Chief, nadia@eng.uerj.br. **ISSN:** 1751-648X. **Subscription Rates:** EUR470 individuals includes surface mail, print only; EUR640 individuals print and online. **URL:** http://www.inderscience.com/browse/index.php?journalCODE=ijica.

952 ■ International Journal of Technological Learning, Innovation & Development
Inderscience Publishers
c/o Prof. Paulo N. Figueiredo, Ed.-in-Ch.
Praia de Botafogo 190, 5th Fl., Rm. 510
22250-900 Rio de Janeiro, Rio de Janeiro, Brazil
Publication E-mail: ijtlid@inderscience.com
Publisher E-mail: editor@inderscience.com

Journal covering issues related to technological learning, innovation, and development. **Founded:** 2007. **Freq:** 4/yr. **Key Personnel:** Paulo N. Figueiredo, Editor-in-Chief, pnf@fgv.br. **ISSN:** 1750-4090. **Subscription Rates:** EUR470 individuals includes surface mail, print only; EUR640 individuals print and online. **URL:** http://www.inderscience.com/browse/index.php?journalCODE=ijtlid.

PEOPLE'S REPUBLIC OF CHINA

BEIJING

953 ■ Chinese Tales and Stories
Beijing Language and Culture University Press
No. 15, Xueyuan Rd.
Haidian District
Beijing 100083, People's Republic of China
Ph: 86 10 82303668
Fax: 86 10 82303668
Publisher E-mail: service@blcup.net

Magazine featuring selected readings from Learning Chinese Magazine. **Founded:** Oct. 2003. **URL:** http://www.blcup.com/en/list_1.asp?id=902.

954 ■ Leaning Chinese
Beijing Language and Culture University Press
No. 15, Xueyuan Rd.
Haidian District
Beijing 100083, People's Republic of China
Ph: 86 10 82303668
Fax: 86 10 82303668
Publisher E-mail: service@blcup.net

Magazine featuring Chinese language. **Trim Size:** 185 x 260 mm. **Subscription Rates:** 300 Yu individuals. **URL:** http://www.blcup.com/en/list_1.asp?id=1931.

955 ■ Life in China
Beijing Language and Culture University Press
No. 15, Xueyuan Rd.
Haidian District
Beijing 100083, People's Republic of China
Ph: 86 10 82303668
Fax: 86 10 82303668
Publisher E-mail: service@blcup.net

Magazine featuring selected readings from Learning Chinese Magazine. **Founded:** Oct. 2003. **Subscription Rates:** 15 Yu individuals. **URL:** http://www.blcup.com/en/list_1.asp?id=900.

956 ■ Science in China Series A
Science in China Press
16, Dong-huang-cheng-gen N St.
Beijing 100717, People's Republic of China
Ph: 86 10 64016350
Publisher E-mail: sys@scichina.org

Journal featuring basic mathematics, applied mathematics, calculation mathematics and science engineering calculation, and statistics. **Subtitle:** Mathematics. **Freq:** Monthly. **Key Personnel:** Yang Lo, Editor-in-Chief. **ISSN:** 1006-9283. **Subscription Rates:** US$1,200 institutions; US$100 individuals. **Remarks:** Accepts advertising. **URL:** http://math.scichina.com/english/en/qkjs.asp. **Circ:** (Not Reported)

957 ■ Science in China Series B
Science in China Press
16, Dong-huang-cheng-gen N St.
Beijing 100717, People's Republic of China
Ph: 86 10 64016350
Publisher E-mail: sys@scichina.org

Journal featuring theoretical chemistry, physical chemistry, organic chemistry, inorganic chemistry, polymer chemistry, biological chemistry, environmental chemistry, and chemical engineering. **Subtitle:** Chemistry. **Freq:** Bimonthly. **Key Personnel:** Xu Guangxian, Editor-in-Chief. **ISSN:** 1006-9291. **Subscription Rates:** US$600 institutions; US$100 individuals. **Remarks:** Accepts advertising. **URL:** http://life.scichina.com/english/en/qkjs.asp. **Circ:** (Not Reported)

958 ■ Science in China Series C
Science in China Press
16, Dong-huang-cheng-gen N St.
Beijing 100717, People's Republic of China
Ph: 86 10 64016350
Publisher E-mail: sys@scichina.org

Journal featuring biology, agriculture, and medicine. **Subtitle:** Life Sciences. **Freq:** 6/yr. **Key Personnel:** Liang Dongcai, Editor-in-Chief; Wu Changxin Wu Zuze, Exec. Ed. **ISSN:** 1006-9305. **Subscription Rates:** US$600 institutions; US$100 individuals. **Remarks:** Accepts advertising. **URL:** http://life.scichina.com/english/en/dqml.asp. **Circ:** (Not Reported)

959 ■ Science in China Series D
Science in China Press
16, Dong-huang-cheng-gen N St.
Beijing 100717, People's Republic of China
Ph: 86 10 64016350
Publisher E-mail: sys@scichina.org

Journal featuring geology, geochemistry, geophysics, geography, atmospheric sciences, and ocean sciences. **Subtitle:** Earth Sciences. **Freq:** Monthly. **Key Personnel:** Sun Shu, Editor-in-Chief; Chen Yong, Exec. Ed.; Ma Zongjin, Exec. Ed. **ISSN:** 1006-9313. **Subscription Rates:** US$1,440 institutions; US$120 individuals. **Remarks:** Accepts advertising. **URL:** http://earth.scichina.com/english/en/qkjs.asp. **Circ:** (Not Reported)

960 ■ Science in China Series E
Science in China Press
16, Dong-huang-cheng-gen N St.
Beijing 100717, People's Republic of China
Ph: 86 10 64016350
Publisher E-mail: sys@scichina.org

Journal featuring mechanical engineering, engineering thermophysics, electronic engineering, architecture, astronomics, civil engineering, nuclear science and technology. **Subtitle:** Technological Sciences. **Freq:** 6/yr. **Key Personnel:** Yan Luguang, Editor-in-Chief. **ISSN:** 1006-9321. **Subscription Rates:** US$1,200 institutions; US$600 individuals. **URL:** http://tech.scichina.com/english/en/qkjs.asp.

961 ■ Science in China Series F
Science in China Press
16, Dong-huang-cheng-gen N St.
Beijing 100717, People's Republic of China
Ph: 86 10 64016350
Publisher E-mail: sys@scichina.org

Journal featuring computer science and technology, control science and technology, communication and information system, electronic science and technology, and bioinformation. **Subtitle:** Information Science. **Freq:** 6/yr. **Key Personnel:** Yang Fuqing, Editor-in-Chief. **ISSN:** 1006-2757. **Subscription Rates:** US$600 institutions; US$100 individuals. **URL:** http://info.scichina.com/english/en/qkjs.asp.

962 ■ Science in China Series G
Science in China Press
16, Dong-huang-cheng-gen N St.
Beijing 100717, People's Republic of China
Ph: 86 10 64016350
Publisher E-mail: sys@scichina.org

Journal featuring basic and applied research in the fields of physics, dynamics, and astronomy. **Subtitle:** Physics, Mechanics & Astronomy. **Freq:** Bimonthly. **Key Personnel:** Bai Yilong, Editor-in-Chief; Chen Jiansheng, Exec. Ed.; Fang Shouxian, Exec. Ed. **ISSN:** 1672-1799. **Subscription Rates:** US$600 institutions; US$100 individuals. **URL:** http://phys.scichina.com/english/en/qkjs.asp.

HONG KONG

963 ■ Asian Anthropology
Chinese University Press
Shatin Galleria, 9th Fl., Unit 1-3 & 18
18-24 Shan Mei St.
Fo Tan, Shatin
Hong Kong, People's Republic of China
Ph: 852 29465300
Fax: 852 26037355
Publisher E-mail: cup@cuhk.edu.hk

Journal covering anthropological research in Asia. **Founded:** June 2002. **Freq:** Annual. **Trim Size:** 152 x 229 mm. **Key Personnel:** Chee-Beng Tan, Editor-in-Chief, cbtan@cuhk.edu.hk; Mathews Gordon, Editor-in-Chief, cmgordon@cuhk.edu.hk. **ISSN:** 1683-478X. **Subscription Rates:** US$11.50 individuals. **URL:** http://www.chineseupress.com/asp/JournalList_en.asp?CatID=1&Lang=E&JournalID=10.

964 ■ Asian Journal of English Language Teaching
Chinese University Press
Shatin Galleria, 9th Fl., Unit 1-3 & 18
18-24 Shan Mei St.
Fo Tan, Shatin
Hong Kong, People's Republic of China
Ph: 852 29465300
Fax: 852 26037355
Publisher E-mail: cup@cuhk.edu.hk

Journal covering English language teaching. **Founded:** Oct. 2001. **Freq:** Annual. **Trim Size:** 152 x 229 mm. **Key Personnel:** Gwendolyn Gong, Editor, ggong@cuhk.edu.hk; Peter Yonggi Gu, Editor, peter.gu@vuw.ac.nz; Lixian Jin, Review Ed., jin@dmu.ac.uk. **ISSN:** 1026-2652. **Subscription Rates:** US$11.50 individuals. **URL:** http://www.

Circulation: ★ = ABC; △ = BPA; ♦ = CAC; • = CCAB; ❑ = VAC; ⊕ = PO Statement; ‡ = Publisher's Report; Boldface figures = sworn; Light figures = estimated.

chineseupress.com/asp/JournalList_en.asp?CatID=1&Lang=E&JournalI D=7.

965 ■ Ching Feng
Chinese University Press
Shatin Galleria, 9th Fl., Unit 1-3 & 18
18-24 Shan Mei St.
Fo Tan, Shatin
Hong Kong, People's Republic of China
Ph: 852 29465300
Fax: 852 26037355
Publisher E-mail: cup@cuhk.edu.hk

Journal covering Chinese religion. **Founded:** 2001. **Freq:** Semiannual. **Trim Size:** 152 x 229 mm. **ISSN:** 0009-4668. **Subscription Rates:** US$22.50 individuals. **URL:** http://www.chineseupress.com/asp/JournalList_en.asp?CatID=1&Lang=E&JournalI D=12.

966 ■ Hong Kong Journal of Sociology
Chinese University Press
Shatin Galleria, 9th Fl., Unit 1-3 & 18
18-24 Shan Mei St.
Fo Tan, Shatin
Hong Kong, People's Republic of China
Ph: 852 29465300
Fax: 852 26037355
Publisher E-mail: cup@cuhk.edu.hk

Journal covering sociology. **Founded:** Nov. 2000. **Freq:** Annual. **Trim Size:** 152 x 229 mm. **Key Personnel:** Lau Siu-kai, Editor-in-Chief. **ISSN:** 1606-8610. **Subscription Rates:** US$12 individuals. **URL:** http://www.chineseupress.com/asp/JournalList_en.asp?CatID=1&Lang=E&JournalI D=8.

967 ■ Journal of Translation Studies
Chinese University Press
Shatin Galleria, 9th Fl., Unit 1-3 & 18
18-24 Shan Mei St.
Fo Tan, Shatin
Hong Kong, People's Republic of China
Ph: 852 29465300
Fax: 852 26037355
Publisher E-mail: cup@cuhk.edu.hk

Journal covering translation studies and Chinese translations of literary works. **Founded:** Mar. 2000. **Freq:** Semiannual. **Trim Size:** 152 x 229 mm. **Key Personnel:** Sin-wai Chan, Editor-in-Chief; Evangeline S.P. Almberg, Editor; Gilbert C.F. Fong, Editor. **ISSN:** 1027-7978. **Subscription Rates:** US$12 individuals. **URL:** http://www.chineseupress.com/asp/JournalList_en.asp?CatID=1&Lang=E&JournalI D=4.

968 ■ Translation Quarterly
Chinese University Press
Shatin Galleria, 9th Fl., Unit 1-3 & 18
18-24 Shan Mei St.
Fo Tan, Shatin
Hong Kong, People's Republic of China
Ph: 852 29465300
Fax: 852 26037355
Publisher E-mail: cup@cuhk.edu.hk

Journal covering translation studies. **Founded:** Mar. 2005. **Freq:** Quarterly. **Trim Size:** 140 x 210 mm. **Key Personnel:** Prof. Leo Tak-hung Chan, Editor-in-Chief. **ISSN:** 1027-8559. **Subscription Rates:** US$20 individuals. **URL:** http://www.chineseupress.com/asp/JournalList_en.asp?CatID=1&Lang=E&JournalI D=14.

SHANGHAI

969 ■ Beauty Home
Shanghai Weekly Culture Media Co., Ltd.
593 Yan'an Rd. W
Shanghai 200050, People's Republic of China
Ph: 86 21 61229100
Fax: 86 21 61229039
Publisher E-mail: rights@shwenyi.com

Magazine featuring Chinese lifestyle. **Founded:** 2000. **Freq:** Semimonthly. **Key Personnel:** Zhang Keping, Licensor, phone 86 21 61229133, fax 86 21 61229129. **Subscription Rates:** 9.80 Yu individuals. **URL:** http://www.wenyigroup.com.cn/ehibition/ehibition_default2.asp?pro_id=253.

970 ■ Calligraphy
Shanghai Fine Arts Publishing House
593 Yan'an Rd. W
Shanghai 200050, People's Republic of China
Ph: 86 21 61229008
Fax: 86 21 61229015

Journal covering Chinese calligraphy. **Freq:** Monthly. **Key Personnel:** Xu Mingsong, Licensor, phone 86 21 61229018, fax 86 21 61229015. **Subscription Rates:** 10 Yu individuals. **URL:** http://www.wenyigroup.com.cn/ehibition/ehibition_default2.asp?pro_id=244.

971 ■ Calligraphy and Painting
Shanghai Fine Arts Publishing House
593 Yan'an Rd. W
Shanghai 200050, People's Republic of China
Ph: 86 21 61229008
Fax: 86 21 61229015

Journal covering calligraphy and painting. **Founded:** 1982. **Freq:** Monthly. **Key Personnel:** Xu Mingsong, Licensor, phone 86 21 61229005, fax 86 21 61229015. **Subscription Rates:** 4.90 Yu individuals. **URL:** http://www.wenyigroup.com.cn/ehibition/ehibition_default2.asp?pro_id=241.

972 ■ Charity Matters
Shanghai Brilliant Books
Changle Rd., Ln. 672, No. 33, Section E
Shanghai 200040, People's Republic of China
Ph: 86 21 54030490
Fax: 86 21 54045466

Magazine covering social charity activities. **Freq:** Monthly. **Key Personnel:** He Sicong, Licensor, phone 86 21 54045981, fax 86 21 54045981. **Subscription Rates:** 3.80 Yu individuals. **URL:** http://www.wenyigroup.com.cn/ehibition/ehibition_default2.asp?pro_id=263.

973 ■ Comic King
Shanghai People's Fine Arts Publishing House
D Bldg., No. 33
Changle Rd., Ln. 672
Shanghai 200040, People's Republic of China
Ph: 86 21 54044520
Fax: 86 21 54032331
Publisher E-mail: mscbs@sh163.net

Magazine featuring cartoon stories. **Founded:** Aug. 1985. **Freq:** Monthly. **Key Personnel:** Le Jian, Licensor, phone 86 21 54031690, fax 86 21 54032331. **Subscription Rates:** 4.20 Yu individuals. **URL:** http://www.wenyigroup.com.cn/ehibition/ehibition_default2.asp?pro_id=280.

974 ■ Fiction World
Shanghai Literature & Arts Publishing House
74 Shaoxing Rd.
Shanghai 200020, People's Republic of China
Ph: 86 21 64336243
Fax: 86 21 64740676

Magazine featuring Chinese literature. **Founded:** May 1981. **Freq:** Bimonthly. **Key Personnel:** Han Ying, Licensor, phone 86 21 64377833, fax 86 21 64740676. **Subscription Rates:** 10 Yu individuals. **URL:** http://www.wenyigroup.com.cn/ehibition/ehibition_default2.asp?pro_id=262.

975 ■ Journal of Editorial Study
Shanghai Using the Right Word Culture Media Co., Ltd.
Jia, 384/11 Jianguo Rd. W
Shanghai 200031, People's Republic of China
Ph: 86 21 64330669
Fax: 86 21 64330669

Journal covering editorial study. **Freq:** Bimonthly. **Key Personnel:** Sun Huan, Licensor, phone 86 21 64311015, fax 86 21 64311015. **Subscription Rates:** 8 Yu individuals. **URL:** http://www.wenyigroup.com.cn/ehibition/ehibition_default2.asp?pro_id=273.

976 ■ Man & Nature
Shanghai Weekly Culture Media Co., Ltd.
593 Yan'an Rd. W
Shanghai 200050, People's Republic of China
Ph: 86 21 61229100
Fax: 86 21 61229039
Publisher E-mail: rights@shwenyi.com

Magazine covering nature and human culture. **Founded:** Sept. 2001. **Freq:** Monthly. **Key Personnel:** Lu Yan, Licensor, phone 86 21 61229241, fax 86 21 61229240. **Subscription Rates:** 16 Yu individuals. **URL:** http://www.wenyigroup.com.cn/ehibition/ehibition_default2.asp?pro_id=247.

977 ■ Oriental Sword
Shanghai Literature & Arts Publishing House
74 Shaoxing Rd.
Shanghai 200020, People's Republic of China
Ph: 86 21 64336243
Fax: 86 21 64740676

Magazine featuring case reports and detective stories. **Founded:** 1993. **Freq:** Monthly. **Key Personnel:** Wang Jian, Licensor, phone 86 21 64723570, fax 86 21 64723570. **Subscription Rates:** 6.50 Yu individuals. **URL:** http://www.wenyigroup.com.cn/ehibition/ehibition_default2.asp?pro_id=261.

978 ■ Shanghai Pictorial
Shanghai Brilliant Books
Changle Rd., Ln. 672, No. 33, Section E
Shanghai 200040, People's Republic of China
Ph: 86 21 54030490
Fax: 86 21 54045466

Magazine covering photography. **Founded:** 1982. **Freq:** Monthly. **Key Personnel:** Zhao Songhua, Licensor, phone 86 21 54045234, fax 86 21 54045234. **Subscription Rates:** 10 Yu individuals. **URL:** http://www.wenyigroup.com.cn/ehibition/ehibition_default2.asp?pro_id=242.

979 ■ Shanghai Residence
Shanghai Fine Arts Publishing House
593 Yan'an Rd. W
Shanghai 200050, People's Republic of China
Ph: 86 21 61229008
Fax: 86 21 61229015

Magazine featuring home decoration. **Freq:** Monthly. **Key Personnel:** Zhang Xiong, Licensor, phone 86 21 54904529, fax 86 21 54904493. **Subscription Rates:** 20 Yu individuals. **URL:** http://www.wenyigroup.com.cn/ehibition/ehibition_default2.asp?pro_id=264. **Circ:** 30,000

980 ■ Shanghai Weekly
Shanghai Weekly Culture Media Co., Ltd.
593 Yan'an Rd. W
Shanghai 200050, People's Republic of China
Ph: 86 21 61229100
Fax: 86 21 61229039
Publisher E-mail: rights@shwenyi.com

Magazine featuring lifestyle in Shanghai. **Founded:** Oct. 2000. **Freq:** Weekly. **Key Personnel:** Wang Xiaolian, Licensor, phone 86 21 61229170, fax 86 21 61229200. **Subscription Rates:** 1 Yu individuals. **URL:** http://www.wenyigroup.com.cn/ehibition/ehibition_default2.asp?pro_id=255.

981 ■ Travelling Scope
Shanghai Stories Culture Media Co., Ltd.
74 Shaoxing Rd.
Shanghai 200020, People's Republic of China
Ph: 86 21 64376635
Fax: 86 21 64376635
Publisher E-mail: f_weien@sohu.com

Magazine covering travel destination in different countries. **Freq:** Monthly. **Key Personnel:** Xia Qinggen, Licensor, phone 86 21 64450298, fax 86 21 64660169. **Subscription Rates:** 16 Yu individuals. **URL:** http://www.wenyigroup.com.cn/ehibition/ehibition_default2.asp?pro_id=256.

982 ■ Using the Right Word
Shanghai Using the Right Word Culture Media Co., Ltd.
Jia, 384/11 Jianguo Rd. W
Shanghai 200031, People's Republic of China
Ph: 86 21 64330669
Fax: 86 21 64330669

Magazine covering Chinese writings. **Founded:** 1995. **Freq:** Monthly. **Key Personnel:** Wang Min, Licensor, phone 86 21 64330669, fax 86 21 64330669. **Subscription Rates:** 2 Yu individuals. **URL:** http://www.wenyigroup.com.cn/ehibition/ehibition_default2.asp?pro_id=288.

983 ■ With
Shanghai Stories Culture Media Co., Ltd.
74 Shaoxing Rd.
Shanghai 200020, People's Republic of China
Ph: 86 21 64376635
Fax: 86 21 64376635
Publisher E-mail: f_weien@sohu.com

Magazine covering women's fashion and lifestyle. **Founded:** Dec. 18, 2002. **Freq:** Monthly. **Key Personnel:** Li Zhenyu, Licensor, phone 86 21 64372608, fax 86 21 64668742. **Subscription Rates:** 16 Yu individuals. **URL:** http://www.wenyigroup.com.cn/ehibition/ehibition_default2.asp?pro_id=279.

984 ■ World Traveller
Shanghai People's Fine Arts Publishing House
D Bldg., No. 33
Changle Rd., Ln. 672
Shanghai 200040, People's Republic of China
Ph: 86 21 54044520
Fax: 86 21 54032331
Publisher E-mail: mscbs@sh163.net

Magazine featuring travel fashion and trend. **Founded:** Oct. 2003. **Freq:** Monthly. **Key Personnel:** Le Jian, Licensor, phone 86 21 54031690, fax 86 21 54032331. **Subscription Rates:** 18 Yu individuals. **URL:** http://www.wenyigroup.com.cn/ehibition/ehibition_default2.asp?pro_id=277.

XI'AN

985 ■ International Journal of Internet Manufacturing & Services
Inderscience Publishers
c/o Prof. Pingyu Jiang, Ed.-in-Ch.
Xi'an Jiaotong University
CAD/CAM Institute
School of Mechanical Engineering
Xi'an 710049, Shaanxi, People's Republic of China
Publisher E-mail: editor@inderscience.com

Journal covering areas of manufacturing science & technologies, services theory and methods, and corresponding information and knowledge management issues. **Founded:** 2007. **Freq:** 4/yr. **Key Personnel:** Prof. Pingyu Jiang, Editor-in-Chief, pjiang@mail.xjtu.edu.cn. **ISSN:** 1751-6048. **Subscription Rates:** EUR470 individuals includes surface mail, print only; EUR640 individuals print and online. **URL:** http://www.inderscience.com/browse/index.php?journalCODE=ijims.

XIANGTAN

986 ■ International Journal of Computing Science and Mathematics
Inderscience Publishers
c/o Prof. Yong Zhou, Ed.-in-Ch.
Xiangtan University
School of Mathematics & Computational Science
Institute for Computational & Applied Mathematics
Xiangtan 411105, Hunan, People's Republic of China
Publication E-mail: ijcsm@inderscience.com
Publisher E-mail: editor@inderscience.com

Journal covering all areas of computing science & mathematics. **Founded:** 2007. **Freq:** 4/yr. **Key Personnel:** Prof. Yong Zhou, Editor-in-Chief, yzhou@xtu.edu.cn. **ISSN:** 1752-5055. **Subscription Rates:** EUR470 individuals includes surface mail, print only; EUR640 individuals print & online. **URL:** http://www.inderscience.com/browse/index.php?journalCODE=ijcsm.

987 ■ International Journal of Dynamical Systems and Differential Equations
Inderscience Publishers
c/o Prof. Yong Zhou, Ed.-in-Ch.
Xiangtan University
School of Mathematics & Computational Science
Institute for Computational & Applied Mathematics
Xiangtan 411105, Hunan, People's Republic of China
Publication E-mail: ijdsde@inderscience.com
Publisher E-mail: editor@inderscience.com

Journal covering dynamical systems & differential equations. **Founded:** 2007. **Freq:** Quarterly. **Key Personnel:** Prof. Yong Zhou, Editor-in-Chief, yzhou@xtu.edu.cn. **ISSN:** 1752-3583. **Subscription Rates:** EUR470 individuals includes surface mail, print only; EUR640 individuals print & online. **URL:** http://www.inderscience.com/browse/index.php?journalCODE=ijdsde.

EGYPT

ALEXANDRIA

988 ■ Research Journal of Cell and Molecular Biology
International Network for Scientific Information
c/o Dr. Elsayed Elsayed Hafez, Ed.-in-Ch.
Nucleic Acid Research Department
Genetic Engineering & Biotechnology Research Institute
Research Area, Borg El Arab
Alexandria 21934, Egypt
Ph: 20 3 4593413
Fax: 20 3 4593423
Publisher E-mail: editor@insinet.net

Journal covering cell and molecular biology. **Freq:** 4/yr. **Key Personnel:** Dr. Elsayed Elsayed Hafez, Editor-in-Chief, elsayed_hafez@yahoo.com. **ISSN:** 1991-8828. **URL:** http://www.insinet.net/rjcmb.html.

FRANCE

AIX-EN-PROVENCE

989 ■ International Journal of Public Sector Performance Management
Inderscience Publishers
c/o Prof. Robert Fouchet, Ed.
Universite Paul-Cezanne Aix-Marseille III
Institute of Public Management & Territorial Governance
21 rue Gaston de Saporta
F-13625 Aix-en-Provence Cedex, France
Publication E-mail: ijpspm@inderscience.com
Publisher E-mail: editor@inderscience.com

Journal covering implementation of performance management in the public sector. **Founded:** 2007. **Freq:** 4/yr. **Key Personnel:** Prof. Robert Fouchet, Editor, robert.fouchet@univ-cezanne.fr. **ISSN:** 1741-1041. **Subscription Rates:** EUR470 individuals includes surface mail, print only; EUR640 individuals print and online. **URL:** http://www.inderscience.com/browse/index.php?journalCODE=ijpspm.

CERGY-PONTOISE

990 ■ International Journal of Teaching & Case Studies
Inderscience Publishers
c/o Prof. David Avison, Ed.-in-Ch.
ESSEC Business School
Av. Bernard Hirsch
BP 95105
F-95105 Cergy-Pontoise Cedex, France
Publisher E-mail: editor@inderscience.com

Journal covering management, management science, computer engineering, computer science, information systems, information technology and software engineering. **Founded:** 2007. **Freq:** 4/yr. **Key Personnel:** Prof. David Avison, Editor-in-Chief, avison@essec.fr; Miltiadis Lytras, Editor-in-Chief, lytras@ceid.upatras.gr; Martin P. Papadatos, Managing Editor, mpp26@cam.ac.uk. **ISSN:** 1750-4090. **Subscription Rates:** EUR470 individuals includes surface mail, print only; EUR640 individuals print and online. **URL:** http://www.inderscience.com/browse/index.php?journalCODE=ijtcs.

FONTENAY-SOUS-BOIS

991 ■ Mer & Bateaux
Boat International Group
118, ave. Marechal de Lattre de Tassigny
F-94120 Fontenay-sous-Bois, France
Ph: 33 1 48774830
Fax: 33 1 48775780
Publication E-mail: mer.et.bateaux@wanadoo.fr
Publisher E-mail: info@boatinternational.co.uk

Magazine covering French yachting market. **Freq:** Bimonthly. **Trim Size:** 223 x 275 mm. **Key Personnel:** Tony Harris, CEO/Publisher; Tony Euden, Publishing Dir., tonye@boatinternational.co.uk; Tim Hartney, Production Mgr., timh@boatinternational.co.uk. **Subscription Rates:** EUR45 two years; EUR81 two years French territories; EUR100 other countries 2 years. **Remarks:** Accepts advertising. **URL:** http://www.boatinternational.com/mags/mag04.htm. . **Ad Rates:** 4C: 3,850 Fr. **Circ:** 13,500

MARSEILLE

992 ■ International Journal of Technoentrepreneurship
Inderscience Publishers
c/o Prof. Francois Therin, Ed.
Euromed Marseille Ecole de Management
Domaine de Luminy, BP 921
F-26222 Marseille Cedex, France
Publisher E-mail: editor@inderscience.com

Journal covering high tech entrepreneurship and intrapreneurship. **Founded:** 2007. **Freq:** 4/yr. **Key Personnel:** Prof. Francois Therin, Editor, francois.therin@euromed-marseille.com. **ISSN:** 1750-4090. **Subscription Rates:** EUR470 individuals includes surface mail; EUR640 individuals print and online. **URL:** http://www.inderscience.com/browse/index.php?journalCODE=ijte.

VILLIERS-SAINT-FREDERIC

993 ■ International Journal of Electric and Hybrid Vehicles
Inderscience Publishers
c/o Benoit Maisseu, Ed.
Renault IDVU, VSF PMB 1 00
42 Rt. de Beynes
F-78640 Villiers-Saint-Frederic, France
Publisher E-mail: editor@inderscience.com

Journal covering field of electric and hybrid automotive systems. **Founded:** 2007. **Freq:** 4/yr. **Key Personnel:** Benoit Maisseu, Editor, benoit.maisseu@renault.com. **ISSN:** 1751-4088. **Subscription Rates:** EUR470 individuals includes surface mail, print only; EUR640 individuals print & online. **URL:** http://www.inderscience.com/browse/index.php?journalCODE=ijehv.

GERMANY

KIEL

994 ■ Meer & Yachten
Boat International Group
Edimer. Flensburger Str. 87
D-24106 Kiel, Germany
Ph: 49 431 336883
Fax: 49 431 331485
Publication E-mail: my@boatinternational.co.uk
Publisher E-mail: info@boatinternational.co.uk

Magazine covering luxury yacht for German speaking readers. **Founded:** 1991. **Freq:** Bimonthly. **Trim Size:** 223 x 275 mm. **Key Personnel:** Tony Harris, CEO/Publisher; Tony Euden, Publishing Dir., tonye@boatinternational.co.uk; Tim Hartney, Production Mgr., timh@boatinternational.co.uk. **Subscription Rates:** EUR36 individuals in Germany; EUR42 individuals in European countries; 72 SFr individuals in Switzerland. **Remarks:** Accepts advertising. **URL:** http://www.boatinternational.com/mags/mag05.htm. . **Ad Rates:** 4C: DM 3,850. **Circ:** 17,000

Circulation: ★ = ABC; △ = BPA; ♦ = CAC; • = CCAB; ❑ = VAC; ⊕ = PO Statement; ‡ = Publisher's Report; Boldface figures = sworn; Light figures = estimated.

LEMGO

995 ■ Contributions to Algebra and Geometry
Heldermann Verlag
Langer Graben 17
D-32657 Lemgo, Germany
Ph: 49 526 110226
Fax: 49 526 115264
Publisher E-mail: mail@heldermann.de

Journal covering areas of algebra, geometry, algebraic geometry and related fields. **Founded:** 1971. **Freq:** 2/yr. **Key Personnel:** H. Martini, Managing Editor; G. Stroth, Managing Editor; J. Stuckland, Managing Editor. **ISSN:** 0138-4821. **Subscription Rates:** EUR130 individuals. **URL:** http://www.heldermann.de/BAG/bagcover.htm.

996 ■ Economic Quality Control
Heldermann Verlag
Langer Graben 17
D-32657 Lemgo, Germany
Ph: 49 526 110226
Fax: 49 526 115264
Publisher E-mail: mail@heldermann.de

Journal covering economic control and maintenance policies for production and inventory. **Subtitle:** International Journal for Quality and Reliability. **Founded:** 1986. **Freq:** 2/yr. **Key Personnel:** E. von Collani, Managing Editor. **ISSN:** 0940-5151. **Subscription Rates:** EUR100 individuals. **URL:** http://www.heldermann.de/EQC/eqccover.htm.

997 ■ Journal of Convex Analysis
Heldermann Verlag
Langer Graben 17
D-32657 Lemgo, Germany
Ph: 49 526 110226
Fax: 49 526 115264
Publisher E-mail: mail@heldermann.de

Journal covering field of convex analysis. **Founded:** 1994. **Freq:** 4/yr. **Key Personnel:** G. Buttazzo, Managing Editor; L. Thibault, Managing Editor. **ISSN:** 0944-6532. **Subscription Rates:** EUR220 individuals. **URL:** http://www.heldermann.de/JCA/jcacover.htm.

GREECE

ATHENS

998 ■ International Journal of Electronic Governance
Inderscience Publishers
c/o Prof. Panagiotis Georgiadis, Ed.-in-Ch.
University of Athens
E-Government Laboratory, TYPA bldg.
Panepistimiopolis Ilission
GR-15784 Athens, Greece
Publisher E-mail: editor@inderscience.com

Journal covering areas of electronic governance. **Founded:** 2007. **Freq:** 4/yr. **Key Personnel:** Prof. Panagiotis Georgiadis, Editor-in-Chief, georgiad@di.uoa.gr; Prof. Dimitris Gouscos, Exec. Ed., gouscos@media.uoa.gr. **ISSN:** 1742-7509. **Subscription Rates:** EUR470 individuals includes surface mail, print only; EUR640 individuals print & online. **URL:** http://www.inderscience.com/browse/index.php?journalCODE=ijeg.

PATRA

999 ■ International Journal of Tourism Policy
Inderscience Publishers
c/o Prof. George Agiomirgianakis, Ed.-in-Ch.
Hellenic Open University
Riga Fereou 169 & Tsamadou
GR-26222 Patra, Greece
Publication E-mail: ijtp@atseap.gr
Publisher E-mail: editor@inderscience.com

Journal covering international tourism development. **Founded:** 2007. **Freq:** 4/yr. **Key Personnel:** Prof. George Agiomirgianakis, Editor-in-Chief; Dr. Konstantinos Andriotis, Editor-in-Chief. **ISSN:** 1750-4090. **Subscription Rates:** EUR470 individuals includes surface mail, print only; EUR640 individuals print and online. **URL:** http://www.inderscience.com/browse/index.php?journalCODE=ijtp.

PIRAEUS

1000 ■ International Journal of Applied Systemic Studies
Inderscience Publishers
c/o Prof. Nikitas Assimakopoulos, Ed.-in-Ch.
University of Piraeus
Department of Informatics
80, Karaoli & Dimitriou Str.
GR-185 34 Piraeus, Greece
Publisher E-mail: editor@inderscience.com

Journal covering applications of systemic methodologies & studies. **Founded:** 2007. **Freq:** 4/yr. **Key Personnel:** Prof. Nikitas Assimakopoulos, Editor-in-Chief, assinik@unipi.gr. **ISSN:** 1751-0589. **Subscription Rates:** EUR470 individuals includes surface mail, print only; EUR640 individuals print & online. **URL:** http://www.inderscience.com/browse/index.php?journalCODE=ijass.

1001 ■ Journal of Applied Systems Studies
Cambridge International Science Publishing
c/o Nikitas A. Assimakopoulos, Ed.-in-Ch.
Department of Informatics
University of Piraeus
80 Karaoli & Dimitriou Str.
GR-185 34 Piraeus, Greece
Fax: 30 1 4179064
Publisher E-mail: cisp@cisp-publishing.com

Journal covering the development of methodologies based on the laws and rules of various sciences. **Founded:** Mar. 2000. **Key Personnel:** Nikitas A. Assimakopoulos, Editor-in-Chief, assinik@unipi.gr; Russell L. Ackoff, Editor. **ISSN:** 1466-7738. **Subscription Rates:** 200 institutions; 150 individuals; 100 members. **URL:** http://www.cisp-publishing.com/jass.html.

GUATEMALA

ALTA VERAPAZ

1002 ■ Emisoras Unidas-AM - 930
4 calle 6-84, zona 13
Guatemala City, Guatemala
Ph: 502 2 4405139

Format: News; Information. **Owner:** Grupo Emisoras Unidas, at above address. **Key Personnel:** Felipe Valenzuela, Dir. Gen. **URL:** http://radio.emisorasunidas.com/frecuencias.php.

1003 ■ Emisoras Unidas-FM - 92.3
4 calle 6-84, zona 13
Guatemala City, Guatemala
Ph: 502 2 4405139

Format: News; Information. **Owner:** Grupo Emisoras Unidas, at above address. **Key Personnel:** Felipe Valenzuela, Dir. Gen. **URL:** http://radio.emisorasunidas.com/frecuencias.php.

1004 ■ Emisoras Unidas-FM - 91.1
4 calle 6-84, zona 13
Guatemala City, Guatemala
Ph: 502 2 4405139

Format: News; Information. **Owner:** Grupo Emisoras Unidas, at above address. **Key Personnel:** Felipe Valenzuela, Dir. Gen. **URL:** http://radio.emisorasunidas.com/frecuencias.php.

ANTIGUA

1005 ■ Emisoras Unidas-FM - 89.7
4 calle 6-84, zona 13
Guatemala City, Guatemala
Ph: 502 2 4405139

Format: News; Information. **Owner:** Grupo Emisoras Unidas, at above address. **Key Personnel:** Felipe Valenzuela, Dir. Gen. **URL:** http://radio.emisorasunidas.com/frecuencias.php.

BAJA VERAPAZ

1006 ■ Emisoras Unidas-FM - 94.3
4 calle 6-84, zona 13
Guatemala City, Guatemala
Ph: 502 2 4405139

Format: News; Information. **Owner:** Grupo Emisoras Unidas, at above address. **Key Personnel:** Felipe Valenzuela, Dir. Gen. **URL:** http://radio.emisorasunidas.com/frecuencias.php.

1007 ■ Emisoras Unidas-FM - 89.7
4 calle 6-84, zona 13
Guatemala City, Guatemala
Ph: 502 2 4405139

Format: News; Information. **Owner:** Grupo Emisoras Unidas, at above address. **Key Personnel:** Felipe Valenzuela, Dir. Gen. **URL:** http://radio.emisorasunidas.com/frecuencias.php.

CHICHICASTENANGO

1008 ■ Emisoras Unidas-AM - 960
4 calle 6-84, zona 13
Guatemala City, Guatemala
Ph: 502 2 4405139

Format: News; Information. **Owner:** Grupo Emisoras Unidas, at above address. **Key Personnel:** Felipe Valenzuela, Dir. Gen. **URL:** http://radio.emisorasunidas.com/frecuencias.php.

1009 ■ Emisoras Unidas-FM - 90.3
4 calle 6-84, zona 13
Guatemala City, Guatemala
Ph: 502 2 4405139

Format: News; Information. **Owner:** Grupo Emisoras Unidas, at above address. **Key Personnel:** Felipe Valenzuela, Dir. Gen. **URL:** http://radio.emisorasunidas.com/frecuencias.php.

CHIMALTENANGO

1010 ■ Emisoras Unidas-FM - 91.1
4 calle 6-84, zona 13
Guatemala City, Guatemala
Ph: 502 2 4405139

Format: News; Information. **Owner:** Grupo Emisoras Unidas, at above address. **Key Personnel:** Felipe Valenzuela, Dir. Gen. **URL:** http://radio.emisorasunidas.com/frecuencias.php.

CHIQUIMULA

1011 ■ Emisoras Unidas-FM - 89.9
4 calle 6-84, zona 13
Guatemala City, Guatemala
Ph: 502 2 4405139

Format: News; Information. **Owner:** Grupo Emisoras Unidas, at above address. **Key Personnel:** Felipe Valenzuela, Dir. Gen. **URL:** http://radio.emisorasunidas.com/frecuencias.php.

COATEPEQUE

1012 ■ Emisoras Unidas-FM - 98.7
4 calle 6-84, zona 13
Guatemala City, Guatemala
Ph: 502 2 4405139

Format: News; Information. **Owner:** Grupo Emisoras Unidas, at above address. **Key Personnel:** Felipe Valenzuela, Dir. Gen. **URL:** http://radio.emisorasunidas.com/frecuencias.php.

EL PETEN

1013 ■ Emisoras Unidas-FM - 98.9
4 calle 6-84, zona 13
Guatemala City, Guatemala
Ph: 502 2 4405139

Format: News; Information. **Owner:** Grupo Emisoras Unidas, at above address. **Key Personnel:** Felipe Valenzuela, Dir. Gen. **URL:** http://radio.emisorasunidas.com/frecuencias.php.

ESCUINTLA

1014 ■ Emisoras Unidas-AM - 600
4 calle 6-84, zona 13
Guatemala City, Guatemala
Ph: 502 2 4405139

Format: News; Information. **Owner:** Grupo Emisoras Unidas, at above address. **Key Personnel:** Felipe Valenzuela, Dir. Gen. **URL:** http://radio.emisorasunidas.com/frecuencias.php.

1015 ■ Emisoras Unidas-FM - 92.3
4 calle 6-84, zona 13
Guatemala City, Guatemala
Ph: 502 2 4405139

Format: News; Information. **Owner:** Grupo Emisoras Unidas, at above address. **Key Personnel:** Felipe Valenzuela, Dir. Gen. **URL:** http://radio.emisorasunidas.com/frecuencias.php.

1016 ■ Emisoras Unidas-FM - 91.9
4 calle 6-84, zona 13
Guatemala City, Guatemala
Ph: 502 2 4405139

Format: News; Information. **Owner:** Grupo Emisoras Unidas, at above address. **Key Personnel:** Felipe Valenzuela, Dir. Gen. **URL:** http://radio.emisorasunidas.com/frecuencias.php.

GUATEMALA CITY

Emisoras Unidas-AM - See Alta Verapaz
Emisoras Unidas-AM - See Chichicastenango
Emisoras Unidas-AM - See Escuintla
Emisoras Unidas-AM - See Quetzaltenango
Emisoras Unidas-AM - See Quiche
Emisoras Unidas-AM - See Retalhuleu
Emisoras Unidas-AM - See San Marcos
Emisoras Unidas-AM - See Suchitepequez

1017 ■ Emisoras Unidas-FM - 89.7
4 calle 6-84, zona 13
Guatemala City, Guatemala
Ph: 502 2 4405139

Format: News; Information. **Owner:** Grupo Emisoras Unidas, at above address. **Key Personnel:** Felipe Valenzuela, Dir. Gen. **URL:** http://radio.emisorasunidas.com/frecuencias.php.

Emisoras Unidas-FM - See Alta Verapaz
Emisoras Unidas-FM - See Alta Verapaz
Emisoras Unidas-FM - See Antigua
Emisoras Unidas-FM - See Baja Verapaz
Emisoras Unidas-FM - See Baja Verapaz
Emisoras Unidas-FM - See Chichicastenango
Emisoras Unidas-FM - See Chimaltenango
Emisoras Unidas-FM - See Chiquimula
Emisoras Unidas-FM - See Coatepeque
Emisoras Unidas-FM - See El Peten
Emisoras Unidas-FM - See Escuintla
Emisoras Unidas-FM - See Escuintla
Emisoras Unidas-FM - See Huehuetenango
Emisoras Unidas-FM - See Occidente
Emisoras Unidas-FM - See Puerto Barrios
Emisoras Unidas-FM - See Quetzaltenango
Emisoras Unidas-FM - See Quiche
Emisoras Unidas-FM - See Retalhuleu
Emisoras Unidas-FM - See Retalhuleu
Emisoras Unidas-FM - See San Marcos
Emisoras Unidas-FM - See Santa Rosa
Emisoras Unidas-FM - See Suchitepequez
Emisoras Unidas-FM - See Suchitepequez
Emisoras Unidas-FM - See Zacapa

HUEHUETENANGO

1018 ■ Emisoras Unidas-FM - 104.1
4 calle 6-84, zona 13
Guatemala City, Guatemala
Ph: 502 2 4405139

Format: News; Information. **Owner:** Grupo Emisoras Unidas, at above address. **Key Personnel:** Felipe Valenzuela, Dir. Gen. **URL:** http://radio.emisorasunidas.com/frecuencias.php.

OCCIDENTE

1019 ■ Emisoras Unidas-FM - 104.3
4 calle 6-84, zona 13
Guatemala City, Guatemala
Ph: 502 2 4405139

Format: News; Information. **Owner:** Grupo Emisoras Unidas, at above address. **Key Personnel:** Felipe Valenzuela, Dir. Gen. **URL:** http://radio.emisorasunidas.com/frecuencias.php.

PUERTO BARRIOS

1020 ■ Emisoras Unidas-FM - 89.9
4 calle 6-84, zona 13
Guatemala City, Guatemala
Ph: 502 2 4405139

Format: News; Information. **Owner:** Grupo Emisoras Unidas, at above address. **Key Personnel:** Felipe Valenzuela, Dir. Gen. **URL:** http://radio.emisorasunidas.com/frecuencias.php.

QUETZALTENANGO

1021 ■ Emisoras Unidas-AM - 1340
4 calle 6-84, zona 13
Guatemala City, Guatemala
Ph: 502 2 4405139

Format: News; Information. **Owner:** Grupo Emisoras Unidas, at above address. **Key Personnel:** Felipe Valenzuela, Dir. Gen. **URL:** http://radio.emisorasunidas.com/frecuencias.php.

1022 ■ Emisoras Unidas-FM - 98.7
4 calle 6-84, zona 13
Guatemala City, Guatemala
Ph: 502 2 4405139

Format: News; Information. **Owner:** Grupo Emisoras Unidas, at above address. **Key Personnel:** Felipe Valenzuela, Dir. Gen. **URL:** http://radio.emisorasunidas.com/frecuencias.php.

QUICHE

1023 ■ Emisoras Unidas-AM - 960
4 calle 6-84, zona 13
Guatemala City, Guatemala
Ph: 502 2 4405139

Format: News; Information. **Owner:** Grupo Emisoras Unidas, at above address. **Key Personnel:** Felipe Valenzuela, Dir. Gen. **URL:** http://radio.emisorasunidas.com/frecuencias.php.

1024 ■ Emisoras Unidas-FM - 90.3
4 calle 6-84, zona 13
Guatemala City, Guatemala
Ph: 502 2 4405139

Format: News; Information. **Owner:** Grupo Emisoras Unidas, at above address. **Key Personnel:** Felipe Valenzuela, Dir. Gen. **URL:** http://radio.emisorasunidas.com/frecuencias.php.

RETALHULEU

1025 ■ Emisoras Unidas-AM - 1130
4 calle 6-84, zona 13
Guatemala City, Guatemala
Ph: 502 2 4405139

Format: News; Information. **Owner:** Grupo Emisoras Unidas, at above address. **Key Personnel:** Felipe Valenzuela, Dir. Gen. **URL:** http://radio.emisorasunidas.com/frecuencias.php.

1026 ■ Emisoras Unidas-FM - 105.1
4 calle 6-84, zona 13
Guatemala City, Guatemala
Ph: 502 2 4405139

Format: News; Information. **Owner:** Grupo Emisoras Unidas, at above address. **Key Personnel:** Felipe Valenzuela, Dir. Gen. **URL:** http://radio.emisorasunidas.com/frecuencias.php.

1027 ■ Emisoras Unidas-FM - 103.1
4 calle 6-84, zona 13
Guatemala City, Guatemala
Ph: 502 2 4405139

Format: News; Information. **Owner:** Grupo Emisoras Unidas, at above address. **Key Personnel:** Felipe Valenzuela, Dir. Gen. **URL:** http://radio.emisorasunidas.com/frecuencias.php.

SAN MARCOS

1028 ■ Emisoras Unidas-AM - 740
4 calle 6-84, zona 13
Guatemala City, Guatemala
Ph: 502 2 4405139

Format: News; Information. **Owner:** Grupo Emisoras Unidas, at above address. **Key Personnel:** Felipe Valenzuela, Dir. Gen. **URL:** http://radio.emisorasunidas.com/frecuencias.php.

1029 ■ Emisoras Unidas-FM - 104.3
4 calle 6-84, zona 13
Guatemala City, Guatemala
Ph: 502 2 4405139

Format: News; Information. **Owner:** Grupo Emisoras Unidas, at above address. **Key Personnel:** Felipe Valenzuela, Dir. Gen. **URL:** http://radio.emisorasunidas.com/frecuencias.php.

SANTA ROSA

1030 ■ Emisoras Unidas-FM - 89.9
4 calle 6-84, zona 13
Guatemala City, Guatemala
Ph: 502 2 4405139

Format: News; Information. **Owner:** Grupo Emisoras Unidas, at above address. **Key Personnel:** Felipe Valenzuela, Dir. Gen. **URL:** http://radio.emisorasunidas.com/frecuencias.php.

SUCHITEPEQUEZ

1031 ■ Emisoras Unidas-AM - 600
4 calle 6-84, zona 13
Guatemala City, Guatemala
Ph: 502 2 4405139

Format: News; Information. **Owner:** Grupo Emisoras Unidas, at above address. **Key Personnel:** Felipe Valenzuela, Dir. Gen. **URL:** http://radio.emisorasunidas.com/frecuencias.php.

1032 ■ Emisoras Unidas-FM - 92.3
4 calle 6-84, zona 13
Guatemala City, Guatemala
Ph: 502 2 4405139

Format: News; Information. **Owner:** Grupo Emisoras Unidas, at above address. **Key Personnel:** Felipe Valenzuela, Dir. Gen. **URL:** http://radio.emisorasunidas.com/frecuencias.php.

1033 ■ Emisoras Unidas-FM - 91.9
4 calle 6-84, zona 13
Guatemala City, Guatemala
Ph: 502 2 4405139

Format: News; Information. **Owner:** Grupo Emisoras Unidas, at above address. **Key Personnel:** Felipe Valenzuela, Dir. Gen. **URL:** http://radio.emisorasunidas.com/frecuencias.php.

ZACAPA

1034 ■ Emisoras Unidas-FM - 89.9
4 calle 6-84, zona 13
Guatemala City, Guatemala
Ph: 502 2 4405139

Format: News; Information. **Owner:** Grupo Emisoras Unidas, at above address. **Key Personnel:** Felipe Valenzuela, Dir. Gen. **URL:** http://radio.emisorasunidas.com/frecuencias.php.

HUNGARY

BUDAPEST

1035 ■ Atrium
Sanoma Budapest Kiadoi Rt.
Montevideo utca 9
H-1037 Budapest, Hungary
Ph: 36 1 4371100
Fax: 36 1 4372303
Publisher E-mail: sanomabp@sanomabp.hu

Magazine featuring architectural design. **Freq:** Bimonthly. **Print Method:** Offset. **Trim Size:** 220 x 300 mm. **Key Personnel:** Csato Edina, Coord., phone 36 1 4373906, fax 36 1 4371180, e.csato@sanomabp.hu. **Subscription Rates:** 780 Ft individuals. **Remarks:** Accepts advertising. **URL:** http://www.sanoma.hu/sajtohirdetes/index.php?action=termek&site=33. **Circ:** 15,000

1036 ■ Baratok Kozt Magazin
Sanoma Budapest Kiadoi Rt.
Montevideo utca 9
H-1037 Budapest, Hungary
Ph: 36 1 4371100
Fax: 36 1 4372303
Publisher E-mail: sanomabp@sanomabp.hu

Magazine featuring information about celebrities, behind-the-scenes stories, and issues in connection with the popular TV series. **Freq:** Quarterly. **Print Method:** Offset. **Trim Size:** 210 x 285 mm. **Key Personnel:** Csato Edina, Coord., phone 36 1 4373906, fax 36 1 4371180, e.csato@sanomabp.hu. **Subscription Rates:** 215 Ft individuals. **Remarks:** Accepts advertising. **URL:** http://www.sanoma.hu/sajtohirdetes/index.php?action=termek&site=21. **Circ:** 60,000

1037 ■ Fakanal
Sanoma Budapest Kiadoi Rt.
Montevideo utca 9
H-1037 Budapest, Hungary
Ph: 36 1 4371100
Fax: 36 1 4372303
Publisher E-mail: sanomabp@sanomabp.hu

Magazine covering gastronomy. **Freq:** 10/yr. **Print Method:** Offset. **Trim Size:** 175 x 255 mm. **Key Personnel:** Kantor Anita, Coord., phone 36 1 4373912, fax 36 1 4371180, a.kantor@sanomabp.hu. **Subscription Rates:** 229 Ft individuals. **Remarks:** Accepts advertising. **URL:** http://www.sanoma.hu/sajtohirdetes/index.php?action=termek&site=26. **Circ:** Paid ‡25,162

1038 ■ Fakanal Recepttar
Sanoma Budapest Kiadoi Rt.
Montevideo utca 9
H-1037 Budapest, Hungary
Ph: 36 1 4371100
Fax: 36 1 4372303
Publisher E-mail: sanomabp@sanomabp.hu

Magazine featuring recipes. **Freq:** Semiannual. **Print Method:** Offset. **Trim Size:** 148 x 210 mm. **Key Personnel:** Kantor Anita, Coord., phone 36 1 4373912, fax 36 1 4371180, a.kantor@sanomabp.hu. **Subscription Rates:** 440 Ft individuals. **Remarks:** Accepts advertising. **URL:** http://www.sanoma.hu/sajtohirdetes/index.php?action=termek&site=7. **Circ:** 32,500

1039 ■ Figyelo
Sanoma Budapest Kiadoi Rt.
Montevideo utca 9
H-1037 Budapest, Hungary
Ph: 36 1 4371100
Fax: 36 1 4372303
Publisher E-mail: sanomabp@sanomabp.hu

Magazine featuring news, analyses, forecasts, and background information about the economy. **Freq:** Weekly (Thurs.). **Print Method:** Offset. **Trim Size:** 200 x 267 mm. **Key Personnel:** Bosanszki Piroska, Coord., phone 36 1 4373905, fax 36 1 4371180, p.bosanszki@sanomabp.hu. **Subscription Rates:** 299 Ft individuals. **Remarks:** Accepts advertising. **URL:** http://www.sanoma.hu/sajtohirdetes/index.php?action=termek&site=28. **Circ:** ‡12,853

1040 ■ Figyelo TOP 200
Sanoma Budapest Kiadoi Rt.
Montevideo utca 9
H-1037 Budapest, Hungary
Ph: 36 1 4371100
Fax: 36 1 4372303
Publisher E-mail: sanomabp@sanomabp.hu

Magazine featuring information about the biggest companies in Hungary. **Freq:** Annual. **Print Method:** Offset. **Trim Size:** 205 x 285 mm. **Key Personnel:** Bosanszki Piroska, Coord., phone 36 1 4373905, fax 36 1 4371180, p.bosanszki@sanomabp.hu. **Subscription Rates:** 5,000 Ft individuals. **Remarks:** Accepts advertising. **URL:** http://www.sanoma.hu/sajtohirdetes/index.php?action=termek&site=29. **Circ:** Paid ‡11,500

1041 ■ Figyelo Trend
Sanoma Budapest Kiadoi Rt.
Montevideo utca 9
H-1037 Budapest, Hungary
Ph: 36 1 4371100
Fax: 36 1 4372303
Publisher E-mail: sanomabp@sanomabp.hu

Magazine featuring in-depth analysis of a given economic sector or area. **Freq:** Quarterly. **Print Method:** Offset. **Trim Size:** 200 x 267 mm. **Key Personnel:** Takacs Zsoka, Coord., phone 36 1 4373927, fax 36 1 4371180, zs.takacs@sanomabp.hu. **Subscription Rates:** 250 Ft individuals. **Remarks:** Accepts advertising. **URL:** http://www.sanoma.hu/sajtohirdetes/index.php?action=termek&site=30. **Circ:** ‡15,000

1042 ■ Fules
Sanoma Budapest Kiadoi Rt.
Montevideo utca 9
H-1037 Budapest, Hungary
Ph: 36 1 4371100
Fax: 36 1 4372303
Publisher E-mail: sanomabp@sanomabp.hu

Puzzle magazine featuring games, entertainment, and quizzes. **Freq:** Weekly (Tues.). **Print Method:** Offset. **Trim Size:** 165 x 236 mm. **Key Personnel:** Takacs Zsoka, Coord., phone 36 1 4373927, fax 36 1 4371180, zs.takacs@sanomabp.hu. **Subscription Rates:** 150 Ft individuals. **Remarks:** Accepts advertising. **URL:** http://www.sanoma.hu/sajtohirdetes/index.php?action=termek&site=6. **Circ:** Paid ‡78,001

1043 ■ Kismama
Sanoma Budapest Kiadoi Rt.
Montevideo utca 9
H-1037 Budapest, Hungary
Ph: 36 1 4371100
Fax: 36 1 4372303
Publisher E-mail: sanomabp@sanomabp.hu

Magazine covering parenthood. **Freq:** Monthly. **Print Method:** Offset. **Trim Size:** 213 x 280 mm. **Key Personnel:** Csato Edina, Coord., phone 36 1 4373906, fax 36 1 4371180, e.csato@sanomabp.hu. **Subscription Rates:** 495 Ft individuals. **Remarks:** Accepts advertising. **URL:** http://www.sanoma.hu/sajtohirdetes/index.php?action=termek&site=31. **Circ:** Paid ‡24,313

1044 ■ Kismama A Baba Elso Eve
Sanoma Budapest Kiadoi Rt.
Montevideo utca 9
H-1037 Budapest, Hungary
Ph: 36 1 4371100
Fax: 36 1 4372303
Publisher E-mail: sanomabp@sanomabp.hu

Magazine featuring information about planning and having a baby. **Freq:** Annual. **Print Method:** Offset. **Key Personnel:** Csato Edina, Coord., phone 36 1 4373906, fax 36 1 4371180, e.csato@sanomabp.hu. **Remarks:** Accepts advertising. **URL:** http://www.sanoma.hu/sajtohirdetes/index.php?action=termek&site=52. **Circ:** 46,000

1045 ■ Kismama Mintaszam
Sanoma Budapest Kiadoi Rt.
Montevideo utca 9
H-1037 Budapest, Hungary
Ph: 36 1 4371100
Fax: 36 1 4372303
Publisher E-mail: sanomabp@sanomabp.hu

Magazine featuring child parenting. **Freq:** Annual. **Print Method:** Offset. **Trim Size:** 213 x 280 mm. **Key Personnel:** Csato Edina, Coord., phone 36 1 4373906, fax 36 1 4371180, e.csato@sanomabp.hu. **Subscription Rates:** Free. **Remarks:** Accepts advertising. **URL:** http://www.sanoma.hu/sajtohirdetes/index.php?action=termek&site=23. **Circ:** Free 130,000

1046 ■ Kismama 9 Honap
Sanoma Budapest Kiadoi Rt.
Montevideo utca 9
H-1037 Budapest, Hungary
Ph: 36 1 4371100
Fax: 36 1 4372303
Publisher E-mail: sanomabp@sanomabp.hu

Magazine featuring series of topics like pregnancy, birth, and taking care of the babies. **Freq:** Annual. **Trim Size:** 200 x 265 mm. **Key Personnel:** Csato Edina, Coord., phone 36 1 4373906, fax 36 1 4371180, e.csato@sanomabp.hu. **Subscription Rates:** 565 Ft individuals. **Remarks:** Accepts advertising. **URL:** http://www.sanoma.hu/sajtohirdetes/index.php?action=termek&site=9. **Circ:** Paid ‡19,534

1047 ■ Market!ng&Media
Sanoma Budapest Kiadoi Rt.
Montevideo utca 9
H-1037 Budapest, Hungary
Ph: 36 1 4371100
Fax: 36 1 4372303
Publisher E-mail: sanomabp@sanomabp.hu

Magazine covering marketing, media, and advertisements. **Freq:** Semimonthly. **Print Method:** Offset. **Trim Size:** 215 x 276 mm. **Key Personnel:** Bosanszki Piroska, Coord., phone 36 1 4373905, fax 36 1 4371180, p.bosanszki@sanomabp.hu. **Subscription Rates:** 1,796 Ft individuals. **Remarks:** Accepts advertising. **URL:** http://www.sanoma.hu/sajtohirdetes/index.php?action=termek&site=19. **Circ:** Combined ‡2,000

1048 ■ Maxima Special
Sanoma Budapest Kiadoi Rt.
Montevideo utca 9
H-1037 Budapest, Hungary
Ph: 36 1 4371100
Fax: 36 1 4372303
Publisher E-mail: sanomabp@sanomabp.hu

Magazine featuring topics that is currently in the focus of young women's interest. **Freq:** Semiannual. **Print Method:** Offset. **Trim Size:** 190 x 250 mm. **Key Personnel:** Takacs Zsoka, Coord., phone 36 1 4373927, fax 36 1 4371180, zs.takacs@sanomabp.hu. **Subscription Rates:** 235 Ft individuals. **Remarks:** Accepts advertising. **URL:** http://www.sanoma.hu/sajtohirdetes/index.php?action=termek&site=51. **Circ:** Paid ‡70,432

1049 ■ Meglepetes
Sanoma Budapest Kiadoi Rt.
Montevideo utca 9
H-1037 Budapest, Hungary
Ph: 36 1 4371100
Fax: 36 1 4372303
Publisher E-mail: sanomabp@sanomabp.hu

Magazine featuring women's interests. **Freq:** Weekly (Thurs.). **Print Method:** Offset. **Trim Size:** 205 x 280 mm. **Key Personnel:** Selmeczi Andrea, Coord., phone 36 1 4373639, fax 36 1 4371180, a.selmeczi@sanomabp.hu. **Subscription Rates:** 135 Ft individuals. **Remarks:** Accepts advertising. **URL:** http://www.sanoma.hu/sajtohirdetes/index.php?action=termek&site=18. **Circ:** Paid ‡97,267

1050 ■ Meglepetes Raadas
Sanoma Budapest Kiadoi Rt.
Montevideo utca 9
H-1037 Budapest, Hungary
Ph: 36 1 4371100
Fax: 36 1 4372303
Publisher E-mail: sanomabp@sanomabp.hu

Magazine covering gastronomy. **Freq:** Semiannual. **Print Method:** Offset. **Trim Size:** 205 x 280 mm. **Key Personnel:** Selmeczi Andrea, Coord., phone 36 1 4373639, fax 36 1 4371180, a.selmeczi@sanomabp.hu. **Subscription Rates:** 194 Ft individuals. **Remarks:** Accepts advertising. **URL:** http://www.sanoma.hu/sajtohirdetes/index.php?action=termek&site=11. **Circ:** ‡109,750

1051 ■ National Geographic Kids
Sanoma Budapest Kiadoi Rt.
Montevideo utca 9
H-1037 Budapest, Hungary
Ph: 36 1 4371100
Fax: 36 1 4372303
Publisher E-mail: sanomabp@sanomabp.hu

Magazine featuring interesting things, inventions, and amazing stories from the world of animals. **Freq:** 10/yr. **Print Method:** Offset. **Key Personnel:** Selmeczi Andrea, Coord., phone 36 1 4373639, fax 36 1 4371180, a.selmeczi@sanomabp.hu. **Subscription Rates:** 355 Ft individuals. **Remarks:** Accepts advertising. **URL:** http://www.sanoma.hu/sajtohirdetes/index.php?action=termek&site=8. **Circ:** ‡35,000

1052 ■ National Geographic Special
Sanoma Budapest Kiadoi Rt.
Montevideo utca 9
H-1037 Budapest, Hungary
Ph: 36 1 4371100
Fax: 36 1 4372303
Publisher E-mail: sanomabp@sanomabp.hu

Magazine featuring interesting stories and photographs. **Freq:** 4/yr. **Print Method:** Offset. **Trim Size:** 229 x 276 mm. **Key Personnel:** Selmeczi Andrea, Coord., phone 36 1 4373639, fax 36 1 4371180, a.selmeczi@sanomabp.hu. **Remarks:** Accepts advertising. **URL:** http://www.sanoma.hu/sajtohirdetes/index.php?action=termek&site=46. **Circ:** ‡32,740

1053 ■ Nok Lapja
Sanoma Budapest Kiadoi Rt.
Montevideo utca 9
H-1037 Budapest, Hungary
Ph: 36 1 4371100
Fax: 36 1 4372303
Publisher E-mail: sanomabp@sanomabp.hu

Magazine featuring women's interests. **Freq:** Weekly (Wed.). **Print Method:** Offset. **Trim Size:** 220 x 300 mm. **Key Personnel:** Mitrovics Aniko, Coord., phone 36 1 4373637, fax 36 1 4371180, a.mitrovics@sanomabp.hu. **Subscription Rates:** 175 Ft individuals. **Remarks:** Accepts advertising. **URL:** http://www.sanoma.hu/sajtohirdetes/index.php?action=termek&site=3. **Circ:** Paid ‡283,425

1054 ■ Nok Lapja Egeszseg
Sanoma Budapest Kiadoi Rt.
Montevideo utca 9
H-1037 Budapest, Hungary
Ph: 36 1 4371100
Fax: 36 1 4372303
Publisher E-mail: sanomabp@sanomabp.hu

Magazine featuring health-related topics. **Founded:** Apr. 2007. **Freq:** Monthly. **Trim Size:** 205 x 275 mm. **Key Personnel:** Kralik Judit, Coord., phone 36 1 4373632, fax 36 1 4371180, j.kralik@sanomabp.hu. **Remarks:** Accepts advertising. **URL:** http://www.sanoma.hu/sajtohirdetes/index.php?action=termek&site=58. **Circ:** ‡65,000

1055 ■ Nok Lapja Eskuvo
Sanoma Budapest Kiadoi Rt.
Montevideo utca 9
H-1037 Budapest, Hungary
Ph: 36 1 4371100
Fax: 36 1 4372303
Publisher E-mail: sanomabp@sanomabp.hu

Magazine featuring wedding. **Freq:** Annual. **Print Method:** Offset. **Trim Size:** 215 x 280 mm. **Key Personnel:** Kantor Anita, Coord., phone 36 1 4373912, fax 36 1 4371180, a.kantor@sanomabp.hu. **Subscription Rates:** 850 Ft individuals. **Remarks:** Accepts advertising. **URL:** http://www.sanoma.hu/sajtohirdetes/index.php?action=termek&site=5. **Circ:** ‡28,000

1056 ■ Nok Lapja Evszakok
Sanoma Budapest Kiadoi Rt.
Montevideo utca 9
H-1037 Budapest, Hungary
Ph: 36 1 4371100
Fax: 36 1 4372303
Publisher E-mail: sanomabp@sanomabp.hu

Magazine featuring women's interests. **Freq:** 10/yr. **Print Method:** Offset. **Trim Size:** 210 x 270 mm. **Key Personnel:** Rakosi Gabriella, Coord., phone 36 1 4373635, fax 36 1 4371180, g.rakosi@sanomabp.hu. **Subscription Rates:** 665 Ft individuals. **Remarks:** Accepts advertising. **URL:** http://www.sanoma.hu/sajtohirdetes/index.php?action=termek&site=37. **Circ:** Paid ‡21,996

1057 ■ Nok Lapja Konyha
Sanoma Budapest Kiadoi Rt.
Montevideo utca 9
H-1037 Budapest, Hungary
Ph: 36 1 4371100
Fax: 36 1 4372303
Publisher E-mail: sanomabp@sanomabp.hu

Magazine covering gastronomy. **Freq:** Quarterly. **Trim Size:** 220 x 277 mm. **Key Personnel:** Kralik Judit, Coord., phone 36 1 4373632, fax 36 1 4371180, j.kralik@sanomabp.hu. **Subscription Rates:** 284 Ft individuals. **Remarks:** Accepts advertising. **URL:** http://www.sanoma.hu/sajtohirdetes/index.php?action=termek&site=10. **Circ:** Paid ‡42,887

1058 ■ Otlet Mozaik
Sanoma Budapest Kiadoi Rt.
Montevideo utca 9
H-1037 Budapest, Hungary
Ph: 36 1 4371100
Fax: 36 1 4372303
Publisher E-mail: sanomabp@sanomabp.hu

Magazine featuring information on construction, interior decoration, and garden settlement. **Freq:** Monthly. **Print Method:** Offset. **Trim Size:** 205 x 295 mm. **Key Personnel:** Krisztina Wilhelm, Coord., phone 36 1 4371612, fax 36 1 4371180, k.wilhelm@sanomabp.hu. **Remarks:** Accepts advertising. **URL:** http://www.sanoma.hu/main.php?temp=E_online_page.pge&id=88. **Circ:** Paid ‡18,217

1059 ■ Otthon
Sanoma Budapest Kiadoi Rt.
Montevideo utca 9
H-1037 Budapest, Hungary
Ph: 36 1 4371100
Fax: 36 1 4372303
Publisher E-mail: sanomabp@sanomabp.hu

Magazine covering lifestyle and interior design. **Freq:** Monthly. **Print Method:** Offset. **Trim Size:** 215 x 300 mm. **Key Personnel:** Rakosi Gabriella, Coord., phone 36 1 4373635, fax 36 1 4371180, g.rakosi@sanomabp.hu. **Subscription Rates:** 470 Ft individuals. **Remarks:** Accepts advertising. **URL:** http://www.sanoma.hu/sajtohirdetes/index.php?action=termek&site=38. **Circ:** Paid ‡39,310

1060 ■ Praktika
Sanoma Budapest Kiadoi Rt.
Montevideo utca 9
H-1037 Budapest, Hungary
Ph: 36 1 4371100
Fax: 36 1 4372303
Publisher E-mail: sanomabp@sanomabp.hu

Magazine featuring women's interests. **Freq:** Monthly. **Print Method:** Offset. **Trim Size:** 200 x 265 mm. **Key Personnel:** Selmeczi Andrea, Coord., phone 36 1 4373639, fax 36 1 4371180, a.selmeczi@sanomabp.hu. **Subscription Rates:** 415 Ft individuals. **Remarks:** Accepts advertising. **URL:** http://www.sanoma.hu/sajtohirdetes/index.php?action=termek&site=39. **Circ:** Paid ‡54,673

1061 ■ RTV Musormagazin
Sanoma Budapest Kiadoi Rt.
Montevideo utca 9
H-1037 Budapest, Hungary
Ph: 36 1 4371100
Fax: 36 1 4372303
Publisher E-mail: sanomabp@sanomabp.hu

Magazine featuring radio and television schedule. **Freq:** Weekly (Fri.). **Print Method:** Cold-set. **Trim Size:** 205 x 270 mm. **Key Personnel:** Rakosi Gabriella, Coord., phone 36 1 4373635, fax 36 1 4371180, g.rakosi@sanomabp.hu. **Subscription Rates:** 69 Ft individuals. **Remarks:** Accepts advertising. **URL:** http://www.sanoma.hu/sajtohirdetes/index.php?action=termek&site=40. **Circ:** Paid ‡74,540

1062 ■ Story Special
Sanoma Budapest Kiadoi Rt.
Montevideo utca 9
H-1037 Budapest, Hungary
Ph: 36 1 4371100
Fax: 36 1 4372303
Publisher E-mail: sanomabp@sanomabp.hu

Magazine featuring events in the life of the popular personalities in Hungary. **Freq:** Semiannual. **Print Method:** Offset. **Trim Size:** 210 x 285 mm. **Key Personnel:** Csato Edina, Coord., phone 36 1 4373906, fax 36 1 4371180, e.csato@sanomabp.hu. **Subscription Rates:** 359 Ft individuals. **Remarks:** Accepts advertising. **URL:** http://www.sanoma.hu/sajtohirdetes/index.php?action=termek&site=43. **Circ:** ‡162,365

1063 ■ Szines RTV
Sanoma Budapest Kiadoi Rt.
Montevideo utca 9
H-1037 Budapest, Hungary
Ph: 36 1 4371100
Fax: 36 1 4372303
Publisher E-mail: sanomabp@sanomabp.hu

Magazine featuring information about programs in television. **Freq:** Weekly (Sat.). **Print Method:** Offset. **Trim Size:** 200 x 285 mm. **Key Personnel:** Gabriella Rakosi, Coord., phone 36 1 4373635, fax 36 1 4371180, g.rakosi@sanomabp.hu. **Subscription Rates:** 149 Ft individuals. **Remarks:** Accepts advertising. **URL:** http://www.sanoma.hu/sajtohirdetes/index.php?action=termek&site=1. **Circ:** Paid ‡240,975

1064 ■ UZLET & SIKER
Sanoma Budapest Kiadoi Rt.
Montevideo utca 9
H-1037 Budapest, Hungary
Ph: 36 1 4371100
Fax: 36 1 4372303
Publisher E-mail: sanomabp@sanomabp.hu

Magazine featuring practical information, analysis, and advice for smalland middle-sized enterprises. **Freq:** 10/yr. **Print Method:** Offset. **Trim Size:** 200 x 267 mm. **Key Personnel:** Torok Marta, Coord., phone 36 1 4371218, fax 36 1 4371180, p.bosanszki@sanomabp.hu. **Subscription Rates:** 350 Ft individuals. **Remarks:** Accepts advertising. **URL:** http://www.sanoma.hu/sajtohirdetes/index.php?action=termek&site=2. **Circ:** ‡20,346

Circulation: ★ = ABC; △ = BPA; ♦ = CAC; • = CCAB; ❑ = VAC; ⊕ = PO Statement; ‡ = Publisher's Report; Boldface figures = sworn; Light figures = estimated.

INDIA

HYDERABAD

1065 ■ Icfai Journal of Accounting Research
ICFAI University Press
Plot. 6-3-354/1, Stellar Sphinx
Rd. 1, Banjara Hills
Panjagutta
Hyderabad 500 082, Andhra Pradesh, India
Ph: 91 40 23430449
Fax: 91 40 23430447
Publisher E-mail: info@iupindia.org

Journal featuring research reports in the field of accounting. **Freq:** Quarterly. **Key Personnel:** E.N. Murthy, Editor; G.R.K. Murty, Managing Editor. **Subscription Rates:** Rs 625 individuals. **URL:** http://www.iupindia.org/IJAR.asp.

1066 ■ The Icfai Journal of Alternative Dispute Resolution
ICFAI University Press
Plot. 6-3-354/1, Stellar Sphinx
Rd. 1, Banjara Hills
Panjagutta
Hyderabad 500 082, Andhra Pradesh, India
Ph: 91 40 23430449
Fax: 91 40 23430447
Publisher E-mail: info@iupindia.org

Journal focusing on arbitration, conciliation, mediation, negotiation, and settlement of disputes. **Freq:** Quarterly. **Key Personnel:** E.N. Murthy, Editor; G.R.K Murty, Managing Editor. **Subscription Rates:** Rs 625 individuals. **URL:** http://www.iupindia.org/ijadr.asp.

1067 ■ The Icfai Journal of Audit Practice
ICFAI University Press
Plot. 6-3-354/1, Stellar Sphinx
Rd. 1, Banjara Hills
Panjagutta
Hyderabad 500 082, Andhra Pradesh, India
Ph: 91 40 23430449
Fax: 91 40 23430447
Publisher E-mail: info@iupindia.org

Journal featuring information about auditing for accountants, finance professionals, and auditors. **Freq:** Quarterly. **Key Personnel:** E.N. Murthy, Editor; G.R.K. Murty, Managing Editor. **Subscription Rates:** Rs 625 individuals. **URL:** http://www.iupindia.org/ijap.asp.

1068 ■ The Icfai Journal of Bank Management
ICFAI University Press
Plot. 6-3-354/1, Stellar Sphinx
Rd. 1, Banjara Hills
Panjagutta
Hyderabad 500 082, Andhra Pradesh, India
Ph: 91 40 23430449
Fax: 91 40 23430447
Publisher E-mail: info@iupindia.org

Journal focusing on the areas of risk management, forex markets, retail banking, HRD & leadership, banking, supervision, convergence of financial services, and e-banking. **Freq:** Quarterly. **Subscription Rates:** Rs 625 individuals. **URL:** http://www.iupindia.org/ijbm.asp.

1069 ■ The Icfai Journal of Behavioral Finance
ICFAI University Press
Plot. 6-3-354/1, Stellar Sphinx
Rd. 1, Banjara Hills
Panjagutta
Hyderabad 500 082, Andhra Pradesh, India
Ph: 91 40 23430449
Fax: 91 40 23430447
Publisher E-mail: info@iupindia.org

Journal focusing on behavioral economics, behavior of markets, behavioral aspects influencing investment decisions of managers and behavioral aspects in corporate finance decision. **Freq:** Quarterly. **Subscription Rates:** Rs 625 individuals. **URL:** http://www.iupindia.org/ijbf.asp.

1070 ■ The Icfai Journal of Corporate and Securities Law
ICFAI University Press
Plot. 6-3-354/1, Stellar Sphinx
Rd. 1, Banjara Hills
Panjagutta
Hyderabad 500 082, Andhra Pradesh, India
Ph: 91 40 23430449
Fax: 91 40 23430447
Publisher E-mail: info@iupindia.org

Journal focusing on capital markets, mutual funds, corporate governance trading and regulatory authorities. **Freq:** Quarterly. **Subscription Rates:** Rs 625 individuals. **URL:** http://www.iupindia.org/ijcsl.asp.

1071 ■ The Icfai Journal of Cyber Law
ICFAI University Press
Plot. 6-3-354/1, Stellar Sphinx
Rd. 1, Banjara Hills
Panjagutta
Hyderabad 500 082, Andhra Pradesh, India
Ph: 91 40 23430449
Fax: 91 40 23430447
Publisher E-mail: info@iupindia.org

Journal focusing on internet, information technology, and international cyber law. **Freq:** Quarterly. **Subscription Rates:** Rs 625 individuals. **URL:** http://www.iupindia.org/ijcl.asp.

1072 ■ The Icfai Journal of Derivatives Markets
ICFAI University Press
Plot. 6-3-354/1, Stellar Sphinx
Rd. 1, Banjara Hills
Panjagutta
Hyderabad 500 082, Andhra Pradesh, India
Ph: 91 40 23430449
Fax: 91 40 23430447
Publisher E-mail: info@iupindia.org

Journal featuring articles dealing with derivative valuation and risk management. **Freq:** Quarterly. **Subscription Rates:** Rs 625 individuals. **URL:** http://www.iupindia.org/ijdm.asp.

1073 ■ The Icfai Journal of Employment Law
ICFAI University Press
Plot. 6-3-354/1, Stellar Sphinx
Rd. 1, Banjara Hills
Panjagutta
Hyderabad 500 082, Andhra Pradesh, India
Ph: 91 40 23430449
Fax: 91 40 23430447
Publisher E-mail: info@iupindia.org

Journal focusing on problems in employment, employer-employee relations, social security, labor welfare legislation, health, safety and welfare of workers. **Freq:** Quarterly. **Subscription Rates:** Rs 625 individuals. **URL:** http://www.iupindia.org/ijeml.asp.

1074 ■ The Icfai Journal of Entrepreneurship Development
ICFAI University Press
Plot. 6-3-354/1, Stellar Sphinx
Rd. 1, Banjara Hills
Panjagutta
Hyderabad 500 082, Andhra Pradesh, India
Ph: 91 40 23430449
Fax: 91 40 23430447
Publisher E-mail: info@iupindia.org

Journal focusing on entrepreneurship mindset, entrepreneurship opportunity, motivation, and case studies. **Freq:** Quarterly. **Subscription Rates:** Rs 625 individuals. **URL:** http://www.iupindia.org/ijed.asp.

1075 ■ The Icfai Journal of Environmental Economics
ICFAI University Press
Plot. 6-3-354/1, Stellar Sphinx
Rd. 1, Banjara Hills
Panjagutta
Hyderabad 500 082, Andhra Pradesh, India
Ph: 91 40 23430449
Fax: 91 40 23430447
Publisher E-mail: info@iupindia.org

Journal featuring issues pertaining to valuation of ecosystem benefits from environmental improvements, environmental planning, role of government and non-governmental organizations and also the various innovations related to environment. **Freq:** Quarterly. **Key Personnel:** E.N. Murthy, Editor; G.R.K. Murty, Managing Editor. **Subscription Rates:** Rs 625 individuals. **URL:** http://www.iupindia.org/ijee.asp.

1076 ■ The Icfai Journal of Financial Economics
ICFAI University Press
Plot. 6-3-354/1, Stellar Sphinx
Rd. 1, Banjara Hills
Panjagutta
Hyderabad 500 082, Andhra Pradesh, India
Ph: 91 40 23430449
Fax: 91 40 23430447
Publisher E-mail: info@iupindia.org

Journal focusing on the issue of economics, finance, statistics, and econometrics. **Freq:** Quarterly. **Subscription Rates:** Rs 625 individuals. **URL:** http://www.iupindia.org/ijfe.asp.

1077 ■ The Icfai Journal of Financial Risk Management
ICFAI University Press
Plot. 6-3-354/1, Stellar Sphinx
Rd. 1, Banjara Hills
Panjagutta
Hyderabad 500 082, Andhra Pradesh, India
Ph: 91 40 23430449
Fax: 91 40 23430447
Publisher E-mail: info@iupindia.org

Journal focusing on identifying financial risk, risk management models, accounting for derivatives, risk-hedging techniques, and asset liability management. **Freq:** Quarterly. **Subscription Rates:** Rs 625 individuals. **URL:** http://www.iupindia.org/ijfrm.asp.

1078 ■ The Icfai Journal of Governance and Public Policy
ICFAI University Press
Plot. 6-3-354/1, Stellar Sphinx
Rd. 1, Banjara Hills
Panjagutta
Hyderabad 500 082, Andhra Pradesh, India
Ph: 91 40 23430449
Fax: 91 40 23430447
Publisher E-mail: info@iupindia.org

Journal focusing on critical evaluation of public policies and governance at the national, state, and local levels. **Freq:** Quarterly. **Key Personnel:** E.N. Murthy, Editor; G.R.K. Murty, Managing Editor. **Subscription Rates:** Rs 625 individuals. **URL:** http://www.iupindia.org/ijgpp.asp.

1079 ■ The Icfai Journal of Healthcare Law
ICFAI University Press
Plot. 6-3-354/1, Stellar Sphinx
Rd. 1, Banjara Hills
Panjagutta
Hyderabad 500 082, Andhra Pradesh, India
Ph: 91 40 23430449
Fax: 91 40 23430447
Publisher E-mail: info@iupindia.org

Journal focusing on legal principles relating to medico-legal, socio-legal, medical-malpractice, tort-claims, malpractice-insurance, drug-safety and efficacy issues, and ethical values with health science. **Freq:** Quarterly. **Subscription Rates:** Rs 625 individuals. **URL:** http://www.iupindia.org/ijhl.asp.

1080 ■ The Icfai Journal of History and Culture
ICFAI University Press
Plot. 6-3-354/1, Stellar Sphinx
Rd. 1, Banjara Hills
Panjagutta
Hyderabad 500 082, Andhra Pradesh, India
Ph: 91 40 23430449
Fax: 91 40 23430447
Publisher E-mail: info@iupindia.org

Journal focusing on the historical developments from the ancient past to the contemporary times. **Freq:** Quarterly. **Key Personnel:** E.N. Murthy, Editor; G.R.K. Murty, Managing Editor. **Subscription Rates:** Rs 625 individuals. **URL:** http://www.iupindia.org/ijhc.asp.

1081 ■ The Icfai Journal of International Business Law
ICFAI University Press
Plot. 6-3-354/1, Stellar Sphinx
Rd. 1, Banjara Hills
Panjagutta
Hyderabad 500 082, Andhra Pradesh, India
Ph: 91 40 23430449
Fax: 91 40 23430447
Publisher E-mail: info@iupindia.org

Journal focusing on international trade agreements and financial law. **Freq:** Quarterly. **Subscription Rates:** Rs 625 individuals. **URL:** http://www.iupindia.org/ijibl.asp.

1082 ■ The Icfai Journal of Knowledge Management
ICFAI University Press
Plot. 6-3-354/1, Stellar Sphinx
Rd. 1, Banjara Hills
Panjagutta
Hyderabad 500 082, Andhra Pradesh, India
Ph: 91 40 23430449
Fax: 91 40 23430447
Publisher E-mail: info@iupindia.org

Journal focusing on product knowledge, services knowledge, process knowledge, customer knowledge, and knowledge assets. **Freq:** Bimonthly. **Subscription Rates:** Rs 625 individuals. **URL:** http://www.iupindia.org/ijkm.asp.

1083 ■ The Icfai Journal of Life Sciences
ICFAI University Press
Plot. 6-3-354/1, Stellar Sphinx
Rd. 1, Banjara Hills
Panjagutta
Hyderabad 500 082, Andhra Pradesh, India
Ph: 91 40 23430449
Fax: 91 40 23430447
Publisher E-mail: info@iupindia.org

Journal focusing on botany, zoology, evolutionary biology, microbiology, and biochemistry. **Freq:** Quarterly. **Subscription Rates:** Rs 625 individuals. **URL:** http://www.iupindia.org/ijls.asp.

1084 ■ The Icfai Journal of Managerial Economics
ICFAI University Press
Plot. 6-3-354/1, Stellar Sphinx
Rd. 1, Banjara Hills
Panjagutta
Hyderabad 500 082, Andhra Pradesh, India
Ph: 91 40 23430449
Fax: 91 40 23430447
Publisher E-mail: info@iupindia.org

Journal focusing on managerial decision-making from all functional areas of economics. **Freq:** Quarterly. **Subscription Rates:** Rs 625 individuals. **URL:** http://www.iupindia.org/ijme.asp.

1085 ■ The Icfai Journal of Mergers & Acquisitions
ICFAI University Press
Plot. 6-3-354/1, Stellar Sphinx
Rd. 1, Banjara Hills
Panjagutta
Hyderabad 500 082, Andhra Pradesh, India
Ph: 91 40 23430449
Fax: 91 40 23430447
Publisher E-mail: info@iupindia.org

Journal focusing on mergers and acquisitions, pre-merger issues, post-merger issues, cross-border mergers, and regulatory aspects. **Freq:** Quarterly. **Subscription Rates:** Rs 625 individuals. **URL:** http://www.iupindia.org/ijma.asp.

1086 ■ The Icfai Journal of Monetary Economics
ICFAI University Press
Plot. 6-3-354/1, Stellar Sphinx
Rd. 1, Banjara Hills
Panjagutta
Hyderabad 500 082, Andhra Pradesh, India
Ph: 91 40 23430449
Fax: 91 40 23430447
Publisher E-mail: info@iupindia.org

Journal focusing on monetary policies and issues. **Freq:** Quarterly. **Subscription Rates:** Rs 625 individuals. **URL:** http://www.iupindia.org/ijmoe.asp.

1087 ■ The Icfai Journal of Operations Management
ICFAI University Press
Plot. 6-3-354/1, Stellar Sphinx
Rd. 1, Banjara Hills
Panjagutta
Hyderabad 500 082, Andhra Pradesh, India
Ph: 91 40 23430449
Fax: 91 40 23430447
Publisher E-mail: info@iupindia.org

Journal focusing on inventory control, supply chain management, enterprise resource planning, total quality management, business process re-engineering, logistics management, and flexible manufacturing systems. **Freq:** Quarterly. **Subscription Rates:** Rs 625 individuals. **URL:** http://www.iupindia.org/ijom.asp.

1088 ■ The Icfai Journal of Organizational Behavior
ICFAI University Press
Plot. 6-3-354/1, Stellar Sphinx
Rd. 1, Banjara Hills
Panjagutta
Hyderabad 500 082, Andhra Pradesh, India
Ph: 91 40 23430449
Fax: 91 40 23430447
Publisher E-mail: info@iupindia.org

Journal focusing on organization design, job, performance, motivation and satisfaction, work-life balance, group dynamics, and leadership. **Freq:** Quarterly. **Key Personnel:** E.N. Murthy, Editor; G.R.K. Murty, Managing Editor. **Subscription Rates:** Rs 625 individuals. **URL:** http://www.iupindia.org/ijob.asp.

1089 ■ The Icfai Journal of Public Finance
ICFAI University Press
Plot. 6-3-354/1, Stellar Sphinx
Rd. 1, Banjara Hills
Panjagutta
Hyderabad 500 082, Andhra Pradesh, India
Ph: 91 40 23430449
Fax: 91 40 23430447
Publisher E-mail: info@iupindia.org

Journal focusing on fiscal policy, economic stabilization, and tax reforms. **Freq:** Quarterly. **Key Personnel:** E.N. Murthy, Editor; G.R.K. Murty, Managing Editor. **Subscription Rates:** Rs 625 individuals. **URL:** http://www.iupindia.org/ijpf.asp.

1090 ■ The Icfai Journal of Risk & Insurance
ICFAI University Press
Plot. 6-3-354/1, Stellar Sphinx
Rd. 1, Banjara Hills
Panjagutta
Hyderabad 500 082, Andhra Pradesh, India
Ph: 91 40 23430449
Fax: 91 40 23430447
Publisher E-mail: info@iupindia.org

Journal focusing on the advancement of knowledge in risk management approaches, tools, practices, and hedging techniques. **Freq:** Quarterly. **Subscription Rates:** Rs 625 individuals. **URL:** http://www.iupindia.org/IJRI.asp.

1091 ■ The Icfai Journal of Science & Technology
ICFAI University Press
Plot. 6-3-354/1, Stellar Sphinx
Rd. 1, Banjara Hills
Panjagutta
Hyderabad 500 082, Andhra Pradesh, India
Ph: 91 40 23430449
Fax: 91 40 23430447
Publisher E-mail: info@iupindia.org

Journal focusing on science, engineering, technology, education and their applications. **Freq:** Semiannual. **Subscription Rates:** Rs 625 individuals. **URL:** http://www.iupindia.org/ijst.asp.

1092 ■ The Icfai Journal of Services Marketing
ICFAI University Press
Plot. 6-3-354/1, Stellar Sphinx
Rd. 1, Banjara Hills
Panjagutta
Hyderabad 500 082, Andhra Pradesh, India
Ph: 91 40 23430449
Fax: 91 40 23430447
Publisher E-mail: info@iupindia.org

Journal featuring latest research, concepts, and managerial implications in the service sector covering financial services, educational services, travel and hospitality services, consultancy services, and many others. **Freq:** Quarterly. **Subscription Rates:** Rs 625 individuals. **URL:** http://www.iupindia.org/ijsem.asp.

1093 ■ The Icfai Journal of Soft Skills
ICFAI University Press
Plot. 6-3-354/1, Stellar Sphinx
Rd. 1, Banjara Hills
Panjagutta
Hyderabad 500 082, Andhra Pradesh, India
Ph: 91 40 23430449
Fax: 91 40 23430447
Publisher E-mail: info@iupindia.org

Journal featuring issues pertaining to soft skills and communication. **Freq:** Quarterly. **Key Personnel:** E.N. Murthy, Editor; G.R.K. Murty, Managing Editor. **Subscription Rates:** Rs 625 individuals. **URL:** http://www.iupindia.org/ijss.asp.

1094 ■ The Icfai Journal of Urban Policy
ICFAI University Press
Plot. 6-3-354/1, Stellar Sphinx
Rd. 1, Banjara Hills
Panjagutta
Hyderabad 500 082, Andhra Pradesh, India
Ph: 91 40 23430449
Fax: 91 40 23430447
Publisher E-mail: info@iupindia.org

Journal featuring the coherent and intelligent debate on managing urban growth and efficient development. **Freq:** Quarterly. **Subscription Rates:** Rs 625 individuals. **URL:** http://www.iupindia.org/ijup.asp.

Circulation: ★ = ABC; △ = BPA; ♦ = CAC; • = CCAB; □ = VAC; ⊕ = PO Statement; ‡ = Publisher's Report; Boldface figures = sworn; Light figures = estimated.

1095 ■ ICFAI Reader
ICFAI University Press
Plot. 6-3-354/1, Stellar Sphinx
Rd. 1, Banjara Hills
Panjagutta
Hyderabad 500 082, Andhra Pradesh, India
Ph: 91 40 23430449
Fax: 91 40 23430447
Publisher E-mail: info@iupindia.org

Magazine featuring issues in finance. **Freq:** Monthly. **Subscription Rates:** Rs 625 individuals. **URL:** http://www.iupindia.org/icfaireader.asp.

1096 ■ Insurance Chronicle
ICFAI University Press
Plot. 6-3-354/1, Stellar Sphinx
Rd. 1, Banjara Hills
Panjagutta
Hyderabad 500 082, Andhra Pradesh, India
Ph: 91 40 23430449
Fax: 91 40 23430447
Publisher E-mail: info@iupindia.org

Magazine featuring insurance business environment. **Freq:** Monthly. **Subscription Rates:** Rs 625 individuals. **URL:** http://www.iupindia.org/magazines.asp.

1097 ■ Porfolio Organizer
ICFAI University Press
Plot. 6-3-354/1, Stellar Sphinx
Rd. 1, Banjara Hills
Panjagutta
Hyderabad 500 082, Andhra Pradesh, India
Ph: 91 40 23430449
Fax: 91 40 23430447
Publisher E-mail: info@iupindia.org

Magazine featuring investment and portfolio management. **Freq:** Monthly. **Subscription Rates:** Rs 625 individuals. **URL:** http://www.iupindia.org/magazines.asp.

1098 ■ Professional Banker
ICFAI University Press
Plot. 6-3-354/1, Stellar Sphinx
Rd. 1, Banjara Hills
Panjagutta
Hyderabad 500 082, Andhra Pradesh, India
Ph: 91 40 23430449
Fax: 91 40 23430447
Publisher E-mail: info@iupindia.org

Magazine featuring banking statistics and banking news. **Freq:** Monthly. **Subscription Rates:** Rs 940 individuals. **URL:** http://www.iupindia.org/magazines.asp.

1099 ■ Projects & Profits
ICFAI University Press
Plot. 6-3-354/1, Stellar Sphinx
Rd. 1, Banjara Hills
Panjagutta
Hyderabad 500 082, Andhra Pradesh, India
Ph: 91 40 23430449
Fax: 91 40 23430447
Publisher E-mail: info@iupindia.org

Magazine featuring cutting edge knowledge pertaining to project management. **Freq:** Monthly. **Subscription Rates:** Rs 625 individuals. **URL:** http://www.iupindia.org/magazines.asp.

JODHPUR

1100 ■ Advances in Horticulture and Forestry
Indian Periodical
5 New Pali Rd.
PO Box 33
Jodhpur 342001, Rajasthan, India
Ph: 91 291 2433323
Fax: 91 291 2512580
Publisher E-mail: info@indianperiodical.in

Journal covering horticulture and forestry. **Key Personnel:** R.K. Khetarpal, Editor-in-Chief; Arun Lal, Exec. Ed.; H.N. Gour, Exec. Ed. **ISSN:** 0971-0507. **URL:** http://www.indianperiodical.in/ip/journalview.aspx?jrnl=ahf&jrnltitle=Advances in Horticulture and Forestry.

1101 ■ Advances in Plant Physiology
Indian Periodical
5 New Pali Rd.
PO Box 33
Jodhpur 342001, Rajasthan, India
Ph: 91 291 2433323
Fax: 91 291 2512580
Publisher E-mail: info@indianperiodical.in

Journal covering the advances in plant physiology. **Key Personnel:** A. Hemantaranjan, Editor-in-Chief; Dr. Anjali Bharti, Exec. Ed. **ISSN:** 0972-9917. **URL:** http://www.indianperiodical.in/ip/journalview.aspx?jrnl=app&jrnltitle=Advan ces in Plant Physiology.

1102 ■ Annual Review of Plant Pathology
Indian Periodical
5 New Pali Rd.
PO Box 33
Jodhpur 342001, Rajasthan, India
Ph: 91 291 2433323
Fax: 91 291 2512580
Publisher E-mail: info@indianperiodical.in

Journal covering plant pathology. **Founded:** Oct. 2002. **Freq:** Annual. **Key Personnel:** R.K. Khetarpal, Editor-in-Chief; H.N. Gour, Exec. Ed.; S.D. Purohit, Exec. Ed. **ISSN:** 0972-9712. **URL:** http://www.indianperiodical.in/ip/journalview.aspx?jrnl=arpp&jrnltitle=Annu al Review of Plant Pathology.

1103 ■ Indian Journal of Applied Entomology
Indian Periodical
5 New Pali Rd.
PO Box 33
Jodhpur 342001, Rajasthan, India
Ph: 91 291 2433323
Fax: 91 291 2512580
Publisher E-mail: info@indianperiodical.in

Journal covering applied entomology. **Key Personnel:** Dr. R.C. Saxena, Editor-in-Chief; Dr. R. Swaminathan, Exec. Ed.; Dr. N.K. Bajpai, Exec. Ed. **ISSN:** 0970-9509. **URL:** http://www.indianperiodical.in/ip/journalview.aspx?jrnl=ijae&jrnltitle=Indi an Journal of Applied Entomology.

1104 ■ Journal of Arid Legumes
Indian Periodical
5 New Pali Rd.
PO Box 33
Jodhpur 342001, Rajasthan, India
Ph: 91 291 2433323
Fax: 91 291 2512580
Publisher E-mail: info@indianperiodical.in

Journal covering legume crops. **Key Personnel:** A. Henry, Editor-in-Chief; D. Kumar, Exec. Ed.; J.V. Singh Hisar, Exec. Ed. **ISSN:** 0973-0907. **URL:** http://www.indianperiodical.in/ip/journalview.aspx?jrnl=jal&jrnltitle=Journ al of Arid Legumes.

1105 ■ Journal of Economic and Taxonomic Botany
Indian Periodical
5 New Pali Rd.
PO Box 33
Jodhpur 342001, Rajasthan, India
Ph: 91 291 2433323
Fax: 91 291 2512580
Publisher E-mail: info@indianperiodical.in

Journal covering economic and taxonomic botany. **Subtitle:** Additional Series. **Key Personnel:** M.K. Jaipuriar, Editor-in-Chief; V. Singh, Exec. Ed.; Pawan Kumar, Managing Editor. **ISSN:** 2050-9768. **URL:** http://www.indianperiodical.in/ip/journalview.aspx?jrnl=jetbad&jrnltitle.

1106 ■ Journal of Phytopharmacotherapy and Natural Products
Indian Periodical
5 New Pali Rd.
PO Box 33
Jodhpur 342001, Rajasthan, India
Ph: 91 291 2433323
Fax: 91 291 2512580
Publisher E-mail: info@indianperiodical.in

Journal covering medicinal plants and natural products. **Freq:** Quarterly. **Key Personnel:** Amritpal Singh Saroya, Editor-in-Chief; V.S. Mathur, Exec. Ed.; Narendra Singh, Exec. Ed. **ISSN:** 0973-5941. **URL:** http://www.indianperiodical.in/ip/journalview.aspx?jrnl=jpnp&jrnltitle.

KANPUR

1107 ■ International Journal of Tomography and Statistics
Indian Society for Development and Environment Research
c/o R.K.S. Rathore, Ed.-in-Ch.
Dept. of Mathematics
Indian Institute of Technology
Kanpur 208016, Uttar Pradesh, India
Publisher E-mail: ceser_isder@yahoo.co.in

Journal covering the latest research and developments in computerized tomography and statistics. **Freq:** 3/yr. **Key Personnel:** R.K.S Rathore, Editor-in-Chief, ijts@isder.ceser.res.in; Tanuja Srivastava, Exec. Ed., tanujfma@yahoo.com; S.C. Sexena, Editor. **ISSN:** 0972-9976. **Subscription Rates:** EUR550 individuals print; EUR300 individuals online; EUR750 individuals print and online; EUR700 institutions print; EUR500 institutions online; EUR1,000 institutions print and online. **URL:** http://www.isder.ceser.res.in/ijts.html.

NEW DELHI

1108 ■ The Indian Journal of Crop Science
Satish Serial Publishing House
115, Express Tower
Commercial Complex
Azadpur
New Delhi 110033, Delhi, India
Ph: 91 11 27672852
Fax: 91 11 27672046
Publication E-mail: ijcs@satishserial.com
Publisher E-mail: info@satishserial.com

Journal focusing on new developments in crop science and their applications. **Key Personnel:** S.P. Singh, Editor-in-Chief. **ISSN:** 0973-4880. **Subscription Rates:** Rs 1,500 individuals; US$100 other countries individual; Rs 2,500 institutions library; US$200 institutions, other countries library. **Remarks:** Accepts advertising. **URL:** http://www.satishserial.com/index.php?p=issn0973-4880. **Circ:** (Not Reported)

1109 ■ Indian Media Studies Journal
Satish Serial Publishing House
115, Express Tower
Commercial Complex
Azadpur
New Delhi 110033, Delhi, India
Ph: 91 11 27672852
Fax: 91 11 27672046
Publisher E-mail: info@satishserial.com

Journal covering field of various mass communication. **Freq:** Semiannual. **Key Personnel:** Dr. Anil K. Rai, Editor-in-Chief. **ISSN:** 0972-9348. **Subscription Rates:** Rs 1,200 individuals; US$90 other countries; Rs 2,000 institutions library; US$150 institutions, other countries library. **Remarks:** Accepts advertising. **URL:** http://www.satishserial.com/index.php?p=issn0972-9348. **Circ:** (Not Reported)

ROORKEE

1110 ■ International Journal of Ecology and Development
Indian Society for Development and Environment Research
PO Box 113
Roorkee 247-667, Uttar Pradesh, India
Publisher E-mail: ceser_isder@yahoo.co.in

Journal featuring the latest research and developments in ecology and development. **Freq;** 3/yr. **Key Personnel:** Dr. Tanuja Srivastava, Editor-in-Chief, tanujfmal@iitr.ernet.in; Dr. Kaushal K. Srivastava, Editor-in-Chief. **ISSN:** 0972-9984. **Subscription Rates:** EUR550 individuals print; EUR300 individuals online; EUR750 individuals online and print; EUR700 institutions print; EUR500 institutions online; EUR1,000 institutions print and online. **URL:** http://www.isder.ceser.res.in/ijed.html.

1111 ■ International Journal of Mathematics and Statistics (IJMS)
Indian Society for Development and Environment Research
PO Box 113
Roorkee 247-667, Uttar Pradesh, India
Publisher E-mail: ceser_isder@yahoo.co.in

Journal focusing on the latest developments in the area of mathematics and statistics. **Freq:** Semiannual. **Key Personnel:** Dr. Tanuja Srivastava, Editor-in-Chief, eic.ijms@yahoo.com. **ISSN:** 0973-8347. **Subscription Rates:** EUR100 individuals; EUR250 institutions. **URL:** http://www.isder.ceser.res.in/ijms.html.

ITALY

MESSINA

1112 ■ International Journal of Economic Policy in Emerging Economies
Inderscience Publishers
c/o Dr. Bruno S. Sergi, Ed.
University of Messina
Via T. Cannizzaro, 278
I-98122 Messina, Italy
Publisher E-mail: editor@inderscience.com

Journal covering all areas of economic policy. **Founded:** 2007. **Freq:** 4/yr. **Key Personnel:** Dr. Bruno S. Sergi, Editor, bsergi@unime.it. **ISSN:** 1752-0452. **Subscription Rates:** EUR470 individuals includes surface mail, print only; EUR640 individuals print & online. **URL:** http://www.inderscience.com/browse/index.php?journalCODE=ijepee.

1113 ■ International Journal of Monetary Economics & Finance
Inderscience Publishers
c/o Dr. Bruno S. Sergi, Ed.
University of Messina
Faculty of Political Science
Via T. Cannizzaro, 278
I-98122 Messina, Italy
Publication E-mail: ijmef@inderscience.com
Publisher E-mail: editor@inderscience.com

Journal covering international monetary economics and finance. **Founded:** 2007. **Freq:** 4/yr. **Key Personnel:** Dr. Bruno S. Sergi, Editor, bsergi@unime.it. **ISSN:** 1752-0479. **Subscription Rates:** EUR470 individuals includes surface mail, print only; EUR640 individuals print and online. **URL:** http://www.inderscience.com/browse/index.php?journalCODE=ijmef.

1114 ■ International Journal of Trade and Global Markets
Inderscience Publishers
c/o Dr. Bruno S. Sergi, Ed.
University of Messina
Faculty of Political Science
Via T. Cannizzaro, 278
I-98122 Messina, Italy
Publication E-mail: ijtgm@inderscience.com
Publisher E-mail: editor@inderscience.com

Journal covering international trade & economic growth. **Founded:** 2007. **Freq:** 4/yr. **Key Personnel:** Dr. Bruno S. Sergi, Editor, bsergi@unime.it. **ISSN:** 1742-7541. **Subscription Rates:** EUR470 individuals surface mail, print only; EUR640 individuals print and online. **URL:** http://www.inderscience.com/browse/index.php?journalCODE=ijtgm.

TRENTO

1115 ■ International Journal of Agent-Oriented Software Engineering
Inderscience Publishers
c/o Paolo Giorgini, Ed.
University of Trento
Department of Information & Communication Technology
Via Sommarive, 14
I-38050 Trento, Italy
Publisher E-mail: editor@inderscience.com

Journal covering software engineering aspects of the use of agent technology for the development of IT systems. **Founded:** 2007. **Freq:** 4/yr. **Key Personnel:** Paolo Giorgini, Editor, paolo.giorgini@unitn.it; Brian Henderson-Sellers, Editor, brian@it.uts.edu.au. **ISSN:** 1746-1375. **Subscription Rates:** EUR470 individuals includes surface mail, print only; EUR640 individuals print & online. **URL:** http://www.inderscience.com/browse/index.php?journalCODE=ijaose.

JAPAN

TOKYO

1116 ■ Big Comic Original
Shogakukan Inc.
2-3-1, Hitotsubashi
Chiyoda-ku
Tokyo 101-8001, Japan

Entertainment comic magazine. **Founded:** Feb. 18, 1974. **Freq:** Bimonthly. **Key Personnel:** Masahiro Oga, President. **Subscription Rates:** 260¥ individuals. **URL:** http://www.shogakukan.co.jp/english/htm/m_mag.htmlbigcomicoriginal.

1117 ■ Big Comic Spirits
Shogakukan Inc.
2-3-1, Hitotsubashi
Chiyoda-ku
Tokyo 101-8001, Japan

Entertainment comic magazine. **Founded:** Oct. 14, 1980. **Freq:** Weekly (Mon.). **Key Personnel:** Masahiro Oga, President. **Subscription Rates:** 260¥ individuals. **URL:** http://www.shogakukan.co.jp/english/htm/m_mag.htmlbigcomicspirits.

1118 ■ CanCam
Shogakukan Inc.
2-3-1, Hitotsubashi
Chiyoda-ku
Tokyo 101-8001, Japan

Fashion magazine for women. **Founded:** Oct. 23, 1981. **Freq:** Monthly. **Key Personnel:** Masahiro Oga, President. **Subscription Rates:** 600¥ individuals. **URL:** http://www.shogakukan.co.jp/english/htm/m_mag.htmlmuffin.

1119 ■ Josei Seven
Shogakukan Inc.
2-3-1, Hitotsubashi
Chiyoda-ku
Tokyo 101-8001, Japan

Lifestyle magazine featuring news, fashion, cooking, money, and health and beauty. **Founded:** Aug. 5, 1969. **Freq:** Weekly (Mon.). **Key Personnel:** Masahiro Oga, President. **Subscription Rates:** 320¥ individuals. **URL:** http://www.shogakukan.co.jp/english/htm/m_mag.htmljoseiseven.

1120 ■ National Geographic Japanese Edition
Nikkei Business Publications Inc.
1-17-3 Shirokane
Minato-ku
Tokyo 108-8646, Japan
Ph: 81 3 68118502
Fax: 81 3 54219058
Publisher E-mail: info@nikkeibp.com

Magazine covering "the unknown facts of Earth" that relate to nature, adventure, history, global environment, science, and culture. **Founded:** Apr. 1995. **Freq:** Monthly. **Print Method:** Four color offset lithography. **Trim Size:** 175 x 254 mm. **Key Personnel:** Yuko Tanaka, International Advertising Sales, phone 81 3 68118311, fax 81 3 54219804, yktanaka@nikkeibp.co.jp. **Remarks:** Accepts advertising. **URL:** http://www.nikkeibp.com/adinfo/printmedia/pm_001007049.html. . **Ad Rates:** BW: 740,000¥, 4C: 1,020,000¥. **Circ:** ★**92,171**

1121 ■ Nikkei Architecture
Nikkei Business Publications Inc.
1-17-3 Shirokane
Minato-ku
Tokyo 108-8646, Japan
Ph: 81 3 68118502
Fax: 81 3 54219058
Publisher E-mail: info@nikkeibp.com

Magazine featuring architects and architectural designs firms, general contractors, and home building and construction company. **Founded:** Apr. 1976. **Freq:** Biweekly. **Print Method:** Four color offset lithography. **Trim Size:** 208 x 280 mm. **Key Personnel:** Yuko Tanaka, International Advertising Sales, phone 81 3 68118311, fax 81 3 54219804, yktanaka@nikkeibp.co.jp. **Remarks:** Accepts advertising. **URL:** http://www.nikkeibp.com/adinfo/printmedia/pm_001006044.html. . **Ad Rates:** BW: 738,000¥, 4C: 1,070,000¥. **Circ:** ★**45,238**

1122 ■ Nikkei Board Guide
Nikkei Business Publications Inc.
1-17-3 Shirokane
Minato-ku
Tokyo 108-8646, Japan
Ph: 81 3 68118502
Fax: 81 3 54219058
Publisher E-mail: info@nikkeibp.com

Magazine featuring information for electronic board development engineers. **Founded:** Apr. 1997. **Freq:** Quarterly. **Print Method:** Four color offset lithography. **Trim Size:** 208 x 280 mm. **Key Personnel:** Yuko Tanaka, International Advertising Sales, phone 81 3 68118311, fax 81 3 54219804, yktanaka@nikkeibp.co.jp. **Remarks:** Accepts advertising. **URL:** http://www.nikkeibp.com/adinfo/printmedia/pm_001002028.html. . **Ad Rates:** 4C: 450,000¥. **Circ:** 60,000

1123 ■ Nikkei BP Government Technology
Nikkei Business Publications Inc.
1-17-3 Shirokane
Minato-ku
Tokyo 108-8646, Japan
Ph: 81 3 68118502
Fax: 81 3 54219058
Publisher E-mail: info@nikkeibp.com

Magazine featuring information technology for government sectors. **Founded:** Sept. 2003. **Freq:** Semiannual. **Print Method:** Four color offset lithography. **Trim Size:** 208 x 280 mm. **Key Personnel:** Yuko Tanaka, International Advertising Sales, phone 81 3 68118311, fax 81 3 54219804, yktanaka@nikkeibp.co.jp. **Remarks:** Accepts advertising. **URL:** http://www.nikkeibp.com/adinfo/printmedia/pm_001003037.html. . **Ad Rates:** 4C: 600,000¥. **Circ:** 11,100

1124 ■ Nikkei Business
Nikkei Business Publications Inc.
1-17-3 Shirokane
Minato-ku
Tokyo 108-8646, Japan
Ph: 81 3 68118502
Fax: 81 3 54219058
Publisher E-mail: info@nikkeibp.com

Magazine covering information about company management and business operations. **Founded:** Sept. 1969. **Freq:** Weekly **Print Method:** Offset lithography. **Trim Size:** 210 x 280 mm. **Key Personnel:** Yuko Tanaka, International Advertising Sales, phone 81 3 68118311, fax 81 3 54219804, yktanaka@nikkeibp.co.jp. **Remarks:** Accepts advertising. **URL:** http://www.nikkeibp.com/adinfo/printmedia/pm_001001010.html; http://business.nikkeibp.co.jp/english/index.html. . **Ad Rates:** BW: 1,670,000¥, 4C: 2,520,000¥. **Circ:** ★**332,828**

Circulation: ★ = ABC; △ = BPA; ♦ = CAC; • = CCAB; ❑ = VAC; ⊕ = PO Statement; ‡ = Publisher's Report; Boldface figures = sworn; Light figures = estimated.

1125 ■ Nikkei Business Associe
Nikkei Business Publications Inc.
1-17-3 Shirokane
Minato-ku
Tokyo 108-8646, Japan
Ph: 81 3 68118502
Fax: 81 3 54219058
Publisher E-mail: info@nikkeibp.com

Magazine covering business for young people. **Founded:** Apr. 2002. **Freq:** Semimonthly. **Print Method:** Four color offset lithography. **Trim Size:** 210 x 280 mm. **Key Personnel:** Yuko Tanaka, International Advertising Sales, phone 81 3 68118311, fax 81 3 54219804, yktanaka@nikkeibp.co.jp. **Remarks:** Accepts advertising. **URL:** http://www.nikkeibp.com/adinfo/printmedia/pm_001001012.html. . **Ad Rates:** BW: 700,000¥, 4C: 1,000,000¥. **Circ:** ★**79,674**

1126 ■ Nikkei Communications
Nikkei Business Publications Inc.
1-17-3 Shirokane
Minato-ku
Tokyo 108-8646, Japan
Ph: 81 3 68118502
Fax: 81 3 54219058
Publisher E-mail: info@nikkeibp.com

Magazine featuring information on communication/network products, services and technologies for people engaged in the planning, building, and operating of corporate networks. **Founded:** Oct. 1985. **Freq:** Semimonthly. **Print Method:** Four color offset lithography. **Trim Size:** 208 x 280 mm. **Key Personnel:** Yuko Tanaka, International Advertising Sales, phone 81 3 68118311, fax 81 3 54219804, yktanaka@nikkeibp.co.jp. **Remarks:** Accepts advertising. **URL:** http://www.nikkeibp.com/adinfo/printmedia/pm_001003035.html. . **Ad Rates:** BW: 540,000¥, 4C: 800,000¥. **Circ:** ★**32,047**

1127 ■ Nikkei Computer
Nikkei Business Publications Inc.
1-17-3 Shirokane
Minato-ku
Tokyo 108-8646, Japan
Ph: 81 3 68118502
Fax: 81 3 54219058
Publisher E-mail: info@nikkeibp.com

Magazine covering information technology. **Founded:** Oct. 1981. **Freq:** Semimonthly. **Print Method:** Four color offset lithography. **Trim Size:** 208 x 280 mm. **Key Personnel:** Yuko Tanaka, International Advertising Sales, phone 81 3 68118311, fax 81 3 54219804, yktanaka@nikkeibp.co.jp. **Remarks:** Accepts advertising. **URL:** http://www.nikkeibp.com/adinfo/printmedia/pm_001003029.html. . **Ad Rates:** BW: 856,000¥, 4C: 1,236,000¥. **Circ:** ★**48,409**

1128 ■ Nikkei Construction
Nikkei Business Publications Inc.
1-17-3 Shirokane
Minato-ku
Tokyo 108-8646, Japan
Ph: 81 3 68118502
Fax: 81 3 54219058
Publisher E-mail: info@nikkeibp.com

Magazine featuring construction industry. **Founded:** Oct. 1989. **Freq:** Semimonthly. **Print Method:** Four color offset lithography. **Trim Size:** 210 x 280 mm. **Key Personnel:** Yuko Tanaka, International Advertising Sales, phone 81 3 68118311, fax 81 3 54219804, yktanaka@nikkeibp.co.jp. **Remarks:** Accepts advertising. **URL:** http://www.nikkeibp.com/adinfo/printmedia/pm_001006045.html. . **Ad Rates:** BW: 459,000¥, 4C: 569,000¥. **Circ:** ★**27,425**

1129 ■ Nikkei Design
Nikkei Business Publications Inc.
1-17-3 Shirokane
Minato-ku
Tokyo 108-8646, Japan
Ph: 81 3 68118502
Fax: 81 3 54219058
Publisher E-mail: info@nikkeibp.com

Magazine covering buildings and stores design. **Founded:** July 1987. **Freq:** Monthly. **Print Method:** Four color offset lithography. **Trim Size:** 210 x 280 mm. **Key Personnel:** Yuko Tanaka, International Advertising Sales, phone 81 3 68118311, fax 81 3 54219804, yktanaka@nikkeibp.co.jp. **Remarks:** Accepts advertising. **URL:** http://www.nikkeibp.com/adinfo/printmedia/pm_001001016.html. . **Ad Rates:** BW: 306,000¥, 4C: 522,000¥. **Circ:** ★**14,084**

1130 ■ Nikkei Drug Information
Nikkei Business Publications Inc.
1-17-3 Shirokane
Minato-ku
Tokyo 108-8646, Japan
Ph: 81 3 68118502
Fax: 81 3 54219058
Publisher E-mail: info@nikkeibp.com

Magazine featuring drug information. **Founded:** Apr. 1998. **Freq:** Monthly. **Print Method:** Four color offset lithography. **Trim Size:** 210 x 297 mm. **Key Personnel:** Yuko Tanaka, International Advertising Sales, phone 81 3 68118311, fax 81 3 54219804, yktanaka@nikkeibp.co.jp. **Remarks:** Accepts advertising. **URL:** http://www.nikkeibp.com/adinfo/printmedia/pm_001005019.html. . **Ad Rates:** BW: 484,000¥, 4C: 704,000¥. **Circ:** ★**64,447**

1131 ■ Nikkei Ecology
Nikkei Business Publications Inc.
1-17-3 Shirokane
Minato-ku
Tokyo 108-8646, Japan
Ph: 81 3 68118502
Fax: 81 3 54219058
Publisher E-mail: info@nikkeibp.com

Magazine covering ecology. **Founded:** June 1999. **Freq:** Monthly. **Print Method:** Four color offset lithography. **Trim Size:** 210 x 280 mm. **Key Personnel:** Yuko Tanaka, International Advertising Sales, phone 81 3 68118311, fax 81 3 54219804, yktanaka@nikkeibp.co.jp. **Remarks:** Accepts advertising. **URL:** http://www.nikkeibp.com/adinfo/printmedia/pm_001001014.html. . **Ad Rates:** BW: 420,000¥, 4C: 580,000¥. **Circ:** ★**16,074**

1132 ■ Nikkei Electronics China
Nikkei Business Publications Inc.
1-17-3 Shirokane
Minato-ku
Tokyo 108-8646, Japan
Ph: 81 3 68118502
Fax: 81 3 54219058
Publisher E-mail: info@nikkeibp.com

Magazine covering electronics. **Founded:** Nov. 2002. **Freq:** Monthly. **Print Method:** Four color offset lithography. **Trim Size:** 205 x 270 mm. **Key Personnel:** Yuko Tanaka, International Advertising Sales, phone 81 3 68118311, fax 81 3 54219804, yktanaka@nikkeibp.co.jp. **Remarks:** Accepts advertising. **URL:** http://www.nikkeibp.com/adinfo/printmedia/pm_001002023.html. . **Ad Rates:** BW: US$1,060, 4C: US$4,150. **Circ:** △**30,220**

1133 ■ Nikkei Entertainment!
Nikkei Business Publications Inc.
1-17-3 Shirokane
Minato-ku
Tokyo 108-8646, Japan
Ph: 81 3 68118502
Fax: 81 3 54219058
Publisher E-mail: info@nikkeibp.com

Magazine featuring entertainment. **Founded:** Mar. 1997. **Freq:** Monthly. **Print Method:** Four color offset lithography. **Trim Size:** 210 x 280 mm. **Key Personnel:** Yuko Tanaka, International Advertising Sales, phone 81 3 68118311, fax 81 3 54219804, yktanaka@nikkeibp.co.jp. **Remarks:** Accepts advertising. **URL:** http://www.nikkeibp.com/adinfo/printmedia/pm_001007047.html. . **Ad Rates:** BW: 760,000¥, 4C: 1,100,000¥. **Circ:** ★**91,146**

1134 ■ Nikkei Health
Nikkei Business Publications Inc.
1-17-3 Shirokane
Minato-ku
Tokyo 108-8646, Japan
Ph: 81 3 68118502
Fax: 81 3 54219058
Publisher E-mail: info@nikkeibp.com

Magazine covering health. **Founded:** Mar. 1998. **Freq:** Monthly. **Print Method:** Four color offset lithography. **Trim Size:** 210 x 280 mm. **Key Personnel:** Yuko Tanaka, International Advertising Sales, phone 81 3 68118311, fax 81 3 54219804, yktanaka@nikkeibp.co.jp. **Remarks:** Accepts advertising. **URL:** http://www.nikkeibp.com/adinfo/printmedia/pm_001005020.html. . **Ad Rates:** BW: 600,000¥, 4C: 800,000¥. **Circ:** ★**88,192**

1135 ■ Nikkei Healthcare
Nikkei Business Publications Inc.
1-17-3 Shirokane
Minato-ku
Tokyo 108-8646, Japan
Ph: 81 3 68118502
Fax: 81 3 54219058
Publisher E-mail: info@nikkeibp.com

Magazine featuring healthcare information. **Founded:** Nov. 1989. **Freq:** Monthly. **Print Method:** Four color offset lithography. **Trim Size:** 208 x 280 mm. **Key Personnel:** Yuko Tanaka, International Advertising Sales, phone 81 3 68118311, fax 81 3 54219804, yktanaka@nikkeibp.co.jp. **Remarks:** Accepts advertising. **URL:** http://www.nikkeibp.com/adinfo/printmedia/pm_001005018.html. . **Ad Rates:** BW: 351,000¥, 4C: 516,000¥. **Circ:** ★**18,556**

1136 ■ Nikkei Home Builder
Nikkei Business Publications Inc.
1-17-3 Shirokane
Minato-ku
Tokyo 108-8646, Japan
Ph: 81 3 68118502
Fax: 81 3 54219058
Publisher E-mail: info@nikkeibp.com

Magazine featuring home design and construction. **Founded:** June 1999. **Freq:** Monthly. **Print Method:** Four color offset lithography. **Trim Size:** 208 x 280 mm. **Key Personnel:** Yuko Tanaka, International Advertising Sales, phone 81 3 68118311, fax 81 3 54219804, yktanaka@nikkeibp.co.jp. **Remarks:** Accepts advertising. **URL:** http://www.nikkeibp.com/adinfo/printmedia/pm_001006046.html. . **Ad Rates:** BW: 300,000¥, 4C: 450,000¥. **Circ:** ★**20,770**

1137 ■ Nikkei Information Strategy
Nikkei Business Publications Inc.
1-17-3 Shirokane
Minato-ku
Tokyo 108-8646, Japan
Ph: 81 3 68118502
Fax: 81 3 54219058
Publisher E-mail: info@nikkeibp.com

Magazine featuring new concepts of computerization and analyses of market trends. **Founded:** Apr. 1992. **Freq:** Monthly. **Print Method:** Four color offset lithography. **Trim Size:** 208 x 280 mm. **Key Personnel:** Yuko Tanaka, International Advertising Sales, phone 81 3 68118311, fax 81 3 54219804, yktanaka@nikkeibp.co.jp. **Remarks:** Accepts advertising. **URL:** http://www.nikkeibp.com/adinfo/printmedia/pm_001003030.html. . **Ad Rates:** BW: 520,000¥, 4C: 760,000¥. **Circ:** ★**20,782**

1138 ■ Nikkei Linux
Nikkei-Business Publications Inc.
1-17-3 Shirokane
Minato-ku
Tokyo 108-8646, Japan
Ph: 81 3 68118502
Fax: 81 3 54219058
Publisher E-mail: info@nikkeibp.com

Magazine covering information technology. **Founded:** Sept. 1999. **Freq:** Monthly. **Print Method:** Four color offset lithography. **Trim Size:** 208 x 280 mm. **Key Personnel:** Yuko Tanaka, International Advertising Sales, phone 81 3 68118311, fax 81 3

54219804, yktanaka@nikkeibp.co.jp. **Remarks:** Accepts advertising. **URL:** http://www.nikkeibp.com/adinfo/printmedia/pm_001003033.html. . **Ad Rates:** BW: 200,000¥, 4C: 450,000¥. **Circ:** ★**14,305**

1139 ■ Nikkei Medical
Nikkei Business Publications Inc.
1-17-3 Shirokane
Minato-ku
Tokyo 108-8646, Japan
Ph: 81 3 68118502
Fax: 81 3 54219058
Publisher E-mail: info@nikkeibp.com

Magazine featuring healthcare information. **Founded:** Apr. 1972. **Freq:** Monthly. **Print Method:** Four color offset lithography. **Trim Size:** 208 x 280 mm. **Key Personnel:** Yuko Tanaka, International Advertising Sales, phone 81 3 68118311, fax 81 3 54219804, yktanaka@nikkeibp.co.jp. **Remarks:** Accepts advertising. **URL:** http://www.nikkeibp.com/adinfo/printmedia/pm_001005017.html. . **Ad Rates:** BW: 1,116,000¥, 4C: 1,564,000¥. **Circ:** ★**111,107**

1140 ■ Nikkei Microdevices
Nikkei Business Publications Inc.
1-17-3 Shirokane
Minato-ku
Tokyo 108-8646, Japan
Ph: 81 3 68118502
Fax: 81 3 54219058
Publisher E-mail: info@nikkeibp.com

Magazine covering microdevices. **Founded:** July 1985. **Freq:** Monthly. **Print Method:** Four color offset lithography. **Trim Size:** 208 x 280 mm. **Key Personnel:** Yuko Tanaka, International Advertising Sales, phone 81 3 68118311, fax 81 3 54219804, yktanaka@nikkeibp.co.jp. **Remarks:** Accepts advertising. **URL:** http://www.nikkeibp.com/adinfo/printmedia/pm_001002025.html. . **Ad Rates:** BW: 407,000¥, 4C: 627,000¥. **Circ:** ★**13,970**

1141 ■ Nikkei Monozukuri
Nikkei Business Publications Inc.
1-17-3 Shirokane
Minato-ku
Tokyo 108-8646, Japan
Ph: 81 3 68118502
Fax: 81 3 54219058
Publisher E-mail: info@nikkeibp.com

Magazine covering information about the manufacturing industry. **Founded:** Apr. 2004. **Freq:** Monthly. **Print Method:** Four color offset lithography. **Trim Size:** 208 x 280 mm. **Key Personnel:** Yuko Tanaka, International Advertising Sales, phone 81 3 68118311, fax 81 3 54219804, yktanaka@nikkeibp.co.jp. **Remarks:** Accepts advertising. **URL:** http://www.nikkeibp.com/adinfo/printmedia/pm_001002026.html. . **Ad Rates:** BW: 478,000¥, 4C: 694,000¥. **Circ:** ★**34,906**

1142 ■ Nikkei Network
Nikkei Business Publications Inc.
1-17-3 Shirokane
Minato-ku
Tokyo 108-8646, Japan
Ph: 81 3 68118502
Fax: 81 3 54219058
Publisher E-mail: info@nikkeibp.com

Magazine featuring network technology. **Founded:** Apr. 2000. **Freq:** Monthly. **Print Method:** Four color offset lithography. **Trim Size:** 208 x 280 mm. **Key Personnel:** Yuko Tanaka, International Advertising Sales, phone 81 3 68118311, fax 81 3 54219804, yktanaka@nikkeibp.co.jp. **Remarks:** Accepts advertising. **URL:** http://www.nikkeibp.com/adinfo/printmedia/pm_001003036.html. . **Ad Rates:** BW: 480,000¥, 4C: 700,000¥. **Circ:** ★**55,594**

1143 ■ Nikkei PC Beginners
Nikkei Business Publications Inc.
1-17-3 Shirokane
Minato-ku
Tokyo 108-8646, Japan
Ph: 81 3 68118502
Fax: 81 3 54219058
Publisher E-mail: info@nikkeibp.com

Magazine featuring information for PC beginners. **Founded:** Sept. 1999. **Freq:** Monthly. **Print Method:** Four color offset lithography. **Trim Size:** 208 x 280 mm. **Key Personnel:** Yuko Tanaka, International Advertising Sales, phone 81 3 68118311, fax 81 3 54219804, yktanaka@nikkeibp.co.jp. **Remarks:** Accepts advertising. **URL:** http://www.nikkeibp.com/adinfo/printmedia/pm_001004040.html. . **Ad Rates:** 4C: 450,000¥. **Circ:** 80,000

1144 ■ Nikkei PC21
Nikkei Business Publications Inc.
1-17-3 Shirokane
Minato-ku
Tokyo 108-8646, Japan
Ph: 81 3 68118502
Fax: 81 3 54219058
Publisher E-mail: info@nikkeibp.com

Magazine featuring personal computing. **Founded:** Mar. 1996. **Freq:** Monthly. **Print Method:** Four color offset lithography. **Trim Size:** 210 x 280 mm. **Key Personnel:** Yuko Tanaka, International Advertising Sales, phone 81 3 68118311, fax 81 3 54219804, yktanaka@nikkeibp.co.jp. **Remarks:** Accepts advertising. **URL:** http://www.nikkeibp.com/adinfo/printmedia/pm_001004039.html. . **Ad Rates:** BW: 410,000¥, 4C: 700,000¥. **Circ:** ★**163,403**

1145 ■ Nikkei Personal Computing
Nikkei Business Publications Inc.
1-17-3 Shirokane
Minato-ku
Tokyo 108-8646, Japan
Ph: 81 3 68118502
Fax: 81 3 54219058
Publisher E-mail: info@nikkeibp.com

Magazine featuring information about computers. **Founded:** Oct. 1983. **Freq:** Semimonthly. **Print Method:** Four color offset lithography. **Trim Size:** 208 x 280 mm. **Key Personnel:** Yuko Tanaka, International Advertising Sales, phone 81 3 68118311, fax 81 3 54219804, yktanaka@nikkeibp.co.jp. **Remarks:** Accepts advertising. **URL:** http://www.nikkeibp.com/adinfo/printmedia/pm_001004038.html. . **Ad Rates:** BW: 1,007,000¥, 4C: 1,460,000¥. **Circ:** ★**242,628**

1146 ■ Nikkei Restaurants
Nikkei Business Publications Inc.
1-17-3 Shirokane
Minato-ku
Tokyo 108-8646, Japan
Ph: 81 3 68118502
Fax: 81 3 54219058
Publisher E-mail: info@nikkeibp.com

Magazine featuring food service industry. **Founded:** Oct. 1988. **Freq:** Monthly. **Print Method:** Four color offset lithography. **Trim Size:** 210 x 280 mm. **Key Personnel:** Yuko Tanaka, International Advertising Sales, phone 81 3 68118311, fax 81 3 54219804, yktanaka@nikkeibp.co.jp. **Remarks:** Accepts advertising. **URL:** http://www.nikkeibp.com/adinfo/printmedia/pm_001001015.html. . **Ad Rates:** BW: 332,000¥, 4C: 504,000¥. **Circ:** ★**21,363**

1147 ■ Nikkei Software
Nikkei Business Publications Inc.
1-17-3 Shirokane
Minato-ku
Tokyo 108-8646, Japan
Ph: 81 3 68118502
Fax: 81 3 54219058
Publisher E-mail: info@nikkeibp.com

Magazine featuring computer software. **Founded:** May 1998. **Freq:** Monthly. **Print Method:** Four color offset lithography. **Trim Size:** 208 x 280 mm. **Key Personnel:** Yuko Tanaka, International Advertising Sales, phone 81 3 68118311, fax 81 3 54219804, yktanaka@nikkeibp.co.jp. **Remarks:** Accepts advertising. **URL:** http://www.nikkeibp.com/adinfo/printmedia/pm_001003034.html. . **Ad Rates:** BW: 300,000¥, 4C: 500,000¥. **Circ:** ★**24,947**

1148 ■ Nikkei Solution Business
Nikkei Business Publications Inc.
1-17-3 Shirokane
Minato-ku
Tokyo 108-8646, Japan
Ph: 81 3 68118502
Fax: 81 3 54219058
Publisher E-mail: info@nikkeibp.com

Magazine covering business solutions. **Founded:** May 1996. **Freq:** Semimonthly. **Print Method:** Offset lithography. **Trim Size:** 210 x 280 mm. **Key Personnel:** Yuko Tanaka, International Advertising Sales, phone 81 3 68118311, fax 81 3 54219804, yktanaka@nikkeibp.co.jp. **Remarks:** Accepts advertising. **URL:** http://www.nikkeibp.com/adinfo/printmedia/pm_001003032.html. . **Ad Rates:** BW: 275,000¥, 4C: 473,000¥. **Circ:** ★**9,589**

1149 ■ Nikkei Systems
Nikkei Business Publications Inc.
1-17-3 Shirokane
Minato-ku
Tokyo 108-8646, Japan
Ph: 81 3 68118502
Fax: 81 3 54219058
Publisher E-mail: info@nikkeibp.com

Magazine covering information technology. **Founded:** Mar. 2006. **Freq:** Monthly. **Print Method:** Four color offset lithography. **Trim Size:** 208 x 280 mm. **Key Personnel:** Yuko Tanaka, International Advertising Sales, phone 81 3 68118311, fax 81 3 54219804, yktanaka@nikkeibp.co.jp. **Remarks:** Accepts advertising. **URL:** http://www.nikkeibp.com/adinfo/printmedia/pm_001003031.html. . **Ad Rates:** BW: 390,000¥, 4C: 650,000¥. **Circ:** ★**49,016**

1150 ■ Nikkei Venture
Nikkei Business Publications Inc.
1-17-3 Shirokane
Minato-ku
Tokyo 108-8646, Japan
Ph: 81 3 68118502
Fax: 81 3 54219058
Publisher E-mail: info@nikkeibp.com

Magazine covering business venture. **Founded:** Oct. 1984. **Freq:** Monthly. **Print Method:** Four color offset lithography. **Trim Size:** 208 x 280 mm. **Key Personnel:** Yuko Tanaka, International Advertising Sales, phone 81 3 68118311, fax 81 3 54219804, yktanaka@nikkeibp.co.jp. **Remarks:** Accepts advertising. **URL:** http://www.nikkeibp.com/adinfo/printmedia/pm_001001013.html. . **Ad Rates:** BW: 500,000¥, 4C: 770,000¥. **Circ:** ★**70,075**

1151 ■ Nikkei WinPC
Nikkei Business Publications Inc.
1-17-3 Shirokane
Minato-ku
Tokyo 108-8646, Japan
Ph: 81 3 68118502
Fax: 81 3 54219058
Publisher E-mail: info@nikkeibp.com

Magazine featuring information about personal computers. **Founded:** Mar. 1995. **Freq:** Monthly. **Print Method:** Four color offset lithography. **Trim Size:** 280 x 280 mm. **Key Personnel:** Yuko Tanaka, International Advertising Sales, phone 81 3 68118311, fax 81 3 54219804, yktanaka@nikkeibp.co.jp. **Remarks:** Accepts advertising. **URL:** http://www.nikkeibp.com/adinfo/printmedia/pm_001004043.html. . **Ad Rates:** BW: 250,000¥, 4C: 550,000¥. **Circ:** ★**51,074**

Circulation: ★ = ABC; △ = BPA; ♦ = CAC; • = CCAB; ❑ = VAC; ⊕ = PO Statement; ‡ = Publisher's Report; Boldface figures = sworn; Light figures = estimated.

1152 ■ Priv.
Nikkei Business Publications Inc.
1-17-3 Shirokane
Minato-ku
Tokyo 108-8646, Japan
Ph: 81 3 68118502
Fax: 81 3 54219058
Publisher E-mail: info@nikkeibp.com

Lifestyle magazine featuring different women's interests. **Founded:** Mar. 2000. **Freq:** Quarterly. **Print Method:** Four color offset lithography. **Trim Size:** 208 x 280 mm. **Key Personnel:** Yuko Tanaka, International Advertising Sales, phone 81 3 68118311, fax 81 3 54219804, yktanaka@nikkeibp.co.jp. **Remarks:** Accepts advertising. **URL:** http://www.nikkeibp.com/adinfo/printmedia/pm_001007048.html. . **Ad Rates:** 4C: 1,850,000¥. **Circ:** ★181,391

1153 ■ Proceedings of the Japan Academy, Series A
The Japan Academy
PO Box 5050
Tokyo 100-3191, Japan
Fax: 81 3 32789256

Journal featuring mathematical sciences. **Key Personnel:** Heisuke Hironaka, Editor-in-Chief; Shigefumi Mori, Exec. Ed. **URL:** http://www.japan-acad.go.jp/english/PJAA/pjasera.htm.

1154 ■ Proceedings of the Japan Academy, Series B
The Japan Academy
PO Box 5050
Tokyo 100-3191, Japan
Fax: 81 3 32789256

Journal featuring scientific and academic progress in all fields of natural sciences. **Founded:** 1912. **Key Personnel:** Tamio Yamakawa, Editor-in-Chief; Masanori Otsuka, Exec. Ed. **URL:** http://www.japan-acad.go.jp/english/PJAB/pjaserb.htm.

1155 ■ Real Simple Japan
Nikkei Business Publications Inc.
1-17-3 Shirokane
Minato-ku
Tokyo 108-8646, Japan
Ph: 81 3 68118502
Fax: 81 3 54219058
Publisher E-mail: info@nikkeibp.com

Magazine featuring information for women who are searching ways to balance their own needs, career, and family. **Founded:** Oct. 2005. **Freq:** Monthly. **Print Method:** Four color offset lithography. **Trim Size:** 228 x 276 mm. **Key Personnel:** Yuko Tanaka, International Advertising Sales, phone 81 3 68118311, fax 81 3 54219804, yktanaka@nikkeibp.co.jp. **Remarks:** Accepts advertising. **URL:** http://www.nikkeibp.com/adinfo/printmedia/pm_001007050.html. . **Ad Rates:** 4C: 1,600,000¥. **Circ:** ★106,805

1156 ■ SARAI
Shogakukan Inc.
2-3-1, Hitotsubashi
Chiyoda-ku
Tokyo 101-8001, Japan

Lifestyle magazine for the older generation. **Founded:** Sept. 7, 1989. **Freq:** Bimonthly. **Key Personnel:** Masahiro Oga, President. **Subscription Rates:** 480¥ individuals. **URL:** http://www.shogakukan.co.jp/english/htm/m_mag.htmlserai.

1157 ■ Weekly Shonen Sunday
Shogakukan Inc.
2-3-1, Hitotsubashi
Chiyoda-ku
Tokyo 101-8001, Japan

Entertainment comic magazine Japanese boys. **Founded:** Mar. 15, 1959. **Freq:** Weekly (Wed.). **Key Personnel:** Masahiro Oga, President. **Subscription Rates:** 240¥ individuals. **URL:** http://www.shogakukan.co.jp/english/htm/m_mag.htmlweeklyshonensunday.

REPUBLIC OF KOREA

SEOUL

1158 ■ International Journal of Grid & Utility Computing
Inderscience Publishers
c/o Dr. Ajith Abraham, Ed.-in-Ch.
Chung-Ang University
Electronic Commerce Lab, School of Computer Science & Engineering
221, Heukseok-Dong, Dongjak-Gu
Seoul 156-756, Republic of Korea
Publisher E-mail: editor@inderscience.com

Journal covering grid and utility computing technology. **Founded:** 2005. **Freq:** 4/yr. **Key Personnel:** Dr. Ajith Abraham, Editor-in-Chief, abraham.ajith@gmail.com; Prof. Vaclav Snasel, Editor-in-Chief, vaclav.snasel@vsb.cz; Dr. Michal Kratky, Managing Editor, michal.kratky@vsb.cz. **ISSN:** 1741-847X. **Subscription Rates:** EUR470 individuals includes surface mail, print only; EUR640 individuals print & online. **URL:** http://www.inderscience.com/browse/index.php?journalCODE=ijguc.

KUWAIT

SAFAT

1159 ■ European Journal of Industrial Engineering
Inderscience Publishers
c/o Ali Allahverdi, Ed.
Kuwait University, Department of Industrial & Management Systems Engineering
College of Engineering & Petroleum
PO Box 5969
Safat 13060, Kuwait
Publication E-mail: ejie@inderscience.com
Publisher E-mail: editor@inderscience.com

Journal covering areas of industrial engineering. **Founded:** 2007. **Freq:** Quarterly. **Key Personnel:** Ali Allahverdi, Editor, allahverdi@kuniv.edu.kw; Jose M. Framinan, Editor, jose@esi.us.es; Ruben Ruiz Garcia, Editor, rruiz@eio.upv.es. **ISSN:** 1751-5254. **Subscription Rates:** EUR470 individuals includes surface mail, print only; EUR640 individuals print and online. **URL:** http://www.inderscience.com/browse/index.php?journalCODE=ejie.

LEBANON

BAALBEK

1160 ■ NBN-TV - 49
Adnan El Hakim St., Jnah
Hala Bldg., Block A
Beirut, Lebanon
Ph: 961 1 841020
Fax: 961 1 841029
E-mail: flash@nbn.com.lb

Owner: National Broadcasting Network, at above address. **URL:** http://www.nbn.com.lb.

BEIRUT

NBN-TV - See Baalbek

NBN-TV - See Beit Mery

NBN-TV - See Nabatieh

NBN-TV - See Zahle

1161 ■ NBN-TV Abay - 49
Adnan El Hakim St., Jnah
Hala Bldg., Block A
Beirut, Lebanon
Ph: 961 1 841020
Fax: 961 1 841029
E-mail: flash@nbn.com.lb

Owner: National Broadcasting Network, at above address. **URL:** http://www.nbn.com.lb.

1162 ■ NBN-TV Akroum - 29
Adnan El Hakim St., Jnah
Hala Bldg., Block A
Beirut, Lebanon
Ph: 961 1 841020
Fax: 961 1 841029
E-mail: flash@nbn.com.lb

Owner: National Broadcasting Network, at above address. **URL:** http://www.nbn.com.lb.

1163 ■ NBN-TV B.A. Haidar - 29
Adnan El Hakim St., Jnah
Hala Bldg., Block A
Beirut, Lebanon
Ph: 961 1 841020
Fax: 961 1 841029
E-mail: flash@nbn.com.lb

Owner: National Broadcasting Network, at above address. **URL:** http://www.nbn.com.lb.

1164 ■ NBN-TV Fatka - 63
Adnan El Hakim St., Jnah
Hala Bldg., Block A
Beirut, Lebanon
Ph: 961 1 841020
Fax: 961 1 841029
E-mail: flash@nbn.com.lb

Owner: National Broadcasting Network, at above address. **URL:** http://www.nbn.com.lb.

1165 ■ NBN-TV Keliat - 44
Adnan El Hakim St., Jnah
Hala Bldg., Block A
Beirut, Lebanon
Ph: 961 1 841020
Fax: 961 1 841029
E-mail: flash@nbn.com.lb

Owner: National Broadcasting Network, at above address. **URL:** http://www.nbn.com.lb.

1166 ■ NBN-TV Maad - 44
Adnan El Hakim St., Jnah
Hala Bldg., Block A
Beirut, Lebanon
Ph: 961 1 841020
Fax: 961 1 841029
E-mail: flash@nbn.com.lb

Owner: National Broadcasting Network, at above address. **URL:** http://www.nbn.com.lb.

1167 ■ NBN-TV Soltanieh - 29
Adnan El Hakim St., Jnah
Hala Bldg., Block A
Beirut, Lebanon
Ph: 961 1 841020
Fax: 961 1 841029
E-mail: flash@nbn.com.lb

Owner: National Broadcasting Network, at above address. **URL:** http://www.nbn.com.lb.

1168 ■ NBN-TV Turbo - 49
Adnan El Hakim St., Jnah
Hala Bldg., Block A
Beirut, Lebanon
Ph: 961 1 841020
Fax: 961 1 841029
E-mail: flash@nbn.com.lb

Owner: National Broadcasting Network, at above address. **URL:** http://www.nbn.com.lb.

BEIT MERY

1169 ■ NBN-TV - 63
Adnan El Hakim St., Jnah
Hala Bldg., Block A
Beirut, Lebanon
Ph: 961 1 841020
Fax: 961 1 841029
E-mail: flash@nbn.com.lb

Owner: National Broadcasting Network, at above address. **URL:** http://www.nbn.com.lb.

NABATIEH

1170 ■ NBN-TV - 63
Adnan El Hakim St., Jnah
Hala Bldg., Block A
Beirut, Lebanon
Ph: 961 1 841020
Fax: 961 1 841029
E-mail: flash@nbn.com.lb

Owner: National Broadcasting Network, at above address. **URL:** http://www.nbn.com.lb.

ZAHLE

1171 ■ NBN-TV - 44
Adnan El Hakim St., Jnah
Hala Bldg., Block A
Beirut, Lebanon
Ph: 961 1 841020
Fax: 961 1 841029
E-mail: flash@nbn.com.lb

Owner: National Broadcasting Network, at above address. **URL:** http://www.nbn.com.lb.

MACEDONIA

BEROVO

1172 ■ Antenna5 - Berovo - 97.9
Tetovska 35
91000 Skopje, Macedonia
Ph: 389 2 3111911

Format: Top 40; World Beat; Full Service. **Owner:** Antenna5, at above address. **Key Personnel:** Zoran Petrov, Gen. Mgr./Music Ed. **URL:** http://www.antenna5.com.mk/about.aspxBroadcast.

BITOLA

1173 ■ Antenna5 - Bitola - 93.9
Tetovska 35
91000 Skopje, Macedonia
Ph: 389 2 3111911

Format: Top 40; World Beat; Full Service. **Owner:** Antenna5, at above address. **Key Personnel:** Zoran Petrov, Gen. Mgr./Music Ed. **URL:** http://www.antenna5.com.mk/about.aspxBroadcast.

1174 ■ Antenna5 - Bitola - 92.9
Tetovska 35
91000 Skopje, Macedonia
Ph: 389 2 3111911

Format: Top 40; World Beat; Full Service. **Owner:** Antenna5, at above address. **Key Personnel:** Zoran Petrov, Gen. Mgr./Music Ed. **URL:** http://www.antenna5.com.mk/about.aspxBroadcast.

BOGDANCI

1175 ■ Antenna5 - Bogdanci - 89.2
Tetovska 35
91000 Skopje, Macedonia
Ph: 389 2 3111911

Format: Top 40; World Beat; Full Service. **Owner:** Antenna5, at above address. **Key Personnel:** Zoran Petrov, Gen. Mgr./Music Ed. **URL:** http://www.antenna5.com.mk/about.aspxBroadcast.

1176 ■ Antenna5 - Bogdanci - 106.3
Tetovska 35
91000 Skopje, Macedonia
Ph: 389 2 3111911

Format: Top 40; World Beat; Full Service. **Owner:** Antenna5, at above address. **Key Personnel:** Zoran Petrov, Gen. Mgr./Music Ed. **URL:** http://www.antenna5.com.mk/about.aspxBroadcast.

DELCEVO

1177 ■ Antenna5 - Delcevo - 97.90
Tetovska 35
91000 Skopje, Macedonia
Ph: 389 2 3111911

Format: Top 40; World Beat; Full Service. **Owner:** Antenna5, at above address. **Key Personnel:** Zoran Petrov, Gen. Mgr./Music Ed. **URL:** http://www.antenna5.com.mk/about.aspxBroadcast.

DEMIR KAPIJA

1178 ■ Antenna5 - Demir Kapija - 91.9
Tetovska 35
91000 Skopje, Macedonia
Ph: 389 2 3111911

Format: Top 40; World Beat; Full Service. **Owner:** Antenna5, at above address. **Key Personnel:** Zoran Petrov, Gen. Mgr./Music Ed. **URL:** http://www.antenna5.com.mk/about.aspxBroadcast.

1179 ■ Antenna5 - Demir Kapija - 88.8
Tetovska 35
91000 Skopje, Macedonia
Ph: 389 2 3111911

Format: Top 40; World Beat; Full Service. **Owner:** Antenna5, at above address. **Key Personnel:** Zoran Petrov, Gen. Mgr./Music Ed. **URL:** http://www.antenna5.com.mk/about.aspxBroadcast.

1180 ■ Antenna5 - Demir Kapija - 104.2
Tetovska 35
91000 Skopje, Macedonia
Ph: 389 2 3111911

Format: Top 40; World Beat; Full Service. **Owner:** Antenna5, at above address. **Key Personnel:** Zoran Petrov, Gen. Mgr./Music Ed. **URL:** http://www.antenna5.com.mk/about.aspxBroadcast.

DOJRAN

1181 ■ Antenna5 - Dojran - 106.3
Tetovska 35
91000 Skopje, Macedonia
Ph: 389 2 3111911

Format: Top 40; World Beat; Full Service. **Owner:** Antenna5, at above address. **Key Personnel:** Zoran Petrov, Gen. Mgr./Music Ed. **URL:** http://www.antenna5.com.mk/about.aspxBroadcast.

GALICNIK

1182 ■ Antenna5 - Galicnik - 92.9
Tetovska 35
91000 Skopje, Macedonia
Ph: 389 2 3111911

Format: Top 40; World Beat; Full Service. **Owner:** Antenna5, at above address. **Key Personnel:** Zoran Petrov, Gen. Mgr./Music Ed. **URL:** http://www.antenna5.com.mk/about.aspxBroadcast.

GEVGELIJA

1183 ■ Antenna5 - Gevgelija - 89.2
Tetovska 35
91000 Skopje, Macedonia
Ph: 389 2 3111911

Format: Top 40; World Beat; Full Service. **Owner:** Antenna5, at above address. **Key Personnel:** Zoran Petrov, Gen. Mgr./Music Ed. **URL:** http://www.antenna5.com.mk/about.aspxBroadcast.

1184 ■ Antenna5 - Gevgelija - 106.3
Tetovska 35
91000 Skopje, Macedonia
Ph: 389 2 3111911

Format: Top 40; World Beat; Full Service. **Owner:** Antenna5, at above address. **Key Personnel:** Zoran Petrov, Gen. Mgr./Music Ed. **URL:** http://www.antenna5.com.mk/about.aspxBroadcast.

GOSTIVAR

1185 ■ Antenna5 - Gostivar - 106.9
Tetovska 35
91000 Skopje, Macedonia
Ph: 389 2 3111911

Format: Top 40; World Beat; Full Service. **Owner:** Antenna5, at above address. **Key Personnel:** Zoran Petrov, Gen. Mgr./Music Ed. **URL:** http://www.antenna5.com.mk/about.aspxBroadcast.

1186 ■ Antenna5 - Gostivar - 105.5
Tetovska 35
91000 Skopje, Macedonia
Ph: 389 2 3111911

Format: Top 40; World Beat; Full Service. **Owner:** Antenna5, at above address. **Key Personnel:** Zoran Petrov, Gen. Mgr./Music Ed. **URL:** http://www.antenna5.com.mk/about.aspxBroadcast.

KAVADARCI

1187 ■ Antenna5 - Kavadarci - 91.9
Tetovska 35
91000 Skopje, Macedonia
Ph: 389 2 3111911

Format: Top 40; World Beat; Full Service. **Owner:** Antenna5, at above address. **Key Personnel:** Zoran Petrov, Gen. Mgr./Music Ed. **URL:** http://www.antenna5.com.mk/about.aspxBroadcast.

1188 ■ Antenna5 - Kavadarci - 104.2
Tetovska 35
91000 Skopje, Macedonia
Ph: 389 2 3111911

Format: Top 40; World Beat; Full Service. **Owner:** Antenna5, at above address. **Key Personnel:** Zoran Petrov, Gen. Mgr./Music Ed. **URL:** http://www.antenna5.com.mk/about.aspxBroadcast.

KICEVO

1189 ■ Antenna5 - Kicevo - 95.5
Tetovska 35
91000 Skopje, Macedonia
Ph: 389 2 3111911

Format: Top 40; World Beat; Full Service. **Owner:** Antenna5, at above address. **Key Personnel:** Zoran Petrov, Gen. Mgr./Music Ed. **URL:** http://www.antenna5.com.mk/about.aspxBroadcast.

KOCANI

1190 ■ Antenna5 - Kocani - 97.9
Tetovska 35
91000 Skopje, Macedonia
Ph: 389 2 3111911

Format: Top 40; World Beat; Full Service. **Owner:** Antenna5, at above address. **Key Personnel:** Zoran Petrov, Gen. Mgr./Music Ed. **URL:** http://www.antenna5.com.mk/about.aspxBroadcast.

1191 ■ Antenna5 - Kocani - 104.8
Tetovska 35
91000 Skopje, Macedonia
Ph: 389 2 3111911

Format: Top 40; World Beat; Full Service. **Owner:** Antenna5, at above address. **Key Personnel:** Zoran Petrov, Gen. Mgr./Music Ed. **URL:** http://www.antenna5.com.mk/about.aspxBroadcast.

KRATOVO

1192 ■ Antenna5 - Kratovo - 106.9
Tetovska 35
91000 Skopje, Macedonia
Ph: 389 2 3111911

Format: Top 40; World Beat; Full Service. **Owner:** Antenna5, at above address. **Key Personnel:** Zoran Petrov, Gen. Mgr./Music Ed. **URL:** http://www.antenna5.com.mk/about.aspxBroadcast.

1193 ■ Antenna5 - Kratovo - 105.5
Tetovska 35
91000 Skopje, Macedonia
Ph: 389 2 3111911

Format: Top 40; World Beat; Full Service. **Owner:** Antenna5, at above address. **Key Personnel:** Zoran Petrov, Gen. Mgr./Music Ed. **URL:** http://www.antenna5.com.mk/about.aspxBroadcast.

KRIVA PALANKA

1194 ■ Antenna5 - Kriva Palanka - 95.5
Tetovska 35
91000 Skopje, Macedonia
Ph: 389 2 3111911
Format: Top 40; World Beat; Full Service. **Owner:** Antenna5, at above address. **Key Personnel:** Zoran Petrov, Gen. Mgr./Music Ed. **URL:** http://www.antenna5.com.mk/about.aspxBroadcast.

1195 ■ Antenna5 - Kriva Palanka - 105.5
Tetovska 35
91000 Skopje, Macedonia
Ph: 389 2 3111911
Format: Top 40; World Beat; Full Service. **Owner:** Antenna5, at above address. **Key Personnel:** Zoran Petrov, Gen. Mgr./Music Ed. **URL:** http://www.antenna5.com.mk/about.aspxBroadcast.

KRUSEVO

1196 ■ Antenna5 - Krusevo - 92.9
Tetovska 35
91000 Skopje, Macedonia
Ph: 389 2 3111911
Format: Top 40; World Beat; Full Service. **Owner:** Antenna5, at above address. **Key Personnel:** Zoran Petrov, Gen. Mgr./Music Ed. **URL:** http://www.antenna5.com.mk/about.aspxBroadcast.

KUMANOVO

1197 ■ Antenna5 - Kumanovo - 106.9
Tetovska 35
91000 Skopje, Macedonia
Ph: 389 2 3111911
Format: Top 40; World Beat; Full Service. **Owner:** Antenna5, at above address. **Key Personnel:** Zoran Petrov, Gen. Mgr./Music Ed. **URL:** http://www.antenna5.com.mk/about.aspxBroadcast.

1198 ■ Antenna5 - Kumanovo - 106.3
Tetovska 35
91000 Skopje, Macedonia
Ph: 389 2 3111911
Format: Top 40; World Beat; Full Service. **Owner:** Antenna5, at above address. **Key Personnel:** Zoran Petrov, Gen. Mgr./Music Ed. **URL:** http://www.antenna5.com.mk/about.aspxBroadcast.

1199 ■ Antenna5 - Kumanovo - 104.8
Tetovska 35
91000 Skopje, Macedonia
Ph: 389 2 3111911
Format: Top 40; World Beat; Full Service. **Owner:** Antenna5, at above address. **Key Personnel:** Zoran Petrov, Gen. Mgr./Music Ed. **URL:** http://www.antenna5.com.mk/about.aspxBroadcast.

MAKEDONSKA KAMENICA

1200 ■ Antenna5 - Makedonska Kamenica - 97.90
Tetovska 35
91000 Skopje, Macedonia
Ph: 389 2 3111911
Format: Top 40; World Beat; Full Service. **Owner:** Antenna5, at above address. **Key Personnel:** Zoran Petrov, Gen. Mgr./Music Ed. **URL:** http://www.antenna5.com.mk/about.aspxBroadcast.

1201 ■ Antenna5 - Makedonska Kamenica - 104.8
Tetovska 35
91000 Skopje, Macedonia
Ph: 389 2 3111911
Format: Top 40; World Beat; Full Service. **Owner:** Antenna5, at above address. **Key Personnel:** Zoran Petrov, Gen. Mgr./Music Ed. **URL:** http://www.antenna5.com.mk/about.aspxBroadcast.

MAVROVO

1202 ■ Antenna5 - Mavrovo - 95.5
Tetovska 35
91000 Skopje, Macedonia
Ph: 389 2 3111911
Format: Top 40; World Beat; Full Service. **Owner:** Antenna5, at above address. **URL:** http://www.antenna5.com.mk.

1203 ■ Antenna5 - Mavrovo - 105.5
Tetovska 35
91000 Skopje, Macedonia
Ph: 389 2 3111911
Format: Top 40; World Beat; Full Service. **Owner:** Antenna5, at above address. **Key Personnel:** Zoran Petrov, Gen. Mgr./Music Ed. **URL:** http://www.antenna5.com.mk/about.aspxBroadcast.

NEGOTINO

1204 ■ Antenna5 - Negotino - 91.9
Tetovska 35
91000 Skopje, Macedonia
Ph: 389 2 3111911
Format: Top 40; World Beat; Full Service. **Owner:** Antenna5, at above address. **Key Personnel:** Zoran Petrov, Gen. Mgr./Music Ed. **URL:** http://www.antenna5.com.mk/about.aspxBroadcast.

1205 ■ Antenna5 - Negotino - 88.8
Tetovska 35
91000 Skopje, Macedonia
Ph: 389 2 3111911
Format: Top 40; World Beat; Full Service. **Owner:** Antenna5, at above address. **Key Personnel:** Zoran Petrov, Gen. Mgr./Music Ed. **URL:** http://www.antenna5.com.mk/about.aspxBroadcast.

1206 ■ Antenna5 - Negotino - 104.2
Tetovska 35
91000 Skopje, Macedonia
Ph: 389 2 3111911
Format: Top 40; World Beat; Full Service. **Owner:** Antenna5, at above address. **Key Personnel:** Zoran Petrov, Gen. Mgr./Music Ed. **URL:** http://www.antenna5.com.mk/about.aspxBroadcast.

OHRID

1207 ■ Antenna5 - Ohrid - 92.0
Tetovska 35
91000 Skopje, Macedonia
Ph: 389 2 3111911
Format: Top 40; World Beat; Full Service. **Owner:** Antenna5, at above address. **Key Personnel:** Zoran Petrov, Gen. Mgr./Music Ed. **URL:** http://www.antenna5.com.mk/about.aspxBroadcast.

1208 ■ Antenna5 - Ohrid - 103.3
Tetovska 35
91000 Skopje, Macedonia
Ph: 389 2 3111911
Format: Top 40; World Beat; Full Service. **Owner:** Antenna5, at above address. **Key Personnel:** Zoran Petrov, Gen. Mgr./Music Ed. **URL:** http://www.antenna5.com.mk/about.aspxBroadcast.

PRILEP

1209 ■ Antenna5 - Prilep - 92.9
Tetovska 35
91000 Skopje, Macedonia
Ph: 389 2 3111911
Format: Top 40; World Beat; Full Service. **Owner:** Antenna5, at above address. **Key Personnel:** Zoran Petrov, Gen. Mgr./Music Ed. **URL:** http://www.antenna5.com.mk/about.aspxBroadcast.

1210 ■ Antenna5 - Prilep - 106.3
Tetovska 35
91000 Skopje, Macedonia
Ph: 389 2 3111911
Format: Top 40; World Beat; Full Service. **Owner:** Antenna5, at above address. **Key Personnel:** Zoran Petrov, Gen. Mgr./Music Ed. **URL:** http://www.antenna5.com.mk/about.aspxBroadcast.

SKOPJE

Antenna5 - Berovo - See Berovo
Antenna5 - Bitola - See Bitola
Antenna5 - Bitola - See Bitola
Antenna5 - Bogdanci - See Bogdanci
Antenna5 - Bogdanci - See Bogdanci
Antenna5 - Delcevo - See Delcevo
Antenna5 - Demir Kapija - See Demir Kapija
Antenna5 - Demir Kapija - See Demir Kapija
Antenna5 - Demir Kapija - See Demir Kapija
Antenna5 - Dojran - See Dojran

1211 ■ Antenna5 - Dusegubica - 103.3
Tetovska 35
91000 Skopje, Macedonia
Ph: 389 2 3111911
Format: Top 40; World Beat; Full Service. **Owner:** Antenna5, at above address. **Key Personnel:** Zoran Petrov, Gen. Mgr./Music Ed. **URL:** http://www.antenna5.com.mk/about.aspxBroadcast.

Antenna5 - Galicnik - See Galicnik
Antenna5 - Gevgelija - See Gevgelija
Antenna5 - Gevgelija - See Gevgelija
Antenna5 - Gostivar - See Gostivar
Antenna5 - Gostivar - See Gostivar
Antenna5 - Kavadarci - See Kavadarci
Antenna5 - Kavadarci - See Kavadarci
Antenna5 - Kicevo - See Kicevo
Antenna5 - Kocani - See Kocani
Antenna5 - Kocani - See Kocani
Antenna5 - Kratovo - See Kratovo
Antenna5 - Kratovo - See Kratovo
Antenna5 - Kriva Palanka - See Kriva Palanka
Antenna5 - Kriva Palanka - See Kriva Palanka
Antenna5 - Krusevo - See Krusevo
Antenna5 - Kumanovo - See Kumanovo
Antenna5 - Kumanovo - See Kumanovo
Antenna5 - Kumanovo - See Kumanovo

1212 ■ Antenna5 - M. Radobil - 97.9
Tetovska 35
91000 Skopje, Macedonia
Ph: 389 2 3111911
Format: Top 40; World Beat; Full Service. **Owner:** Antenna5, at above address. **Key Personnel:** Zoran Petrov, Gen. Mgr./Music Ed. **URL:** http://www.antenna5.com.mk/about.aspxBroadcast.

Antenna5 - Makedonska Kamenica - See Makedonska Kamenica
Antenna5 - Makedonska Kamenica - See Makedonska Kamenica
Antenna5 - Mavrovo - See Mavrovo
Antenna5 - Mavrovo - See Mavrovo
Antenna5 - Negotino - See Negotino
Antenna5 - Negotino - See Negotino
Antenna5 - Negotino - See Negotino
Antenna5 - Ohrid - See Ohrid
Antenna5 - Ohrid - See Ohrid

1213 ■ Antenna5 - Pesocan - 101.9
Tetovska 35
91000 Skopje, Macedonia
Ph: 389 2 3111911
Format: Top 40; World Beat; Full Service. **Owner:** Antenna5, at above address. **Key Personnel:** Zoran Petrov, Gen. Mgr./Music Ed. **URL:** http://www.antenna5.com.mk/about.aspxBroadcast.

1214 ■ Antenna5 - Popova Sapka - 106.9
Tetovska 35
91000 Skopje, Macedonia
Ph: 389 2 3111911
Format: Top 40; World Beat; Full Service. **Owner:** Antenna5, at above address. **Key Personnel:** Zoran Petrov, Gen. Mgr./Music Ed. **URL:** http://www.antenna5.com.mk/about.aspxBroadcast.

Antenna5 - Prilep - See Prilep

Antenna5 - Prilep - See Prilep

Antenna5 - Stip - See Stip

Antenna5 - Stip - See Stip

1215 ■ Antenna5 - Stracin - 105.5
Tetovska 35
91000 Skopje, Macedonia
Ph: 389 2 3111911

Format: Top 40; World Beat; Full Service. **Owner:** Antenna5, at above address. **Key Personnel:** Zoran Petrov, Gen. Mgr./Music Ed. **URL:** http://www.antenna5.com.mk/about.aspxBroadcast.

Antenna5 - Struga - See Struga

Antenna5 - Struga - See Struga

Antenna5 - Strumica - See Strumica

Antenna5 - Strumica - See Strumica

Antenna5 - Sveti Nikole - See Sveti Nikole

Antenna5 - Tetovo - See Tetovo

Antenna5 - Tetovo - See Tetovo

Antenna5 - Valandovo - See Valandovo

Antenna5 - Veles - See Veles

Antenna5 - Vinica - See Vinica

Antenna5 - Vinica - See Vinica

STIP

1216 ■ Antenna5 - Stip - 91.9
Tetovska 35
91000 Skopje, Macedonia
Ph: 389 2 3111911

Format: Top 40; World Beat; Full Service. **Owner:** Antenna5, at above address. **Key Personnel:** Zoran Petrov, Gen. Mgr./Music Ed. **URL:** http://www.antenna5.com.mk/about.aspxBroadcast.

1217 ■ Antenna5 - Stip - 104.8
Tetovska 35
91000 Skopje, Macedonia
Ph: 389 2 3111911

Format: Top 40; World Beat; Full Service. **Owner:** Antenna5, at above address. **Key Personnel:** Zoran Petrov, Gen. Mgr./Music Ed. **URL:** http://www.antenna5.com.mk/about.aspxBroadcast.

STRUGA

1218 ■ Antenna5 - Struga - 92.0
Tetovska 35
91000 Skopje, Macedonia
Ph: 389 2 3111911

Format: Top 40; World Beat; Full Service. **Owner:** Antenna5, at above address. **Key Personnel:** Zoran Petrov, Gen. Mgr./Music Ed. **URL:** http://www.antenna5.com.mk/about.aspxBroadcast.

1219 ■ Antenna5 - Struga - 103.3
Tetovska 35
91000 Skopje, Macedonia
Ph: 389 2 3111911

Format: Top 40; World Beat; Full Service. **Owner:** Antenna5, at above address. **Key Personnel:** Zoran Petrov, Gen. Mgr./Music Ed. **URL:** http://www.antenna5.com.mk/about.aspxBroadcast.

STRUMICA

1220 ■ Antenna5 - Strumica - 91.9
Tetovska 35
91000 Skopje, Macedonia
Ph: 389 2 3111911

Format: Top 40; World Beat; Full Service. **Owner:** Antenna5, at above address. **Key Personnel:** Zoran Petrov, Gen. Mgr./Music Ed. **URL:** http://www.antenna5.com.mk/about.aspxBroadcast.

1221 ■ Antenna5 - Strumica - 100.5
Tetovska 35
91000 Skopje, Macedonia
Ph: 389 2 3111911

Format: Top 40; World Beat; Full Service. **Owner:** Antenna5, at above address. **Key Personnel:** Zoran Petrov, Gen. Mgr./Music Ed. **URL:** http://www.antenna5.com.mk/about.aspxBroadcast.

SVETI NIKOLE

1222 ■ Antenna5 - Sveti Nikole - 91.9
Tetovska 35
91000 Skopje, Macedonia
Ph: 389 2 3111911

Format: Top 40; World Beat; Full Service. **Owner:** Antenna5, at above address. **Key Personnel:** Zoran Petrov, Gen. Mgr./Music Ed. **URL:** http://www.antenna5.com.mk/about.aspxBroadcast.

TETOVO

1223 ■ Antenna5 - Tetovo - 106.9
Tetovska 35
91000 Skopje, Macedonia
Ph: 389 2 3111911

Format: Top 40; World Beat; Full Service. **Owner:** Antenna5, at above address. **Key Personnel:** Zoran Petrov, Gen. Mgr./Music Ed. **URL:** http://www.antenna5.com.mk/about.aspxBroadcast.

1224 ■ Antenna5 - Tetovo - 105.5
Tetovska 35
91000 Skopje, Macedonia
Ph: 389 2 3111911

Format: Top 40; World Beat; Full Service. **Owner:** Antenna5, at above address. **Key Personnel:** Zoran Petrov, Gen. Mgr./Music Ed. **URL:** http://www.antenna5.com.mk/about.aspxBroadcast.

VALANDOVO

1225 ■ Antenna5 - Valandovo - 106.3
Tetovska 35
91000 Skopje, Macedonia
Ph: 389 2 3111911

Format: Top 40; World Beat; Full Service. **Owner:** Antenna5, at above address. **Key Personnel:** Zoran Petrov, Gen. Mgr./Music Ed. **URL:** http://www.antenna5.com.mk/about.aspxBroadcast.

VELES

1226 ■ Antenna5 - Veles - 91.9
Tetovska 35
91000 Skopje, Macedonia
Ph: 389 2 3111911

Format: Top 40; World Beat; Full Service. **Owner:** Antenna5, at above address. **Key Personnel:** Zoran Petrov, Gen. Mgr./Music Ed. **URL:** http://www.antenna5.com.mk/about.aspxBroadcast.

VINICA

1227 ■ Antenna5 - Vinica - 97.9
Tetovska 35
91000 Skopje, Macedonia
Ph: 389 2 3111911

Format: Top 40; World Beat; Full Service. **Owner:** Antenna5, at above address. **Key Personnel:** Zoran Petrov, Gen. Mgr./Music Ed. **URL:** http://www.antenna5.com.mk/about.aspxBroadcast.

1228 ■ Antenna5 - Vinica - 104.8
Tetovska 35
91000 Skopje, Macedonia
Ph: 389 2 3111911

Format: Top 40; World Beat; Full Service. **Owner:** Antenna5, at above address. **Key Personnel:** Zoran Petrov, Gen. Mgr./Music Ed. **URL:** http://www.antenna5.com.mk/about.aspxBroadcast.

MALAYSIA

BESUT

1229 ■ Minnal FM - Besut - 95.3
Tingkat 4, Wisma Radio
Peti Surat 11272
Angkasapuri
50740 Kuala Lumpur, Malaysia
Ph: 60 3 22887279
Fax: 60 3 22849137
E-mail: mail@minnalfm.com

Format: Classical; News; Information. **Owner:** Radio Televisyen Malaysia, Jabatan Penyiaran Malaysia, Angkasapuri, 50614 Kuala Lumpur, Malaysia. **Operating Hours:** Continuous. **Key Personnel:** Raja Sekaran, Director; S. Santa, Supvr. **URL:** http://www.minnalfm.com/english/frequency.html.

BINTULU

1230 ■ Traxx FM - Bintulu - 98.5
2nd Fl., Wisma Radio
Angkasapuri
PO Box 11272
50740 Kuala Lumpur, Malaysia
Ph: 60 3 22887663
Fax: 60 3 22845750
E-mail: feedback@traxxfm.net

Format: Contemporary Hit Radio (CHR); Talk; Information. **Owner:** Radio Televisyen Malaysia, Jabatan Penyiaran Malaysia, Angkasapuri, 50614 Kuala Lumpur, Malaysia. **Operating Hours:** Continuous. **Key Personnel:** Norbaiyah Baharudin, Controller, norabaha@rtm.net.my; Kak Salmiah, Admin. **URL:** http://www.traxxfm.net/frequencies.php.

BUKIT TINGGI

1231 ■ Traxx FM - Bukit Tinggi - 92.9
2nd Fl., Wisma Radio
Angkasapuri
PO Box 11272
50740 Kuala Lumpur, Malaysia
Ph: 60 3 22887663
Fax: 60 3 22845750
E-mail: feedback@traxxfm.net

Format: Contemporary Hit Radio (CHR); Talk; Information. **Owner:** Radio Televisyen Malaysia, Jabatan Penyiaran Malaysia, Angkasapuri, 50614 Kuala Lumpur, Malaysia. **Operating Hours:** Continuous. **Key Personnel:** Norbaiyah Baharudin, Controller, norabaha@rtm.net.my; Kak Salmiah, Admin. **URL:** http://www.traxxfm.net/frequencies.php.

GURUN

1232 ■ Traxx FM - Gurun - 98.7
2nd Fl., Wisma Radio
Angkasapuri
PO Box 11272
50740 Kuala Lumpur, Malaysia
Ph: 60 3 22887663
Fax: 60 3 22845750
E-mail: feedback@traxxfm.net

Format: Contemporary Hit Radio (CHR); Talk; Information. **Owner:** Radio Televisyen Malaysia, Jabatan Penyiaran Malaysia, Angkasapuri, 50614 Kuala Lumpur, Malaysia. **Operating Hours:** Continuous. **Key Personnel:** Norbaiyah Baharudin, Controller, norabaha@rtm.net.my; Kak Salmiah, Admin. **URL:** http://www.traxxfm.net/frequencies.php.

IPOH

1233 ■ Traxx FM - Ipoh - 90.1
2nd Fl., Wisma Radio
Angkasapuri
PO Box 11272
50740 Kuala Lumpur, Malaysia
Ph: 60 3 22887663
Fax: 60 3 22845750
E-mail: feedback@traxxfm.net

Format: Contemporary Hit Radio (CHR); Talk;

Information. **Owner:** Radio Televisyen Malaysia, Jabatan Penyiaran Malaysia, Angkasapuri, 50614 Kuala Lumpur, Malaysia. **Operating Hours:** Continuous. **Key Personnel:** Norbaiyah Baharudin, Controller, norabaha@rtm.net.my; Kak Salmiah, Admin. **URL:** http://www.traxxfm.net/frequencies.php.

JELI

1234 ■ Minnal FM - Jeli - 92.2
Tingkat 4, Wisma Radio
Peti Surat 11272
Angkasapuri
50740 Kuala Lumpur, Malaysia
Ph: 60 3 22887279
Fax: 60 3 22849137
E-mail: mail@minnalfm.com

Format: Classical; News; Information. **Owner:** Radio Televisyen Malaysia, Jabatan Penyiaran Malaysia, Angkasapuri, 50614 Kuala Lumpur, Malaysia. **Operating Hours:** Continuous. **Key Personnel:** Raja Sekaran, Director; S. Santa, Supvr. **URL:** http://www.minnalfm.com/english/frequency.html.

1235 ■ Traxx FM - Jeli - 90.8
2nd Fl., Wisma Radio
Angkasapuri
PO Box 11272
50740 Kuala Lumpur, Malaysia
Ph: 60 3 22887663
Fax: 60 3 22845750
E-mail: feedback@traxxfm.net

Format: Contemporary Hit Radio (CHR); Talk; Information. **Owner:** Radio Televisyen Malaysia, Jabatan Penyiaran Malaysia, Angkasapuri, 50614 Kuala Lumpur, Malaysia. **Operating Hours:** Continuous. **Key Personnel:** Norbaiyah Baharudin, Controller, norabaha@rtm.net.my; Kak Salmiah, Admin. **URL:** http://www.traxxfm.net/frequencies.php.

JERANTUT

1236 ■ Minnal FM - Jerantut - 91.9
Tingkat 4, Wisma Radio
Peti Surat 11272
Angkasapuri
50740 Kuala Lumpur, Malaysia
Ph: 60 3 22887279
Fax: 60 3 22849137
E-mail: mail@minnalfm.com

Format: Classical; News; Information. **Owner:** Radio Televisyen Malaysia, Jabatan Penyiaran Malaysia, Angkasapuri, 50614 Kuala Lumpur, Malaysia. **Operating Hours:** Continuous. **Key Personnel:** Raja Sekaran, Director; S. Santa, Supvr. **URL:** http://www.minnalfm.com/english/frequency.html.

JOHOR BAHARU

1237 ■ Traxx FM - Johor Baharu - 102.9
2nd Fl., Wisma Radio
Angkasapuri
PO Box 11272
50740 Kuala Lumpur, Malaysia
Ph: 60 3 22887663
Fax: 60 3 22845750
E-mail: feedback@traxxfm.net

Format: Contemporary Hit Radio (CHR); Talk; Information. **Owner:** Radio Televisyen Malaysia, Jabatan Penyiaran Malaysia, Angkasapuri, 50614 Kuala Lumpur, Malaysia. **Operating Hours:** Continuous. **Key Personnel:** Norbaiyah Baharudin, Controller, norabaha@rtm.net.my; Kak Salmiah, Admin. **URL:** http://www.traxxfm.net/frequencies.php.

KOTA BARU

1238 ■ Traxx FM - Kota Baru - 104.7
2nd Fl., Wisma Radio
Angkasapuri
PO Box 11272
50740 Kuala Lumpur, Malaysia
Ph: 60 3 22887663
Fax: 60 3 22845750
E-mail: feedback@traxxfm.net

Format: Contemporary Hit Radio (CHR); Talk; Information. **Owner:** Radio Televisyen Malaysia, Jabatan Penyiaran Malaysia, Angkasapuri, 50614 Kuala Lumpur, Malaysia. **Operating Hours:** Continuous. **Key Personnel:** Norbaiyah Baharudin, Controller, norabaha@rtm.net.my; Kak Salmiah, Admin. **URL:** http://www.traxxfm.net/frequencies.php.

KOTA BELUD

1239 ■ Traxx FM - Kota Belud - 102.5
2nd Fl., Wisma Radio
Angkasapuri
PO Box 11272
50740 Kuala Lumpur, Malaysia
Ph: 60 3 22887663
Fax: 60 3 22845750
E-mail: feedback@traxxfm.net

Format: Contemporary Hit Radio (CHR); Talk; Information. **Owner:** Radio Televisyen Malaysia, Jabatan Penyiaran Malaysia, Angkasapuri, 50614 Kuala Lumpur, Malaysia. **Operating Hours:** Continuous. **Key Personnel:** Norbaiyah Baharudin, Controller, norabaha@rtm.net.my; Kak Salmiah, Admin. **URL:** http://www.traxxfm.net/frequencies.php.

KOTA KINABALU

1240 ■ Biotechnology & Molecular Biology Reviews
Academic Journals
c/o P. Ravindra, PhD, Ed.-in-Ch.
School of Engineering & IT
University Malaysia Sabah
88999 Kota Kinabalu, Malaysia
Publication E-mail: bmbr_acadjourn@yahoo.com
Publisher E-mail: service@academicjournals.org

Journal covering reviews in biotechnology and molecular biology. **Freq:** Quarterly. **Key Personnel:** P. Ravindra, PhD, Editor-in-Chief, dr_ravindra@hotmail.com; David Maina Menge, PhD, Editor, dmenge@uci.edu; Evans Kaimoyo, PhD, Editor, ek246@cornell.edu. **ISSN:** 1538-2273. **Remarks:** Accepts advertising. **URL:** http://www.academicjournals.org/bmbr/index.htm. **Circ:** (Not Reported)

1241 ■ Traxx FM - Kota Kinabalu - 90.7
2nd Fl., Wisma Radio
Angkasapuri
PO Box 11272
50740 Kuala Lumpur, Malaysia
Ph: 60 3 22887663
Fax: 60 3 22845750
E-mail: feedback@traxxfm.net

Format: Contemporary Hit Radio (CHR); Talk; Information. **Owner:** Radio Televisyen Malaysia, Jabatan Penyiaran Malaysia, Angkasapuri, 50614 Kuala Lumpur, Malaysia. **Operating Hours:** Continuous. **Key Personnel:** Norbaiyah Baharudin, Controller, norabaha@rtm.net.my; Kak Salmiah, Admin. **URL:** http://www.traxxfm.net/frequencies.php.

KUALA BESUT

1242 ■ Traxx FM - Kuala Besut - 97.0
2nd Fl., Wisma Radio
Angkasapuri
PO Box 11272
50740 Kuala Lumpur, Malaysia
Ph: 60 3 22887663
Fax: 60 3 22845750
E-mail: feedback@traxxfm.net

Format: Contemporary Hit Radio (CHR); Talk; Information. **Owner:** Radio Televisyen Malaysia, Jabatan Penyiaran Malaysia, Angkasapuri, 50614 Kuala Lumpur, Malaysia. **Operating Hours:** Continuous. **Key Personnel:** Norbaiyah Baharudin, Controller, norabaha@rtm.net.my; Kak Salmiah, Admin. **URL:** http://www.traxxfm.net/frequencies.php.

KUALA DUNGUN

1243 ■ Traxx FM - Kuala Dungun - 98.8
2nd Fl., Wisma Radio
Angkasapuri
PO Box 11272
50740 Kuala Lumpur, Malaysia
Ph: 60 3 22887663
Fax: 60 3 22845750
E-mail: feedback@traxxfm.net

Format: Contemporary Hit Radio (CHR); Talk; Information. **Owner:** Radio Televisyen Malaysia, Jabatan Penyiaran Malaysia, Angkasapuri, 50614 Kuala Lumpur, Malaysia. **Operating Hours:** Continuous. **Key Personnel:** Norbaiyah Baharudin, Controller, norabaha@rtm.net.my; Kak Salmiah, Admin. **URL:** http://www.traxxfm.net/frequencies.php.

KUALA LUMPUR

Minnal FM - Besut - See Besut
Minnal FM - Jeli - See Jeli
Minnal FM - Jerantut - See Jerantut

1244 ■ Minnal FM - Kuala Lumpur - 92.3
Tingkat 4, Wisma Radio
Peti Surat 11272
Angkasapuri
50740 Kuala Lumpur, Malaysia
Ph: 60 3 22887279
Fax: 60 3 22849137
E-mail: mail@minnalfm.com

Format: Classical; News; Information. **Owner:** Radio Televisyen Malaysia, Jabatan Penyiaran Malaysia, Angkasapuri, 50614 Kuala Lumpur, Malaysia. **Operating Hours:** Continuous. **Key Personnel:** Raja Sekaran, Director; S. Santa, Supvr. **URL:** http://www.minnalfm.com/english/frequency.html.

Minnal FM - Kuala Terengganu - See Kuala Terengganu
Minnal FM - Kuantan - See Kuantan
Minnal FM - Mersing - See Mersing
Minnal FM - Muar - See Muar
Minnal FM - Taiping - See Taiping
Traxx FM - Bintulu - See Bintulu
Traxx FM - Bukit Tinggi - See Bukit Tinggi
Traxx FM - Gurun - See Gurun
Traxx FM - Ipoh - See Ipoh
Traxx FM - Jeli - See Jeli
Traxx FM - Johor Baharu - See Johor Baharu
Traxx FM - Kota Baru - See Kota Baru
Traxx FM - Kota Belud - See Kota Belud
Traxx FM - Kota Kinabalu - See Kota Kinabalu
Traxx FM - Kuala Besut - See Kuala Besut
Traxx FM - Kuala Dungun - See Kuala Dungun
Traxx FM - Kuala Terengganu - See Kuala Terengganu
Traxx FM - Kuantan - See Kuantan
Traxx FM - Kuching - See Kuching
Traxx FM - Miri - See Miri
Traxx FM - Sandakan - See Sandakan
Traxx FM - Taiping - See Taiping
Traxx FM - Tangkak - See Tangkak

KUALA TERENGGANU

1245 ■ Minnal FM - Kuala Terengganu - 87.9
Tingkat 4, Wisma Radio
Peti Surat 11272
Angkasapuri
50740 Kuala Lumpur, Malaysia
Ph: 60 3 22887279
Fax: 60 3 22849137
E-mail: mail@minnalfm.com

Format: Classical; News; Information. **Owner:** Radio Televisyen Malaysia, Jabatan Penyiaran Malaysia, Angkasapuri, 50614 Kuala Lumpur, Malaysia. **Operating Hours:** Continuous. **Key Personnel:** Raja Sekaran, Director; S. Santa, Supvr. **URL:** http://www.minnalfm.com/english/frequency.html.

1246 ■ Traxx FM - Kuala Terengganu - 89.7
2nd Fl., Wisma Radio
Angkasapuri
PO Box 11272
50740 Kuala Lumpur, Malaysia
Ph: 60 3 22887663
Fax: 60 3 22845750
E-mail: feedback@traxxfm.net

Format: Contemporary Hit Radio (CHR); Talk; Information. **Owner:** Radio Televisyen Malaysia, Jabatan Penyiaran Malaysia, Angkasapuri, 50614 Kuala Lumpur, Malaysia. **Operating Hours:** Continuous. **Key Personnel:** Norbaiyah Baharudin, Controller, norabaha@rtm.net.my; Kak Salmiah, Admin. **URL:** http://www.traxxfm.net/frequencies.php.

KUANTAN

1247 ■ Minnal FM - Kuantan - 103.3
Tingkat 4, Wisma Radio
Peti Surat 11272
Angkasapuri
50740 Kuala Lumpur, Malaysia
Ph: 60 3 22887279
Fax: 60 3 22849137
E-mail: mail@minnalfm.com

Format: Classical; News; Information. **Owner:** Radio Televisyen Malaysia, Jabatan Penyiaran Malaysia, Angkasapuri, 50614 Kuala Lumpur, Malaysia. **Operating Hours:** Continuous. **Key Personnel:** Raja Sekaran, Director; S. Santa, Supvr. **URL:** http://www.minnalfm.com/english/frequency.html.

1248 ■ Traxx FM - Kuantan - 105.3
2nd Fl., Wisma Radio
Angkasapuri
PO Box 11272
50740 Kuala Lumpur, Malaysia
Ph: 60 3 22887663
Fax: 60 3 22845750
E-mail: feedback@traxxfm.net

Format: Contemporary Hit Radio (CHR); Talk; Information. **Owner:** Radio Televisyen Malaysia, Jabatan Penyiaran Malaysia, Angkasapuri, 50614 Kuala Lumpur, Malaysia. **Operating Hours:** Continuous. **Key Personnel:** Norbaiyah Baharudin, Controller, norabaha@rtm.net.my; Kak Salmiah, Admin. **URL:** http://www.traxxfm.net/frequencies.php.

KUCHING

1249 ■ Traxx FM - Kuching - 89.9
2nd Fl., Wisma Radio
Angkasapuri
PO Box 11272
50740 Kuala Lumpur, Malaysia
Ph: 60 3 22887663
Fax: 60 3 22845750
E-mail: feedback@traxxfm.net

Format: Contemporary Hit Radio (CHR); Talk; Information. **Owner:** Radio Televisyen Malaysia, Jabatan Penyiaran Malaysia, Angkasapuri, 50614 Kuala Lumpur, Malaysia. **Operating Hours:** Continuous. **Key Personnel:** Norbaiyah Baharudin, Controller, norabaha@rtm.net.my; Kak Salmiah, Admin. **URL:** http://www.traxxfm.net/frequencies.php.

MELAKA

1250 ■ International Journal of Precision Technology
Inderscience Publishers
c/o Prof. V.C. Venkatesh, Ed.-in-Ch.
Multimedia University
Faculty of Enginooring & Technology
Jalan Ayer Keroh Lama
75450 Melaka, Malaysia
Publication E-mail: ijptech@inderscience.com
Publisher E-mail: editor@inderscience.com

Journal covering technological advances in precision engineering. **Founded:** 2007. **Freq:** 4/yr. **Key Personnel:** Prof. V. C. Venkatesh, Editor-in-Chief, v.c. venkatesh@mmu.edu.my. **ISSN:** 1755-2060. **Subscription Rates:** EUR470 individuals includes surface mail, print only; EUR640 individuals print and online. **URL:** http://www.inderscience.com/browse/index.php?journalCODE=ijptech.

MERSING

1251 ■ Minnal FM - Mersing - 101.1
Tingkat 4, Wisma Radio
Peti Surat 11272
Angkasapuri
50740 Kuala Lumpur, Malaysia
Ph: 60 3 22887279
Fax: 60 3 22849137
E-mail: mail@minnalfm.com

Format: Classical; News; Information. **Owner:** Radio Televisyen Malaysia, Jabatan Penyiaran Malaysia, Angkasapuri, 50614 Kuala Lumpur, Malaysia. **Operating Hours:** Continuous. **Key Personnel:** Raja Sekaran, Director; S. Santa, Supvr. **URL:** http://www.minnalfm.com/english/frequency.html.

MIRI

1252 ■ Traxx FM - Miri - 104.5
2nd Fl., Wisma Radio
Angkasapuri
PO Box 11272
50740 Kuala Lumpur, Malaysia
Ph: 60 3 22887663
Fax: 60 3 22845750
E-mail: feedback@traxxfm.net

Format: Contemporary Hit Radio (CHR); Talk; Information. **Owner:** Radio Televisyen Malaysia, Jabatan Penyiaran Malaysia, Angkasapuri, 50614 Kuala Lumpur, Malaysia. **Operating Hours:** Continuous. **Key Personnel:** Norbaiyah Baharudin, Controller, norabaha@rtm.net.my; Kak Salmiah, Admin. **URL:** http://www.traxxfm.net/frequencies.php.

MUAR

1253 ■ Minnal FM - Muar - 103.3
Tingkat 4, Wisma Radio
Peti Surat 11272
Angkasapuri
50740 Kuala Lumpur, Malaysia
Ph: 60 3 22887279
Fax: 60 3 22849137
E-mail: mail@minnalfm.com

Format: Classical; News; Information. **Owner:** Radio Televisyen Malaysia, Jabatan Penyiaran Malaysia, Angkasapuri, 50614 Kuala Lumpur, Malaysia. **Operating Hours:** Continuous. **Key Personnel:** Raja Sekaran, Director; S. Santa, Supvr. **URL:** http://www.minnalfm.com/english/frequency.html.

SANDAKAN

1254 ■ Traxx FM - Sandakan - 94.3
2nd Fl., Wisma Radio
Angkasapuri
PO Box 11272
50740 Kuala Lumpur, Malaysia
Ph: 60 3 22887663
Fax: 60 3 22845750
E-mail: feedback@traxxfm.net

Format: Contemporary Hit Radio (CHR); Talk; Information. **Owner:** Radio Televisyen Malaysia, Jabatan Penyiaran Malaysia, Angkasapuri, 50614 Kuala Lumpur, Malaysia. **Operating Hours:** Continuous. **Key Personnel:** Norbaiyah Baharudin, Controller, norabaha@rtm.net.my; Kak Salmiah, Admin. **URL:** http://www.traxxfm.net/frequencies.php.

TAIPING

1255 ■ Minnal FM - Taiping - 107.9
Tingkat 4, Wisma Radio
Peti Surat 11272
Angkasapuri
50740 Kuala Lumpur, Malaysia
Ph: 60 3 22887279
Fax: 60 3 22849137
E-mail: mail@minnalfm.com

Format: Classical; News; Information. **Owner:** Radio Televisyen Malaysia, Jabatan Penyiaran Malaysia, Angkasapuri, 50614 Kuala Lumpur, Malaysia. **Operating Hours:** Continuous. **Key Personnel:** Raja Sekaran, Director; S. Santa, Supvr. **URL:** http://www.minnalfm.com/english/frequency.html.

1256 ■ Traxx FM - Taiping - 105.3
2nd Fl., Wisma Radio
Angkasapuri
PO Box 11272
50740 Kuala Lumpur, Malaysia
Ph: 60 3 22887663
Fax: 60 3 22845750
E-mail: feedback@traxxfm.net

Format: Contemporary Hit Radio (CHR); Talk; Information. **Owner:** Radio Televisyen Malaysia, Jabatan Penyiaran Malaysia, Angkasapuri, 50614 Kuala Lumpur, Malaysia. **Operating Hours:** Continuous. **Key Personnel:** Norbaiyah Baharudin, Controller, norabaha@rtm.net.my; Kak Salmiah, Admin. **URL:** http://www.traxxfm.net/frequencies.php.

TANGKAK

1257 ■ Traxx FM - Tangkak - 97.4
2nd Fl., Wisma Radio
Angkasapuri
PO Box 11272
50740 Kuala Lumpur, Malaysia
Ph: 60 3 22887663
Fax: 60 3 22845750
E-mail: feedback@traxxfm.net

Format: Contemporary Hit Radio (CHR); Talk; Information. **Owner:** Radio Televisyen Malaysia, Jabatan Penyiaran Malaysia, Angkasapuri, 50614 Kuala Lumpur, Malaysia. **Operating Hours:** Continuous. **Key Personnel:** Norbaiyah Baharudin, Controller, norabaha@rtm.net.my; Kak Salmiah, Admin. **URL:** http://www.traxxfm.net/frequencies.php.

MALTA

SAINT JULIAN

1258 ■ The Malta Business Weekly
Standard Publications Ltd.
Standard House
Birkirkara Hill
Saint Julian STJ 09, Malta
Ph: 356 21345888

Newspaper featuring business. **Freq:** Weekly. **Key Personnel:** Christopher Sultana, Editor, csultana@independent.com.mt; Noel Azzopardi, Managing Editor, nazzopardi@independent.com.mt. **Remarks:** Accepts advertising. **URL:** http://www.maltabusinessweekly.com.mt/mainpage.asp. **Circ:** (Not Reported)

1259 ■ The Malta Independent
Standard Publications Ltd.
Standard House
Birkirkara Hill
Saint Julian STJ 09, Malta
Ph: 356 21345888

Newspaper featuring latest news, information, and events. **Key Personnel:** Stephen Calleja, Editor, scalleja@independent.com.mt; Noel Azzopardi, Managing Editor, nazzopardi@independent.com.mt. **Remarks:** Accepts advertising. **URL:** http://www.independent.com.mt/mainpage.asp. **Circ:** (Not Reported)

Circulation: ★ = ABC; △ = BPA; ♦ = CAC; • = CCAB; ❑ = VAC; ⊕ = PO Statement; ‡ = Publisher's Report; Boldface figures = sworn; Light figures = estimated.

1260 ■ The Malta Independent on Sunday
Standard Publications Ltd.
Standard House
Birkirkara Hill
Saint Julian STJ 09, Malta
Ph: 356 21345888

Newspaper featuring latest news, information, and events. **Freq:** Weekly (Sun.). **Key Personnel:** Noel Grima, Editor, ngrima@independent.com.mt; Noel Azzopardi, Managing Editor, nazzopardi@independent.com.mt. **Remarks:** Accepts advertising. **URL:** http://www.independent.com.mt/. **Circ:** (Not Reported)

MEXICO

MEXICO CITY

1261 ■ Stereo Joya-FM - 93.7
Constituyentes 1154, 7 Piso
Col. Lomas Altas
11950 Mexico City, Federal District, Mexico
Ph: 52 55 7284800
Fax: 52 55 7284875

Format: Hispanic. **Owner:** Grupo Radio Centro, at above address. **Key Personnel:** Mariano Osorio, Contact; Jesus Zuniga, Contact. **URL:** http://www.stereojoya.com.mx/.

1262 ■ Universal Stereo-FM - 92.1
Constituyentes 1154, 7 Piso
Col. Lomas Altas
11950 Mexico City, Federal District, Mexico
Ph: 52 55 7284800
Fax: 52 55 7284875
E-mail: universal@grc.com.mx

Format: Classic Rock. **Owner:** Grupo Radio Centro, at above address. **Key Personnel:** Adolfo Fernandez, Contact; Manuel Guerrero, Contact. **URL:** http://www.universalstereo.com.mx/.

NETHERLANDS

AMSTERDAM

1263 ■ Babel
John Benjamins Publishing Co.
Klaprozenweg 105
PO Box 36224
NL-1033 NN Amsterdam, Netherlands
Ph: 31 206 304747
Fax: 31 206 739773
Publisher E-mail: benjamins@presswarehouse.com

Journal designed primarily for translators and interpreters. **Freq:** Quarterly. **Key Personnel:** Rene Haeseryn, Editor-in-Chief. **Subscription Rates:** EUR178 institutions; EUR80 individuals. **URL:** http://www.benjamins.nl/; http://www.benjamins.com/cgi-bin/t_seriesview.cgi?series=Babel.

1264 ■ English Text Construction
John Benjamins Publishing Co.
Klaprozenweg 105
PO Box 36224
NL-1033 NN Amsterdam, Netherlands
Ph: 31 206 304747
Fax: 31 206 739773
Publisher E-mail: benjamins@presswarehouse.com

Journal focusing on English text construction. **Freq:** Semiannual. **Key Personnel:** An Laffut, Editor; Keith Carlon, Managing Editor. **ISSN:** 1874-8767. **Subscription Rates:** EUR148 institutions; EUR70 individuals. **URL:** http://www.benjamins.nl/; http://www.benjamins.com/cgi-bin/t_seriesview.cgi?series=ETC.

1265 ■ Information Design Journal
John Benjamins Publishing Co.
Klaprozenweg 105
PO Box 36224
NL-1033 NN Amsterdam, Netherlands
Ph: 31 206 304747
Fax: 31 206 739773
Publisher E-mail: benjamins@presswarehouse.com

Journal featuring research and practice in information design. **Freq:** 3/yr. **Key Personnel:** Jan Renkema, Gen. Ed.; Carel Jansen, Editor. **ISSN:** 0142-5471. **Subscription Rates:** EUR190 institutions; EUR75 individuals. **URL:** http://www.benjamins.nl/; http://www.benjamins.com/cgi-bin/t_seriesview.cgi?series=IDJ.

1266 ■ Interpreting
John Benjamins Publishing Co.
Klaprozenweg 105
PO Box 36224
NL-1033 NN Amsterdam, Netherlands
Ph: 31 206 304747
Fax: 31 206 739773
Publisher E-mail: benjamins@presswarehouse.com

Journal featuring research and practice in interpreting. **Freq:** Semiannual. **Key Personnel:** Miriam Shlesinger, Editor. **Subscription Rates:** EUR185 institutions; EUR75 institutions. **URL:** http://www.benjamins.nl/; http://www.benjamins.com/cgi-bin/t_seriesview.cgi?series=INTP.

1267 ■ Languages in Contrast
John Benjamins Publishing Co.
Klaprozenweg 105
PO Box 36224
NL-1033 NN Amsterdam, Netherlands
Ph: 31 206 304747
Fax: 31 206 739773
Publisher E-mail: benjamins@presswarehouse.com

Journal featuring contrastive studies of two or more languages. **Freq:** Semiannual. **Key Personnel:** Silvia Bernardini, Editor. **ISSN:** 1387-6759. **Subscription Rates:** EUR70 individuals; EUR165 institutions. **URL:** http://www.benjamins.nl/; http://www.benjamins.com/cgi-bin/t_seriesview.cgi?series=LiC.

1268 ■ Narrative Inquiry
John Benjamins Publishing Co.
Klaprozenweg 105
PO Box 36224
NL-1033 NN Amsterdam, Netherlands
Ph: 31 206 304747
Fax: 31 206 739773
Publisher E-mail: benjamins@presswarehouse.com

Journal featuring a forum for theoretical, empirical, and methodological work on narrative. **Freq:** Semiannual. **Key Personnel:** Michael Bamberg, Editor. **ISSN:** 1387-6740. **Subscription Rates:** EUR80 individuals; EUR279 institutions. **URL:** http://www.benjamins.nl/; http://www.benjamins.com/cgi-bin/t_seriesview.cgi?series=NI.

MAASTRICHT

1269 ■ Legisprudence
Hart Publishing Ltd.
c/o Jaap Hage, Gen. Ed.
Department of Metajuridica
University of Maastricht
PO Box 616
NL-6200 Maastricht, Netherlands
Ph: 31 43 3883053
Publisher E-mail: mail@hartpub.co.uk

Journal covering legislation. **Freq:** 3/yr. **Key Personnel:** Jaap Hage, Gen. Ed., jaap.hage@metajur.unimaas.nl; Luc Wintgens, Gen. Ed., luc.wintgens@kubrussel.ac.be. **Subscription Rates:** 85 individuals standard, United Kingdom and Europe; 95 other countries standard; 35 individuals reduced, United Kingdom and Europe; 42 other countries reduced. **Remarks:** Accepts advertising. **URL:** http://www.hartjournals.co.uk/legisprudence/. **Circ:** (Not Reported)

NEW ZEALAND

ALEXANDRA

1270 ■ Newstalk ZB-FM Central Otago - 95.1
54 Cook St.
Private Bag 92198
Auckland, New Zealand
Ph: 64 9 3730000

Format: News; Talk. **Owner:** The Radio Network, at above address. **Key Personnel:** Paul Holmes, Contact, grumpy@newstalkzb.co.nz; Leighton Smith, Contact, leighton@newstalkzb.co.nz; Danny Watson, Contact, danny@newstalkzb.co.nz. **URL:** http://www.newstalkzb.co.nz/featdetail.asp?recnumber=1.

AUCKLAND

Newstalk ZB-AM Northland - See Whangarei
Newstalk ZB-AM Northland - See Whangarei
Newstalk ZB-AM Southland - See Invercargill
Newstalk ZB-AM Waikato - See Hamilton
Newstalk ZB-FM Central Otago - See Alexandra
Newstalk ZB-FM Central Otago - See Wanaka
Radio Sport-AM Southland - See Invercargill

BLENHEIM

1271 ■ Classic Hits-FM Marlborough - 96.9
12a Kinross St.
PO Box 225
Blenheim, New Zealand
Ph: 64 3 5780129
Fax: 64 3 5780981

Format: Adult Contemporary. **Owner:** The Radio Network, 54 Cook St., Private Bag 92198, Auckland, New Zealand, Fax: 64 9 3674797. **Key Personnel:** Glen Kirby, Contact, glennkirby@radionetwork.co.nz; Jenny Laws, Contact. **Ad Rates:** Advertising accepted; rates available upon request. **URL:** http://www.classichits.co.nz/.

DUNEDIN

1272 ■ Otago Daily Times
Allied Press Ltd.
52 Stuart St.
PO Bcx 517
Dunedin, New Zealand
Ph: 64 3 4774760
Publisher E-mail: corporate@alliedpress.co.nz

Newspaper featuring news and events in New Zealand. **Freq:** Mon.-Sat. **Key Personnel:** Paul Dwyer, Advertising Mgr., phone 64 3 4793565, fax 64 3 4747421, paul.dwyer@alliedpress.co.nz. **Subscription Rates:** NZ$245 individuals digital; NZ$300 individuals print. **Remarks:** Accepts advertising. **URL:** http://www.alliedpress.co.nz/. **Circ:** Combined 45,004

HAMILTON

1273 ■ New Zealand Journal of Zoology
Royal Society of New Zealand
c/o Dr C.M. King, Ed.
Department of Biological Sciences
University of Waikato
Private Bag 3105
Hamilton, New Zealand
Ph: 64 7 8562889
Fax: 64 7 8384324

Journal covering fields of zoological science concerning New Zealand, the Pacific Basin, and Antarctica. **Freq:** Quarterly. **Key Personnel:** Dr C.M. King, Editor, nzjz@rsnz.org. **ISSN:** 0301-4223. **Subscription Rates:** NZ$455 institutions print and online; NZ$215 individuals print and online; NZ$172 members print and online; US$325 institutions print and online; US$145 individuals print and online; NZ$385 institutions online; NZ$180 individuals online. **URL:** http://www.rsnz.org/publish/nzjz/.

1274 ■ Classic Hits-FM Waikato - 98.6
PO Box 489
Hamilton, New Zealand
Ph: 64 7 8580700
Fax: 64 7 8580730

Format: Adult Contemporary. **Owner:** The Radio Network, 54 Cook St., Private Bag 92198, Auckland, New Zealand, Fax: 64 9 3674797. **Key Personnel:** Jason Reeves, Contact. **URL:** http://www.classichits.co.nz.

1275 ■ Newstalk ZB-AM Waikato - 1296
54 Cook St.
Private Bag 92198
Auckland, New Zealand
Ph: 64 9 3730000

Format: News; Talk. **Owner:** The Radio Network, at above address. **Key Personnel:** Paul Holmes, Contact, grumpy@newstalkzb.co.nz; Leighton Smith,

Contact, leighton@newstalkzb.co.nz; Danny Watson, Contact, danny@newstalkzb.co.nz. **URL:** http://www.newstalkzb.co.nz/featdetail.asp?recnumber=1.

INVERCARGILL

1276 ■ Classic Hits-FM Southland - 98.8
Radio Network House, 1st Fl.
PO Box 802
Invercargill, New Zealand
Ph: 64 3 2187209
Fax: 64 3 2111532

Format: Adult Contemporary. **Owner:** The Radio Network, 54 Cook St., Private Bag 92198, Auckland, New Zealand, Fax: 64 9 3674797. **Key Personnel:** John McDowell, Contact, boggy@classichits.co.nz. **Ad Rates:** Advertising accepted; rates available upon request. **URL:** http://www.classichits.co.nz.

1277 ■ Hauraki-FM Southland - 93.2
Radio Network House, 1st Fl.
PO Box 802
Invercargill, New Zealand
Ph: 64 3 2111503
Fax: 64 3 2111532
E-mail: trn.invercargill@radionetwork.co.nz

Format: Classic Rock. **Owner:** The Radio Network, 54 Cook St., Private Bag 92198, Auckland, New Zealand, Fax: 64 9 3674802. **Key Personnel:** Lee Piper, General Mgr., phone 64 3 2111501, leepiper@radionetwork.co.nz. **URL:** http://www.radiohauraki.co.nz/.

1278 ■ Newstalk ZB-AM Southland - 864
54 Cook St.
Private Bag 92198
Auckland, New Zealand
Ph: 64 9 3730000

Format: News; Talk. **Owner:** The Radio Network, at above address. **Key Personnel:** Paul Holmes, Contact, grumpy@newstalkzb.co.nz; Leighton Smith, Contact, leighton@newstalkzb.co.nz; Danny Watson, Contact, danny@newstalkzb.co.nz. **URL:** http://www.newstalkzb.co.nz/featdetail.asp?recnumber=1.

1279 ■ Radio Sport-AM Southland - 558
Private Bag 92198
Auckland, New Zealand
Ph: 64 9 3730000
Fax: 64 9 3674644
E-mail: sport@radiosport.co.nz

Format: Sports. **Owner:** The Radio Network, at above address. **Key Personnel:** Peter Everatt, Program Dir., phone 64 9 3674705, peter@radionetwork.co.nz; Malcolm Jordan, Prog. Ed., phone 64 9 3674672, malcolm@radiosport.co.nz; Deidre Bailey, Sales & Marketing Dir., phone 64 9 3674641, deidrebailey@radionetwork.co.nz. **URL:** http://www.radiosport.co.nz/.

NEW PLYMOUTH

1280 ■ Classic Hits-FM Taranaki - 90.7
PO Box 141
New Plymouth, New Zealand
Ph: 64 6 7592460
Fax: 64 6 7592440

Format: Adult Contemporary. **Owner:** The Radio Network, 54 Cook St., Private Bag 92198, Auckland, New Zealand, Fax: 64 9 3674797. **Key Personnel:** Charlotte Butler, Contact, gumboots@classichits.co.nz. **URL:** http://www.classichits.co.nz.

1281 ■ Classic Hits-FM Taranaki - 90.0
PO Box 141
New Plymouth, New Zealand
Ph: 64 6 7592460
Fax: 64 6 7592440

Format: Adult Contemporary. **Owner:** The Radio Network, 54 Cook St., Private Bag 92198, Auckland, New Zealand, Fax: 64 9 3674797. **Key Personnel:** Charlotte Butler, Contact, gumboots@classichits.co.nz. **URL:** http://www.classichits.co.nz.

1282 ■ Hauraki-FM Taranaki - 90.8
PO Box 141
New Plymouth, New Zealand
Ph: 64 6 7592460
Fax: 64 6 7592440
E-mail: taranaki@radionetwork.co.n

Format: Classic Rock. **Owner:** The Radio Network, 54 Cook St., Private Bag 92198, Auckland, New Zealand, Fax: 64 9 3674802. **Key Personnel:** Richard Williams, General Mgr., phone 64 6 7592469, fax 64 6 7592442, richardwilliams@radionetwork.co.nz. **URL:** http://www.radiohauraki.co.nz.

PALMERSTON NORTH

1283 ■ Classic Hits-FM Manawatu - 97.8
619 Main St.
PO Box 1045
Palmerston North, New Zealand
Ph: 64 6 3503550
Fax: 64 6 3503580
E-mail: manawatu@classichits.co.nz

Format: Adult Contemporary. **Owner:** The Radio Network, 54 Cook St., Private Bag 92198, Auckland, New Zealand, Fax: 64 9 3674797. **Key Personnel:** Robbie Walker, Contact, mikeandrobbie@classichits.co.nz; Mike Roke, Contact. **URL:** http://www.classichits.co.nz/.

PARAPARAUMU

1284 ■ Classic Hits-FM Kapiti - 92.7
Shop 5, Coastlands Mall
PO Box 596
Paraparaumu, New Zealand
Ph: 64 4 2983086
Fax: 64 4 2983086

Format: Adult Contemporary. **Owner:** The Radio Network, 54 Cook St., Private Bag 92198, Auckland, New Zealand, Fax: 64 9 3674797. **Key Personnel:** Phil Costello, Contact, phil@classichits.co.n. **URL:** http://www.classichits.co.nz/.

TAURANGA

1285 ■ Classic Hits-FM Bay of Plenty - 95.0
4th Fl., Harrington House
Harrington St.
PO Box 642
Tauranga, New Zealand
Ph: 64 7 5789139
Fax: 64 7 5778522

Format: Adult Contemporary. **Owner:** The Radio Network, 54 Cook St., Private Bag 92198, Auckland, New Zealand, Fax: 64 9 3674797. **Key Personnel:** Louise Dean, Contact. **URL:** http://www.classichits.co.nz/.

WANAKA

1286 ■ Newstalk ZB-FM Central Otago - 90.6
54 Cook St.
Private Bag 92198
Auckland, New Zealand
Ph: 64 9 3730000

Format: News; Talk. **Owner:** The Radio Network, at above address. **Key Personnel:** Paul Holmes, Contact, grumpy@newstalkzb.co.nz; Leighton Smith, Contact, leighton@newstalkzb.co.nz. **URL:** http://www.newstalkzb.co.nz/featdetail.asp?recnumber=1.

WELLINGTON

1287 ■ Journal of the Royal Society of New Zealand
Royal Society of New Zealand
4 Halswell St.
PO Box 598
Wellington 6011, New Zealand
Ph: 64 4 4727421
Fax: 64 4 4731841

Journal covering research papers on the science of New Zealand and the Pacific region. **Freq:** Quarterly. **Key Personnel:** Dr. Anna Meyer, Editor. **ISSN:** 0303-6758. **Subscription Rates:** NZ$455 institutions print and online; NZ$215 individuals print and online; NZ$172 members print and online; US$325 institutions print and online; US$145 individuals print and online; NZ$385 institutions online; NZ$180 individuals online. **URL:** http://www.rsnz.org/publish/jrsnz/.

1288 ■ Kotuitui
Royal Society of New Zealand
4 Halswell St.
PO Box 598
Wellington 6011, New Zealand
Ph: 64 4 4727421
Fax: 64 4 4731841
Publication E-mail: kotuitui@rsnz.org

Journal covering all social science disciplines. **Key Personnel:** Prof. Stephen Levine, Editor; Prof. Paul Spoonley, Editor, kotuitui@rsnz.org. **ISSN:** 1177-083X. **URL:** http://www.rsnz.org/publish/kotuitui/.

1289 ■ New Zealand Journal of Agricultural Research
Royal Society of New Zealand
4 Halswell St.
PO Box 598
Wellington 6011, New Zealand
Ph: 64 4 4727421
Fax: 64 4 4731841

Journal covering aspects of animal and pastoral science relevant to temperate and subtropical regions. **Freq:** Quarterly. **Key Personnel:** David Swain, Editor, nzjar@rsnz.org. **ISSN:** 0028-8233. **Subscription Rates:** NZ$455 institutions print and online; NZ$215 individuals print and online; NZ$172 members print and online; US$325 institutions print and online; US$145 individuals print and online; NZ$385 institutions online; NZ$180 individuals online. **URL:** http://www.rsnz.org/publish/nzjar/.

1290 ■ New Zealand Journal of Botany
Royal Society of New Zealand
4 Halswell St.
PO Box 598
Wellington 6011, New Zealand
Ph: 64 4 4727421
Fax: 64 4 4731841

Journal covering all aspects of the botany, mycology, and phycology of the South Pacific, Australia, South America, southern Africa, and Antarctica. **Subtitle:** The International Journal Of Austral Botany. **Freq:** Quarterly. **Key Personnel:** Dr. Frances Kell, Editor, nzjb@rsnz.org. **ISSN:** 0028-825X. **Subscription Rates:** NZ$475 institutions print and online; NZ$225 individuals print and online; NZ$180 members print and online; US$340 institutions print and online; US$150 individuals print and online; NZ$400 institutions online; NZ$190 individuals online; NZ$152 members online. **URL:** http://www.rsnz.org/publish/nzjb/.

1291 ■ New Zealand Journal of Crop and Horticultural Science
Royal Society of New Zealand
4 Halswell St.
PO Box 598
Wellington 6011, New Zealand
Ph: 64 4 4727421
Fax: 64 4 4731841

Journal covering aspects of crop and horticultural science. **Freq:** Quarterly. **Key Personnel:** Sandra Stanislawek, Editor, nzjchs@rsnz.org. **ISSN:** 0114-0671. **Subscription Rates:** NZ$455 institutions print and online; NZ$215 individuals print and online; NZ$172 members print and online; US$325 institutions print and online; US$145 individuals print and online; NZ$385 institutions online; NZ$180 individuals online. **URL:** http://www.rsnz.org/publish/nzjchs/.

Circulation: ★ = ABC; △ = BPA; ♦ = CAC; • = CCAB; ❑ = VAC; ⊕ = PO Statement; ‡ = Publisher's Report; Boldface figures = sworn; Light figures = estimated.

1292 ■ New Zealand Journal of Geology and Geophysics
Royal Society of New Zealand
4 Halswell St.
PO Box 598
Wellington 6011, New Zealand
Ph: 64 4 4727421
Fax: 64 4 4731841

Journal covering all aspects of the earth sciences relevant to New Zealand, the Pacific, and Antarctica. **Subtitle:** An international journal of Pacific Rim Geosciences. **Freq:** Quarterly. **Key Personnel:** Robert Lynch, Editor, nzjgg@rsnz.org. **ISSN:** 0028-8306. **Subscription Rates:** NZ$475 institutions print and online; NZ$225 individuals print and online; NZ$180 members print and online; US$340 institutions print and online; US$150 individuals print and online; NZ$400 institutions online; NZ$190 individuals online; NZ$152 members online. **URL:** http://www.rsnz.org/publish/nzjgg/.

1293 ■ New Zealand Journal of Marine and Freshwater Research
Royal Society of New Zealand
4 Halswell St.
PO Box 598
Wellington 6011, New Zealand
Ph: 64 4 4727421
Fax: 64 4 4731841

Journal covering aspects of aquatic science. **Freq:** Quarterly. **Key Personnel:** Dr. Katrin Berkenbusch, Editor, katrin.berkenbusch@rsnz.org. **ISSN:** 0028-8330. **Subscription Rates:** NZ$475 institutions print and online; NZ$225 individuals print and online; NZ$180 members print and online; US$340 institutions print and online; US$150 individuals print and online; NZ$400 institutions online; NZ$190 individuals online; NZ$152 members online. **URL:** http://www.rsnz.org/publish/nzjmfr/.

WHANGAREI

1294 ■ Newstalk ZB-AM Northland - 1215
54 Cook St.
Private Bag 92198
Auckland, New Zealand
Ph: 64 9 3730000

Format: News; Talk. **Owner:** The Radio Network, at above address. **Key Personnel:** Paul Holmes, Contact, grumpy@newstalkzb.co.nz; Leighton Smith, Contact, leighton@newstalkzb.co.nz; Danny Watson, Contact, danny@newstalkzb.co.nz. **URL:** http://www.newstalkzb.co.nz/featdetail.asp?recnumber=1.

1295 ■ Newstalk ZB-AM Northland - 1026
54 Cook St.
Private Bag 92198
Auckland, New Zealand
Ph: 64 9 3730000

Format: News; Talk. **Owner:** The Radio Network, at above address. **Key Personnel:** Paul Holmes, Contact, grumpy@newstalkzb.co.nz; Leighton Smith, Contact, leighton@newstalkzb.co.nz; Danny Watson, Contact, danny@newstalkzb.co.nz. **URL:** http://www.newstalkzb.co.nz/featdetail.asp?recnumber=1.

NIGERIA

LAGOS

1296 ■ African Journal of Agricultural Research
Academic Journals
73023 Victoria Island
Lagos, Lagos, Nigeria
Publication E-mail: ajar@academicjournals.org
Publisher E-mail: service@academicjournals.org

Journal covering all areas of agriculture in Africa. **Freq:** Monthly. **Key Personnel:** Prof. N.A. Amusa, Acting Ed. **ISSN:** 1991-637X. **Remarks:** Accepts advertising. **URL:** http://www.academicjournals.org/AJAR/. **Circ:** (Not Reported)

1297 ■ African Journal of Biochemistry Research
Academic Journals
73023 Victoria Island
Lagos, Lagos, Nigeria
Publication E-mail: ajbr@academicjournals.org
Publisher E-mail: service@academicjournals.org

Journal covering in all areas of biochemistry. **Freq:** Monthly. **Key Personnel:** Prof. Johnson Lin, Acting Ed.; Carlos H. I. Ramos, Assoc. Ed.; Gregory Lloyd Blatch, Assoc. Ed. **ISSN:** 1996-0778. **Remarks:** Accepts advertising. **URL:** http://www.academicjournals.org/AJBR/. **Circ:** (Not Reported)

1298 ■ African Journal of Food Science
Academic Journals
73023 Victoria Island
Lagos, Lagos, Nigeria
Publication E-mail: ajfs.acadjourn@gmail.com
Publisher E-mail: service@academicjournals.org

Journal covering in all areas of food science. **Freq:** Monthly. **Key Personnel:** Mamoudou H. Dicko, Acting Ed.; Joan O. Amarteifio, PhD, Assoc. Ed. **ISSN:** 1996-0794. **Remarks:** Accepts advertising. **URL:** http://www.academicjournals.org/AJFS/index.htm. **Circ:** (Not Reported)

1299 ■ African Journal of Microbiology Research
Academic Journals
73023 Victoria Island
Lagos, Lagos, Nigeria
Publication E-mail: ajmr@academicjournals.org
Publisher E-mail: service@academicjournals.org

Journal covering in all areas of microbiology. **Freq:** Monthly. **Key Personnel:** Jacques Theron, Acting Ed.; Mamadou Gueye, Assoc. Ed.; Carolina Mary Knox, Assoc. Ed. **ISSN:** 1996-0808. **Remarks:** Accepts advertising. **URL:** http://www.academicjournals.org/AJMR/. **Circ:** (Not Reported)

1300 ■ African Journal of Political Science & International Relations
Academic Journals
73023 Victoria Island
Lagos, Lagos, Nigeria
Publication E-mail: ajpsir@academicjournals.org
Publisher E-mail: service@academicjournals.org

Journal covering Africa's political science and international relations. **Freq:** Monthly. **Key Personnel:** Ayandiji Daniel Aina, Assoc. Ed.; Cyril K. Daddieh, Assoc. Ed.; Prof. F.J. Kolapo, Assoc. Ed. **ISSN:** 1996-0832. **Remarks:** Accepts advertising. **URL:** http://www.academicjournals.org/AJPSIR/. **Circ:** (Not Reported)

1301 ■ African Journal of Pure & Applied Chemistry
Academic Journals
73023 Victoria Island
Lagos, Nigeria
Publication E-mail: ajpac@academicjournals.org
Publisher E-mail: service@academicjournals.org

Journal covering research analysis on applied chemistry. **Freq:** Monthly. **Key Personnel:** Prof. Tebello Nyokong, Acting Ed.; Prof. F. Tafesse, Assoc. Ed. **ISSN:** 1996-0840. **Remarks:** Accepts advertising. **URL:** http://www.academicjournals.org/AJPAC/index.htm. **Circ:** (Not Reported)

1302 ■ Educational Research & Reviews
Academic Journals
73023 Victoria Island
Lagos, Lagos, Nigeria
Publication E-mail: err_acadjourn@yahoo.com
Publisher E-mail: service@academicjournals.org

Journal covering all areas of education including education policies and management. **Freq:** Monthly. **Key Personnel:** Peter Mayo, Assoc. Ed.; Malcom Vick, Assoc. Ed. **ISSN:** 1990-3839. **Remarks:** Accepts advertising. **URL:** http://www.academicjournals.org/ERR/index.htm. **Circ:** (Not Reported)

1303 ■ Journal of Cell & Animal Biology
Academic Journals
73023 Victoria Island
Lagos, Lagos, Nigeria
Publication E-mail: jcab@academicjournals.org
Publisher E-mail: service@academicjournals.org

Journal covering in all areas of cell and animal biology. **Freq:** Monthly. **Key Personnel:** Jacob Tavern Ross, Editor; Hamada Nohamed Mahmoud, Co-Ed. **ISSN:** 1996-0867. **Remarks:** Accepts advertising. **URL:** http://www.academicjournals.org/JCAB/index.htm. **Circ:** (Not Reported)

1304 ■ Scientific Research & Essays
Academic Journals
73023 Victoria Island
Lagos, Lagos, Nigeria
Publication E-mail: sre@academicjournals.org
Publisher E-mail: service@academicjournals.org

Journal covering applied research in science, medicine, agriculture and engineering. **Freq:** Bimonthly. **Key Personnel:** Dr. NJ Tonukari, Editor-in-Chief, sre@academicjournals.org. **ISSN:** 1992-2248. **Remarks:** Accepts advertising. **URL:** http://www.academicjournals.org/SRE/. **Circ:** (Not Reported)

PAKISTAN

FAISALABAD

1305 ■ Academic Journal of Cancer Research
International Digital Organization for Scientific Information
P-100, St. 7
Sohailabad
Peoples Colony No. 2
Faisalabad, Pakistan
Ph: 92 41 8542906
Fax: 92 333 4006789
Publisher E-mail: idosi@idosi.org

Journal covering cancer research. **Freq:** Quarterly. **Key Personnel:** Muhammad Zeeshan, Managing Editor. **ISSN:** 1992-6197. **Subscription Rates:** EUR400 individuals; EUR500 institutions; EUR100 single issue. **URL:** http://www.idosi.org/ajcr/ajcr.htm.

1306 ■ Academic Journal of Financial Management
International Digital Organization for Scientific Information
P-100, St. 7
Sohailabad
Peoples Colony No. 2
Faisalabad, Pakistan
Ph: 92 41 8542906
Fax: 92 333 4006789
Publisher E-mail: idosi@idosi.org

Journal covering financial management. **Freq:** Semiannual. **Key Personnel:** Muhammad Zeeshan, Managing Editor. **ISSN:** 1992-6197. **Subscription Rates:** EUR200 individuals; EUR250 institutions; EUR100 single issue. **URL:** http://www.idosi.org/ajfm/ajfm.htm.

1307 ■ Advances in Biological Research
International Digital Organization for Scientific Information
P-100, St. 7
Sohailabad
Peoples Colony No. 2
Faisalabad, Pakistan
Ph: 92 41 8542906
Fax: 92 333 4006789
Publisher E-mail: idosi@idosi.org

Journal covering biological research. **Freq:** Bimonthly. **Key Personnel:** Dr. Ramesh Katam, Editor-in-Chief; Dr. Shaban Sharaf El-Deen, Editor-in-Chief; Muhammad Zeeshan, Managing Editor. **ISSN:** 1992-0067. **Subscription Rates:** EUR480 individuals; EUR600 institutions; EUR80 single issue. **URL:** http://www.idosi.org/abr/abr.htm.

1308 ■ American-Eurasian Journal of Agricultural & Environmental Sciences
International Digital Organization for Scientific Information
P-100, St. 7
Sohailabad
Peoples Colony No. 2
Faisalabad, Pakistan
Ph: 92 41 8542906
Fax: 92 333 4006789
Publisher E-mail: idosi@idosi.org

Journal covering agricultural and environmental sciences. **Freq:** Bimonthly. **Key Personnel:** Dr. Muhammad Saleem, Editor-in-Chief; Dr. Jafar Nouri, Editor-in-Chief; Dr. Wenju Liang, Editor-in-Chief. **ISSN:** 1818-6769. **Subscription Rates:** EUR720 individuals; EUR900 institutions. **URL:** http://www.idosi.org/aejaes/aejaes.htm.

1309 ■ American-Eurasian Journal of Botany
International Digital Organization for Scientific Information
P-100, St. 7
Sohailabad
Peoples Colony No. 2
Faisalabad, Pakistan
Ph: 92 41 8542906
Fax: 92 333 4006789
Publisher E-mail: idosi@idosi.org

Journal covering botany. **Freq:** Quarterly. **Key Personnel:** Dr. Ramesh Katam, Editor-in-Chief; Muhammad Zeeshan, Managing Editor. **ISSN:** 1992-6197. **Subscription Rates:** EUR400 individuals; EUR500 institutions; EUR100 single issue. **URL:** http://www.idosi.org/aejb/aejb.htm.

1310 ■ American-Eurasian Journal of Scientific Research
International Digital Organization for Scientific Information
P-100, St. 7
Sohailabad
Peoples Colony No. 2
Faisalabad, Pakistan
Ph: 92 41 8542906
Fax: 92 333 4006789
Publisher E-mail: idosi@idosi.org

Journal covering scientific research. **Freq:** Semiannual. **Key Personnel:** Dr. Mokhtar Abdel-Kader, Editor-in-Chief; Dr. Muhammad Aslam, Editor-in-Chief; Dr. Duponnois Robin, Editor-in-Chief. **ISSN:** 1818-6785. **Subscription Rates:** EUR160 individuals; EUR200 institutions; EUR80 single issue. **URL:** http://www.idosi.org/aejsr/aejsr.htm.

1311 ■ Australian Journal of Basic and Applied Sciences
International Network for Scientific Information
Haseeb Shaheed Colony, P-112, No. 10
Hilal Rd.
Faisalabad, Pakistan
Ph: 92 333 6616624
Publisher E-mail: editor@insinet.net

Journal covering basic and applied sciences. **Freq:** Quarterly. **Key Personnel:** Dr. Deborah Mooney, Editorial Board Member; Dr. Kadambot Siddique, Editorial Board Member. **ISSN:** 1991-8178. **URL:** http://www.insinet.net/ajbas.html.

1312 ■ Global Journal of Biotechnology & Biochemistry
International Digital Organization for Scientific Information
P-100, St. 7
Sohailabad
Peoples Colony No. 2
Faisalabad, Pakistan
Ph: 92 41 8542906
Fax: 92 333 4006789
Publisher E-mail: idosi@idosi.org

Journal covering biotechnology and biochemistry. **Freq:** Semiannual. **Key Personnel:** Dr. Hamed M. Eel-Shora, Editor-in-Chief; Dr. Mohammad Anis, Editor-in-Chief; Muhammad Zeeshan, Managing Editor. **ISSN:** 1990-925X. **Subscription Rates:** EUR200 individuals; EUR250 institutions; EUR100 single issue. **URL:** http://www.idosi.org/gjbb/gjbb.htm.

1313 ■ Global Journal of Environmental Research
International Digital Organization for Scientific Information
P-100, St. 7
Sohailabad
Peoples Colony No. 2
Faisalabad, Pakistan
Ph: 92 41 8542906
Fax: 92 333 4006789
Publisher E-mail: idosi@idosi.org

Journal covering environment research. **Freq:** Quarterly. **Key Personnel:** Dr. Atef Ali Al Kharabshe, Editor-in-Chief; Dr. Ramesh Katam, Editor-in-Chief; Dr. Ismail Bin Sahid, Editor-in-Chief. **ISSN:** 1992-0075. **Subscription Rates:** EUR240 individuals; EUR300 institutions; EUR80 single issue. **URL:** http://www.idosi.org/gjer/gjer.htm.

1314 ■ Global Journal of Molecular Sciences
International Digital Organization for Scientific Information
P-100, St. 7
Sohailabad
Peoples Colony No. 2
Faisalabad, Pakistan
Ph: 92 41 8542906
Fax: 92 333 4006789
Publisher E-mail: idosi@idosi.org

Journal covering molecular sciences. **Freq:** Semiannual. **Key Personnel:** Dr. Abdelrahim Hunaiti, Editor-in-Chief; Dr. Yehiam Salts, Editor-in-Chief; Muhammad Zeeshan, Managing Editor. **ISSN:** 1990-9241. **Subscription Rates:** EUR200 individuals; EUR250 institutions; EUR100 single issue. **URL:** http://www.idosi.org/gjms/gjms.htm.

1315 ■ Global Journal of Pharmacology
International Digital Organization for Scientific Information
P-100, St. 7
Sohailabad
Peoples Colony No. 2
Faisalabad, Pakistan
Ph: 92 41 8542906
Fax: 92 333 4006789
Publisher E-mail: idosi@idosi.org

Journal covering pharmacology. **Freq:** 3/yr. **Key Personnel:** Muhammad Zeeshan, Managing Editor. **ISSN:** 1992-0083. **Subscription Rates:** EUR240 individuals; EUR300 institutions; EUR80 single issue. **URL:** http://www.idosi.org/gjp/gjp.htm.

1316 ■ Global Veterinaria
International Digital Organization for Scientific Information
P-100, St. 7
Sohailabad
Peoples Colony No. 2
Faisalabad, Pakistan
Ph: 92 41 8542906
Fax: 92 333 4006789
Publisher E-mail: idosi@idosi.org

Journal covering all aspects of veterinary sciences. **Freq:** Quarterly. **Key Personnel:** Dr. Wahid Mohamed Ahmed, Editor-in-Chief; Muhammad Zeeshan, Managing Editor. **ISSN:** 1992-6197. **Subscription Rates:** EUR600 individuals; EUR500 institutions; EUR100 single issue. **URL:** http://www.idosi.org/gv/gv.htm.

1317 ■ Humanity & Social Sciences Journal
International Digital Organization for Scientific Information
P-100, St. 7
Sohailabad
Peoples Colony No. 2
Faisalabad, Pakistan
Ph: 92 41 8542906
Fax: 92 333 4006789
Publisher E-mail: idosi@idosi.org

Journal covering humanity and social sciences. **Freq:** Quarterly. **Key Personnel:** Dr. Sefika Sule Ercetin, Editor-in-Chief; Dr. Alay Ahmad, Editor-in-Chief; Dr. Mohammed Subbarini, Editor-in-Chief. **ISSN:** 1818-4960. **Subscription Rates:** EUR400 individuals; EUR320 institutions; EUR80 single issue. **URL:** http://www.idosi.org/hssj/hss.htm.

1318 ■ International Journal of Planetary and Space Research
International Digital Organization for Scientific Information
P-100, St. 7
Sohailabad
Peoples Colony No. 2
Faisalabad, Pakistan
Ph: 92 41 8542906
Fax: 92 333 4006789
Publisher E-mail: idosi@idosi.org

Journal covering planetary and space research. **Freq:** Semiannual. **Key Personnel:** Muhammad Zeeshan, Managing Editor. **ISSN:** 1993-145X. **Subscription Rates:** US$300 individuals; US$150 single issue. **URL:** http://www.idosi.org/ijpsr/ijpsr.htm.

1319 ■ Journal of Applied Sciences Research
International Network for Scientific Information
Haseeb Shaheed Colony, P-112, No. 10
Hilal Rd.
Faisalabad, Pakistan
Ph: 92 333 6616624
Publisher E-mail: editor@insinet.net

Journal covering research in applied sciences. **Freq:** Monthly. **Key Personnel:** Aadel Chaudhuri, Technical Ed.; Dr. Deborah Mooney, Editorial Board Member. **ISSN:** 1816-157X. **URL:** http://www.insinet.net/jasr.html.

1320 ■ Middle East Journal of Scientific Research
International Digital Organization for Scientific Information
P-100, St. 7
Sohailabad
Peoples Colony No. 2
Faisalabad, Pakistan
Ph: 92 41 8542906
Fax: 92 333 4006789
Publisher E-mail: idosi@idosi.org

Journal covering scientific research. **Freq:** Quarterly. **Key Personnel:** Dr. Hamed M. Eel-Shora, Editor-in-Chief; Dr. Munir Aziz Noah Turk, Editor-in-Chief; Muhammad Zeeshan, Managing Editor. **ISSN:** 1990-9233. **Subscription Rates:** EUR320 individuals; EUR400 institutions; EUR80 single issue. **URL:** http://www.idosi.org/mejsr/mejsr.htm.

1321 ■ Research Journal of Agriculture and Biological Sciences
International Network for Scientific Information
Haseeb Shaheed Colony, P-112, No. 10
Hilal Rd.
Faisalabad, Pakistan
Ph: 92 333 6616624
Publisher E-mail: editor@insinet.net

Journal covering agriculture and biological sciences. **Freq:** Bimonthly. **Key Personnel:** Dr. Kyung Dong Lee, Editorial Board Member; Dr. Richard A. Heckmann, Editorial Board Member. **ISSN:** 1816-1561. **URL:** http://www.insinet.net/rjabs.html.

1322 ■ Research Journal of Animal and Veterinary Sciences
International Network for Scientific Information
Haseeb Shaheed Colony, P-112, No. 10
Hilal Rd.
Faisalabad, Pakistan
Ph: 92 333 6616624
Publisher E-mail: editor@insinet.net

Journal covering animal and veterinary sciences. **Freq:** Annual. **Key Personnel:** Dr. Joseph M. Erwin, Editorial Board Member; Dr. Bahy Ahmed Ali, Editorial Board Member. **ISSN:** 1816-2746. **URL:** http://www.insinet.net/rjavs.html.

Circulation: ★ = ABC; △ = BPA; ♦ = CAC; • = CCAB; ❑ = VAC; ⊕ = PO Statement; ‡ = Publisher's Report; Boldface figures = sworn; Light figures = estimated.

1323 ■ Research Journal of Fisheries and Hydrobiology
International Network for Scientific Information
Haseeb Shaheed Colony, P-112, No. 10
Hilal Rd.
Faisalabad, Pakistan
Ph: 92 333 6616624
Publisher E-mail: editor@insinet.net

Journal covering fisheries and hydrobiology. **Freq:** Semiannual. **Key Personnel:** Dr. Ibrahim El-Shishtawy Hassan Belal, Editorial Board Member; Dr. Mohsen Abdel-Tawwab, Editorial Board Member. **ISSN:** 1816-9112. **URL:** http://www.insinet.net/rjfh.html.

1324 ■ Research Journal of Medicine and Medical Sciences
International Network for Scientific Information
Haseeb Shaheed Colony, P-112, No. 10
Hilal Rd.
Faisalabad, Pakistan
Ph: 92 333 6616624
Publisher E-mail: editor@insinet.net

Journal covering medicine and medical sciences. **Freq:** Semiannual. **Key Personnel:** Dr. Neal Davies, Editorial Board Member; Dr. Ugur Cavlak, Editorial Board Member. **ISSN:** 1816-272X. **URL:** http://www.insinet.net/rjmms.html.

1325 ■ Research Journal Telecommunication and Information Technology
International Network for Scientific Information
Haseeb Shaheed Colony, P-112, No. 10
Hilal Rd.
Faisalabad, Pakistan
Ph: 92 333 6616624
Publisher E-mail: editor@insinet.net

Journal covering telecommunication and information technology. **Freq:** Annual. **Key Personnel:** Dr. Ahmad T. Al-Taani, Editorial Board Member; Dr. Kevin Curran, Editorial Board Member. **ISSN:** 1816-2738. **URL:** http://www.insinet.net/rjtit.html.

1326 ■ Universal Science and Engineering for Marine Environment
International Digital Organization for Scientific Information
P-100, St. 7
Sohailabad
Peoples Colony No. 2
Faisalabad, Pakistan
Ph: 92 41 8542906
Fax: 92 333 4006789
Publisher E-mail: idosi@idosi.org

Journal covering coastal environment. **Freq:** 3/yr. **Key Personnel:** Dr. Sarwoko Mangkoedihardjo, Editor-in-Chief. **ISSN:** 1992-0083. **Subscription Rates:** EUR300 individuals; EUR375 institutions; EUR125 single issue. **URL:** http://www.idosi.org/useme/useme.htm.

1327 ■ World Applied Sciences Journal
International Digital Organization for Scientific Information
P-100, St. 7
Sohailabad
Peoples Colony No. 2
Faisalabad, Pakistan
Ph: 92 41 8542906
Fax: 92 333 4006789
Publisher E-mail: idosi@idosi.org

Journal covering world applied sciences. **Freq:** Bimonthly. **Key Personnel:** Dr. Ahmad K. Hegazy, Editor-in-Chief; Dr. Sarwoko Mangkoedihardjo, Editor-in-Chief; Dr. Jesus Martinez-Frias, Editor-in-Chief. **ISSN:** 1818-4952. **Subscription Rates:** EUR600 individuals; EUR750 institutions. **URL:** http://www.idosi.org/wasj/wasj.htm.

1328 ■ World Information Technology Journal
International Digital Organization for Scientific Information
P-100, St. 7
Sohailabad
Peoples Colony No. 2
Faisalabad, Pakistan
Ph: 92 41 8542906
Fax: 92 333 4006789
Publisher E-mail: idosi@idosi.org

Journal covering world information technology. **Freq:** Semiannual. **Key Personnel:** Dr. Omer M. Al-Jarrah, Regional Ed.-in.Ch.; Muhammad Zeeshan, Managing Editor. **ISSN:** 1818-4944. **Subscription Rates:** US$150 individuals; US$200 institutions. **URL:** http://www.idosi.org/witj/witj.htm.

1329 ■ World Journal of Agricultural Sciences
International Digital Organization for Scientific Information
P-100, St. 7
Sohailabad
Peoples Colony No. 2
Faisalabad, Pakistan
Ph: 92 41 8542906
Fax: 92 333 4006789
Publisher E-mail: idosi@idosi.org

Journal covering agricultural sciences. **Freq:** Bimonthly. **Key Personnel:** Dr. Muhammad Saleem, Editor-in-Chief; Dr. A.M. Al-Tawaha, Editor-in-Chief; Dr. Quazi Mesbahul Alam, Editor-in-Chief. **ISSN:** 1817-3047. **Subscription Rates:** EUR720 individuals; EUR900 institutions. **URL:** http://www.idosi.org/wjas/wjas.htm.

1330 ■ World Journal of Chemical
International Digital Organization for Scientific Information
P-100, St. 7
Sohailabad
Peoples Colony No. 2
Faisalabad, Pakistan
Ph: 92 41 8542906
Fax: 92 333 4006789
Publisher E-mail: idosi@idosi.org

Journal covering chemistry. **Freq:** Semiannual. **Key Personnel:** Dr. Sultan T. Abu-Orabi, Editor-in-Chief; Dr. Miguel Yus, Editor-in-Chief; Muhammad Zeeshan, Managing Editor. **ISSN:** 1817-3071. **Subscription Rates:** US$250 individuals; US$200 institutions. **URL:** http://www.idosi.org/wjc/wjc.htm.

1331 ■ World Journal of Dairy & Food Sciences
International Digital Organization for Scientific Information
P-100, St. 7
Sohailabad
Peoples Colony No. 2
Faisalabad, Pakistan
Ph: 92 41 8542906
Fax: 92 333 4006789
Publisher E-mail: idosi@idosi.org

Journal covering daily and food sciences. **Freq:** Semiannual. **Key Personnel:** Dr. Mahmut Dogan, Editor-in-Chief; Dr. Vikas Nanda, Editor-in-Chief; Dr. Anita Rani Jindal, Editor-in-Chief. **ISSN:** 1817-308X. **Subscription Rates:** EUR150 individuals; EUR200 institutions. **URL:** http://www.idosi.org/wjdfs/wjdfs.htm.

1332 ■ World Journal of Medical Sciences
International Digital Organization for Scientific Information
P-100, St. 7
Sohailabad
Peoples Colony No. 2
Faisalabad, Pakistan
Ph: 92 41 8542906
Fax: 92 333 4006789
Publisher E-mail: idosi@idosi.org

Journal covering medical sciences. **Freq:** Semiannual. **Key Personnel:** Dr. Gulzar A. Niazi, Editor-in-Chief, raiwind786@hotmail.com; Dr. Mahmoud Al-Sheyyab, Assoc. Ed.-in-Ch. **ISSN:** 1817-3055. **Subscription Rates:** EUR160 individuals; EUR200 institutions. **URL:** http://www.idosi.org/wjms/wjms.htm.

1333 ■ World Journal of Zoology
International Digital Organization for Scientific Information
P-100, St. 7
Sohailabad
Peoples Colony No. 2
Faisalabad, Pakistan
Ph: 92 41 8542906
Fax: 92 333 4006789
Publisher E-mail: idosi@idosi.org

Journal covering zoology. **Freq:** Semiannual. **Key Personnel:** Dr. Wahid Mohamed Ahmed, Editor-in-Chief; Dr. Marwan M. Muwalla, Editor-in-Chief; Muhammad Zeeshan, Managing Editor. **ISSN:** 1817-3098. **Subscription Rates:** EUR160 individuals; EUR200 institutions. **URL:** http://www.idosi.org/wjz/zoology.htm.

PHILIPPINES

BACOLOD CITY

1334 ■ DYBM-FM - 99.1
Cordova Ave.
Mountain View Subd.
Mandalagan
Bacolod City 6100, Philippines
Ph: 63 34 4339991
Fax: 63 34 4410235

Format: Adult Contemporary. **URL:** http://www.crossover.com.ph/.

1335 ■ DYIF-FM - 95.5 MHz
3rd Fl., CBS Development
Corporation Bldg.
Lacson St.
Mandalagan
Bacolod City 6100, Philippines
Ph: 63 34 4411670
Fax: 63 34 7090766
E-mail: starfm_bacolod@bomboradyo.com

Format: Adult Contemporary. **Owner:** Bombo Radyo Philippines, 2406 Florete Bldg., Edison cor. Nobel St., Makati City 1200, Philippines, Fax: 63 2 8173631. **Key Personnel:** Errol Emil Adrian V. Ledesma, Station Mgr. **Wattage:** 10,000. **URL:** http://www.bomboradyo.com/archive/new/stationprofile/starfmbacolod/index.ht m.

BAGUIO CITY

1336 ■ DWHB-FM - 103.9
4 Fl., State Condo. Bldg.
1 Salcedo St.
Legaspi Village
Makati City 1229, Philippines
Ph: 63 2 8120529
Fax: 63 2 8108362

Format: Adult Contemporary. **Owner:** Radio Mindanao Network, at above address. **Ad Rates:** Advertising accepted; rates available upon request. **URL:** http://www.rmn.ph/fmstations/dwhb.

1337 ■ DWIM-FM - 89.5 MHz
87 Lourdes Subd.
Baguio City, Philippines
Ph: 63 74 4427926
Fax: 63 74 6192897
E-mail: starfmbaguio@mountainview.com.ph

Format: Adult Contemporary. **Owner:** Bombo Radyo Philippines, 2406 Florete Bldg., Edison cor. Nobel St., Makati City 1200, Philippines, Fax: 63 2 8173631. **Key Personnel:** Mrs. Floribel Caja Sales, Station Mgr. **Wattage:** 4500. **URL:** http://www.bomboradyo.com/archive/new/stationprofile/starfmbaguio/index.htm.

1338 ■ DZBM-FM - 105.1
6 Tirad Pass
Sta. Mesa Hts.
Quezon City, Philippines
Ph: 63 2 7311667
Fax: 63 2 7124213

Format: Adult Contemporary. **Owner:** Mareco Broadcasting Network Inc., at above address. **URL:** http://www.crossover.com.ph/.

CEBU CITY

1339 ■ DYAC-FM - 90.7
Rm. 210, Dona Luisa Bldg.
Fuente Osmena
Cebu City, Philippines
Ph: 63 32 4125043

Format: Adult Contemporary. **URL:** http://www.crossover.com.ph/.

1340 ■ DYMX-FM - 95.5 MHz
140 M. Velez St.
Cebu City, Philippines
Ph: 63 32 2430340
Fax: 63 32 2549143
E-mail: starfm_cebu@bomboradyo.com

Format: Adult Contemporary. **Owner:** Bombo Radyo Philippines, 2406 Florete Bldg., Edison cor. Nobel St., Makati City 1200, Philippines, Fax: 63 2 8173631. **Founded:** 1995. **Key Personnel:** Ernesto F. Yap, Jr., Station Mgr. **Wattage:** 25,000. **URL:** http://www.bomboradyo.com/archive/new/stationprofile/starfmcebu/index.htm.

COTABATO CITY

1341 ■ DXFD-FM - 93.7 MHz
5th St., Don E. Sero, RH-5
Cotabato City, Philippines
Ph: 63 64 3902989
Fax: 63 64 3902989
E-mail: starfm_cot@yahoo.com.ph

Format: Adult Contemporary. **Owner:** Bombo Radyo Philippines, 2406 Florete Bldg., Edison cor. Nobel St., Makati City 1200, Philippines, Fax: 63 2 8173631. **Key Personnel:** Mr. Garry A. Aragoncillo, Station Mgr. **Wattage:** 5000. **URL:** http://www.bomboradyo.com/archive/new/stationprofile/starfmcotabato/index.h tm.

DAGUPAN CITY

1342 ■ DWHT-FM - 107.9
4 Fl., State Condo. Bldg.
1 Salcedo St.
Legaspi Village
Makati City 1229, Philippines
Ph: 63 2 8120529
Fax: 63 2 8108362

Format: Oldies; Talk; News. **Owner:** Radio Mindanao Network, at above address. **Ad Rates:** Advertising accepted; rates available upon request. **URL:** http://www.rmn.ph/fmstations/dwht.

1343 ■ DWHY-FM - 100.7 MHz
Bombo Radyo Broadcast Ctr.
Maramba Bankers Village
Bonuan Catacdang
Dagupan City 2400, Philippines
Ph: 63 75 5229908
E-mail: starfmdagupan@bomboradyo.com

Format: Adult Contemporary. **Owner:** Bombo Radyo Philippines, 2406 Florete Bldg., Edison cor. Nobel St., Makati City 1200, Philippines, Fax: 63 2 8173631. **Founded:** June 12, 1993. **Key Personnel:** Christian E. Queyquep, Station Mgr. **Wattage:** 10,000. **URL:** http://www.bomboradyo.com/archive/new/stationprofile/starfmdagupan/index.ht m.

1344 ■ DWON-FM - 104.7
4 Fl., State Condo. Bldg.
1 Salcedo St.
Legaspi Village
Makati City 1229, Philippines
Ph: 63 2 8120529
Fax: 63 2 8108362

Format: Adult Contemporary. **Owner:** Radio Mindanao Network, at above address. **Ad Rates:** Advertising accepted; rates available upon request. **URL:** http://www.rmn.ph/fmstations/dwon.

DAVAO CITY

1345 ■ DXFX-FM - 96.3 MHz
Bombo Radyo Broadcast Ctr.
San Pedro St.
Davao City 8000, Philippines
Ph: 63 82 2225924
Fax: 63 82 2226007
E-mail: starfm_davao@bomboradyo.com

Format: Adult Contemporary. **Owner:** Bombo Radyo Philippines, 2406 Florete Bldg., Edison cor. Nobel St., Makati City 1200, Philippines, Fax: 63 2 8173631. **Key Personnel:** Rogelio G. Caballo, Station Mgr. **Wattage:** 10,000. **URL:** http://www.bomboradyo.com/archive/new/stationprofile/starfmdavao/index.htm.

1346 ■ DXLR-FM - 93.1
251-F Valrose Bldg.
C.M. Recto Ave.
Davao City, Philippines
Ph: 63 82 2220931
Fax: 63 82 2222931

Format: Adult Contemporary. **URL:** http://www.crossover.com.ph/.

DIPOLOG CITY

1347 ■ DXFB-FM - 93.3 MHz
2nd Fl., Lordel Bldg.
Gen. Luna cor. Osmena Sts.
Dipolog City 7100, Philippines
Ph: 63 65 2126596
Fax: 63 65 2126596
E-mail: stardip@mozcom.com

Format: Adult Contemporary. **Owner:** Bombo Radyo Philippines, 2406 Florete Bldg., Edison cor. Nobel St., Makati City 1200, Philippines, Fax: 63 2 8173631. **Key Personnel:** Harriet Ybanez-Paquibot, Station Mgr. **Wattage:** 5000. **URL:** http://www.bomboradyo.com/archive/new/stationprofile/starfmdipolog/index.ht m.

ILOILO CITY

1348 ■ DYRF-FM - 99.5 MHz
3rd Fl., R. Florete Bldg.
cor. Rizal Fermin Caram Ave.
Iloilo City 5000, Philippines
Ph: 63 33 3379087
Fax: 63 33 3379777
E-mail: starfm_iloilo@bomboradyo.com

Format: Adult Contemporary. **Owner:** Bombo Radyo Philippines, 2406 Florete Bldg., Edison cor. Nobel St., Makati City 1200, Philippines, Fax: 63 2 8173631. **Key Personnel:** Mr. Alquin Rubidy, Station Mgr. **Wattage:** 10,000. **URL:** http://www.bomboradyo.com/archive/new/stationprofile/starfmiloilo/index.htm.

LAOAG CITY

1349 ■ DWHP-FM - 99.5
4 Fl., State Condo. Bldg.
1 Salcedo St.
Legaspi Village
Makati City 1229, Philippines
Ph: 63 2 8120529
Fax: 63 2 8108362

Format: Adult Contemporary. **Owner:** Radio Mindanao Network, at above address. **Ad Rates:** Advertising accepted; rates available upon request. **URL:** http://www.rmn.ph/fmstations/dwhp.

MAKATI CITY

DWHB-FM - See Baguio City
DWHP-FM - See Laoag City
DWHT-FM - See Dagupan City

1350 ■ DWKC-FM - 93.9
4th Fl., Guadalupe Commercial Complex Bldg.
Edsa, Guadalupe
Makati City 1212, Philippines
Ph: 63 2 8822375
Fax: 63 2 8822374
E-mail: ifmmanila@rmn.ph

Format: Adult Contemporary. **Owner:** Radio Mindanao Network, at above address. **Ad Rates:** Advertising accepted; rates available upon request. **URL:** http://www.rmn.ph/fmstations/dwkc.

DWON-FM - See Dagupan City

PASAY CITY

1351 ■ DWSM-FM - 102.7 MHz
10th Fl., EGI Rufino Bldg.
Taft cor. Gil Puyat Sts.
Pasay City 1300, Philippines
Ph: 63 2 5520391
Fax: 63 2 5520393
E-mail: starfm_manila@bomboradyo.com

Format: Oldies. **Owner:** Bombo Radyo Philippines, 2406 Florete Bldg., Edison cor. Nobel St., Makati City 1200, Philippines, Fax: 63 2 8173631. **Key Personnel:** Rey Dela Cruz, Station Mgr. **Wattage:** 25,000. **URL:** http://www.bomboradyo.com/archive/new/stationprofile/starfmmanila/index.htm.

PASIG CITY

1352 ■ DWAV-FM - 89.1
Unit 201, Strata 2000 Bldg.
F. Ortigas Jr. Ave.
Ortigas Ctr.
Pasig City 1605, Philippines
Fax: 63 2 6874483

Format: Adult Contemporary; Hip Hop. **Owner:** Blockbuster Broadcasting Systems, Inc., at above address. **URL:** http://www.wave891.fm/.

1353 ■ DWKX-FM - 103.5
Unit 1508, Jollibee Plz.
F. Ortigas Jr. Rd.
Ortigas Ctr.
Pasig City 1605, Philippines
Ph: 63 2 6382529
Fax: 63 2 6363394
E-mail: info@1035max.fm

Format: Adult Contemporary. **Owner:** Advanced Media Broadcasting System, Inc., at above address. **Key Personnel:** Patt Perado, Sales Dir., patt@1035max.fm; Bing Bayron-Galano, Sen. Account Mgr., bing@1035max.fm; Reden Pamintuan, Account Mgr., reden@1035max.fm. **URL:** http://1035max.fm/.

1354 ■ DWQZ-FM - 97.9
5/F Citystate Ctr.
709 Shaw Blvd.
Pasig City 1600, Philippines
Ph: 63 2 6373965
Fax: 63 2 6339162

Format: Easy Listening. **Owner:** Aliw Broadcasting Corporation, at above address. **URL:** http://www.homeradiofm.net/.

QUEZON CITY

1355 ■ DWBM-FM - 105.1
6 Tirad Pass
Sta. Mesa Hts.
Quezon City, Philippines
Ph: 63 2 7311667
Fax: 63 2 7124213

Format: Jazz. **Owner:** Mareco Broadcasting Network Inc., at above address. **URL:** http://www.crossover.com.ph/.

1356 ■ DWLS-FM - 97.1
14/F & 15/F GMA Network Ctr.
EDSA corner Timog Ave.
Diliman
Quezon City 1104, Philippines
Ph: 63 2 9827777
Fax: 63 2 9282024

Format: Adult Contemporary. **Owner:** GMA Network Inc., at above address. **URL:** http://www.gmanetwork.com/about.

DZBM-FM - See Baguio City

ROXAS CITY

1357 ■ DYRX-FM - 103.7 MHz
Arnaldo Blvd.
Roxas City, Philippines
Ph: 63 36 6210119
Fax: 63 36 6214967
E-mail: star_roxas@bomboradyo.com

Format: Adult Contemporary. **Owner:** Bombo Radyo Philippines, 2406 Florete Bldg., Edison cor. Nobel St., Makati City 1200, Philippines, Fax: 63 2 8173631. **Key Personnel:** Michael B. Loja, Station Mgr. **Wattage:** 5000. **URL:** http://www.bomboradyo.com/archive/new/stationprofile/starfmroxas/index.htm.

TACLOBAN CITY

1358 ■ DYTX-FM - 95.1 MHz
YPL Bldg.
Sto. Nino St. cor. Imelda Ave.
Tacloban City, Philippines
Ph: 63 32 2430340
Fax: 63 32 2549143
E-mail: star_tacfm@yahoo.com

Format: Adult Contemporary; News; Information. **Owner:** Bombo Radyo Philippines, 2406 Florete Bldg., Edison cor. Nobel St., Makati City 1200, Philippines, Fax: 63 2 8173631. **Key Personnel:** Virn I. Villagracia, Station Mgr. **Wattage:** 10,000. **URL:** http://www.bomboradyo.com/archive/new/stationprofile/starfmtacloban/index.h tm.

ZAMBOANGA CITY

1359 ■ DXCB-FM - 93.9 MHz
4th Fl., AJS Bldg.
Valderosa St.
Zamboanga City 7000, Philippines
Ph: 63 2 9932099
E-mail: starfm_zamboanga@bomboradyo.com

Format: Adult Contemporary. **Owner:** Bombo Radyo Philippines, 2406 Florete Bldg., Edison cor. Nobel St., Makati City 1200, Philippines, Fax: 63 2 8173631. **Key Personnel:** Mr. Gil Jay Lazo, Station Mgr. **Wattage:** 10,000. **URL:** http://www.bomboradyo.com/archive/new/stationprofile/starfmzamboanga/index. htm.

POLAND

BYDGOSZCZ

1360 ■ Acta Angiologica
Via Medica
ul. K. Ujejskiego 75
PL-85-168 Bydgoszcz, Poland
Ph: 48 52 3715482
Fax: 48 52 3715782
Publisher E-mail: viamedica@viamedica.pl

Journal covering vascular disorders. **Freq:** Quarterly. **Key Personnel:** Arkadiusz Jawien, Editor-in-Chief, ajawien@ceti.com.pl. **ISSN:** 1234-950X. **Subscription Rates:** 230 Zl individuals; 680 Zl institutions; EUR58 individuals; EUR170 institutions. **Remarks:** Accepts advertising. **URL:** http://www.viamedica.pl/en/gazety/xgazAang/index.phtml. **Circ:** (Not Reported)

1361 ■ Advances in Palliative Medicine
Via Medica
ul. Marii Sklodowskiej-Curie 9
PL-85-094 Bydgoszcz, Poland
Ph: 48 52 5853461
Publisher E-mail: viamedica@viamedica.pl

Journal focusing on palliative medicine. **Freq:** Quarterly. **Key Personnel:** Zbigniew Zylicz, Editor-in-Chief, b.zylicz@dovehouse.org.uk. **ISSN:** 1898-3863. **Remarks:** Accepts advertising. **URL:** http://www.viamedica.pl/en/gazety/xgazEang/stopka.phtml. **Circ:** (Not Reported)

1362 ■ Psychiatry in General Practice
Via Medica
ul. Kurpinskiego 19
PL-85-096 Bydgoszcz, Poland
Ph: 48 52 5854039
Fax: 48 52 5853766
Publication E-mail: kikpsych@amb.bydgoszcz.pl
Publisher E-mail: viamedica@viamedica.pl

Journal covering psychiatry. **Freq:** Quarterly. **Key Personnel:** Aleksander Araszkiewicz, Editor-in-Chief. **ISSN:** 1643-0956. **Subscription Rates:** 195 Zl individuals; 390 Zl institutions; EUR49 individuals; EUR98 individuals. **URL:** http://www.viamedica.pl/en/gazety/xgaz8ang/index.phtml.

GDANSK

1363 ■ Arterial Hypertension
Via Medica
ul. Swietokrzyska 73
PL-80-180 Gdansk, Poland
Ph: 48 58 3209494
Fax: 48 58 3209460
Publication E-mail: redakcja@viamedica.pl
Publisher E-mail: viamedica@viamedica.pl

Journal covering field of arterial hypertension. **Founded:** 1997. **Freq:** Bimonthly. **Key Personnel:** Andrzej Tykarski, Editor-in-Chief. **ISSN:** 1428-5851. **Subscription Rates:** 230 Zl individuals; 680 Zl institutions; EUR58 individuals; EUR170 institutions. **Remarks:** Accepts advertising. **URL:** http://www.viamedica.pl/en/gazety/xgaz5ang/index.phtml. **Circ:** (Not Reported)

1364 ■ Endocrinology, Obesity and Metabolic Disorders
Via Medica
ul. Swietokrzyska 73
PL-80-180 Gdansk, Poland
Ph: 48 58 3209494
Fax: 48 58 3209460
Publisher E-mail: viamedica@viamedica.pl

Journal covering fields of metabolic disorders, endocrinology, and obesity. **Freq:** Quarterly. **Key Personnel:** Prof. Ewa Malecka-Tendera, Editor-in-Chief. **ISSN:** 1734-3321. **Subscription Rates:** 240 Zl individuals; 480 Zl institutions; EUR60 individuals; EUR120 institutions. **Remarks:** Accepts advertising. **URL:** http://www.viamedica.pl/en/gazety/xgazMang/index.phtml. **Circ:** (Not Reported)

1365 ■ Folia Morphologica
Via Medica
ul. Debinki 1
PL-80-211 Gdansk, Poland
Ph: 48 58 3491401
Fax: 48 58 3491421
Publisher E-mail: viamedica@viamedica.pl

Journal covering morphology. **Freq:** Quarterly. **Key Personnel:** Janusz Morys, Editor-in-Chief, jmorys@amg.gda.pl. **ISSN:** 0015-5659. **Subscription Rates:** 312 Zl individuals; 624 Zl institutions; EUR78 individuals; EUR156 institutions. **Remarks:** Accepts advertising. **URL:** http://www.viamedica.pl/en/gazety/gazetax1ang/index.phtml. **Circ:** (Not Reported)

1366 ■ Nuclear Medicine Review
Via Medica
ul. Debinki 7
PL-80-211 Gdansk, Poland
Ph: 48 58 3492204
Fax: 48 58 3492204
Publisher E-mail: viamedica@viamedica.pl

Journal covering all nuclear medicine topics. **Freq:** Semiannual. **Key Personnel:** Julian Liniecki, Editor-in-Chief. **ISSN:** 1506-9680. **Subscription Rates:** 114.55 Zl individuals; 232 Zl institutions; EUR29 individuals; EUR58 institutions. **Remarks:** Accepts advertising. **URL:** http://www.viamedica.pl/en/gazety/gazetax2ang/index.phtml. **Circ:** (Not Reported)

1367 ■ Nursing Topics
Via Medica
ul. Do Studzienki 38
PL-80-227 Gdansk, Poland
Ph: 48 58 3491292
Fax: 48 58 3491292
Publisher E-mail: viamedica@viamedica.pl

Journal covering field of nursing. **Founded:** 1993. **Key Personnel:** Dr. Aleksandra Gaworska-Krzeminska, Editor-in-Chief, a.gawor@friend.pl. **ISSN:** 1233-9989. **Subscription Rates:** 62 Zl individuals; 124 Zl institutions; EUR16 individuals; EUR31 institutions. **Remarks:** Accepts advertising. **URL:** http://www.viamedica.pl/en/gazety/xgazUang/index.phtml. **Circ:** (Not Reported)

1368 ■ Psychooncology
Via Medica
ul. Pomorska 68
PL-80-343 Gdansk, Poland
Ph: 48 58 5571414
Fax: 48 58 5571414
Publisher E-mail: viamedica@viamedica.pl

Journal covering psychological aspects of oncological diseases. **Freq:** Semiannual. **Key Personnel:** Mikolaj Majkowicz, Editor-in-Chief, mmajk@amg.gda.pl. **ISSN:** 1429-8538. **Subscription Rates:** 32 Zl individuals; 64 Zl institutions; EUR8 individuals; EUR16 institutions. **URL:** http://www.viamedica.pl/en/gazety/xgazRang/index.phtml.

GLIWICE

1369 ■ International Journal of Computational Materials Science & Surface Engineering
Inderscience Publishers
c/o Dr. Leszek A. Dobrzanski, Ed.-in-Ch.
Silesian University of Technology
Institute of Engineering Materials & Biomaterials
Ul. Konarskiego 18A
PL-44-100 Gliwice, Poland
Publication E-mail: ijcmsse@inderscience.com
Publisher E-mail: editor@inderscience.com

Journal covering computational materials science & surface engineering. **Founded:** 2007. **Freq:** 6/yr. **Key Personnel:** Dr. Leszek A. Dobrzanski, Editor-in-Chief, leszek.dobrzanski@polsl.pl. **ISSN:** 1753-3465. **Subscription Rates:** EUR565 individuals includes surface mail, print only; EUR790 individuals print & online. **URL:** http://www.inderscience.com/browse/index.php?journalCODE=ijcmsse.

KATOWICE

1370 ■ Polish Surgery
Via Medica
ul. Ziolowa 45/47
PL-40-635 Katowice, Poland
Ph: 48 32 2029577
Fax: 48 32 2061728
Publisher E-mail: viamedica@viamedica.pl

Journal covering Polish surgery. **Freq:** Quarterly. **Key Personnel:** Krzysztof Ziaja, Editor-in-Chief. **ISSN:** 1507-5524. **Subscription Rates:** 230 Zl individuals; 680 Zl institutions; EUR58 individuals; EUR170 institutions. **Remarks:** Accepts advertising. **URL:** http://www.viamedica.pl/en/gazety/xgazCang/index.phtml. **Circ:** (Not Reported)

KRAKOW

1371 ■ Polish Gerontology
Via Medica
ul. Sniadeckich 10
31-531 Krakow, Poland
Ph: 48 12 4248800
Fax: 48 12 4248854
Publisher E-mail: viamedica@viamedica.pl

Journal covering field of biological processes of aging. **Freq:** Quarterly. **Key Personnel:** Prof. Josef Kocemba, Editor-in-Chief. **ISSN:** 1425-4956. **Subscription Rates:** 155 Zl individuals; 450 Zl institutions; EUR39 individuals; EUR113 institutions. **Re-**

marks: Accepts advertising. **URL:** http://www.viamedica.pl/en/gazety/xgazKang/index.phtml. **Circ:** (Not Reported)

LODZ

1372 ■ Cardiovascular Forum
Via Medica
ul. Kniaziewicza 1/5
PL-91-347 Lodz, Poland
Ph: 48 42 2516015
Fax: 48 42 2516015
Publisher E-mail: viamedica@viamedica.pl

Journal covering developments in cardiology, diagnostic and treatment in cardiovascular medicine. **Founded:** 1996. **Freq:** Quarterly. **Key Personnel:** Jaroslaw Drozdz, Editor-in-Chief, drozdz@ptkardio.pl. **ISSN:** 1425-3674. **Remarks:** Accepts advertising. **URL:** http://www.viamedica.pl/en/gazety/xgaz3ang/index.phtml. **Circ:** (Not Reported)

WARSAW

1373 ■ Polish Pneumology and Allergology
Via Medica
ul. Plocka 26
PL-01-138 Warsaw, Poland
Ph: 48 22 4312144
Fax: 48 22 4312454
Publisher E-mail: viamedica@viamedica.pl

Journal covering pneumology and allergology. **Freq:** Quarterly. **Key Personnel:** Prof. Dorota Gorecka, Editor-in-Chief. **ISSN:** 0867-7077. **Subscription Rates:** 62 Zl individuals; 124 Zl institutions; EUR16 individuals; EUR31 institutions. **Remarks:** Accepts advertising. **URL:** http://www.viamedica.pl/en/gazety/xgazSang/index.phtml. **Circ:** (Not Reported)

1374 ■ Polish Sexology
Via Medica
ul. Koszykowa 1/28
PL-00-564 Warsaw, Poland
Publication E-mail: redakcja@seksuologia.med.pl
Publisher E-mail: viamedica@viamedica.pl

Journal covering all branches of sexology. **Founded:** 2003. **Freq:** Semiannual. **Key Personnel:** Slawomir Jakima, Editor-in-Chief. **ISSN:** 1731-6677. **Subscription Rates:** 108 Zl individuals; 216 Zl institutions; EUR27 individuals; EUR54 institutions. **Remarks:** Accepts advertising. **URL:** http://www.viamedica.pl/en/gazety/xgazFang/index.phtml. **Circ:** (Not Reported)

1375 ■ Suicidology
Via Medica
Sobieski St. 9
PL-02-957 Warsaw, Poland
Ph: 48 22 4582659
Publisher E-mail: viamedica@viamedica.pl

Journal covering field of suicidology. **Freq:** Annual. **Key Personnel:** Prof. Brunon Holyst, Editor-in-Chief. **ISSN:** 1895-3786. **Subscription Rates:** 49 Zl individuals; 98 Zl institutions; EUR12 individuals; EUR25 institutions. **URL:** http://www.viamedica.pl/en/gazety/xgazOang/index.phtml.

WROCLAW

1376 ■ Interdisciplinary Problems of Stroke
Via Medica
ul. Traugutta 118
PL-50-420 Wroclaw, Poland
Ph: 48 71 3427021
Fax: 48 71 3424919
Publisher E-mail: viamedica@viamedica.pl

Journal covering problems of stroke. **Freq:** Semiannual. **Key Personnel:** Ryszard Podemski, Editor-in-Chief. **ISSN:** 1505-6740. **Subscription Rates:** 95 Zl individuals; 195 Zl institutions; EUR24 individuals; EUR49 institutions. **Remarks:** Accepts advertising. **URL:** http://www.viamedica.pl/en/gazety/xgazDang/index.phtml. **Circ:** (Not Reported)

1377 ■ International Journal of Intelligent Information and Database Systems
Inderscience Publishers
c/o Prof. Ngoc Thanh Nguyen, Ed.-in-Ch.
Wroclaw University of Technology
Institute of Information Science & Engineering
Str. Janiszewskiego 11/17
50-370 Wroclaw, Poland
Publisher E-mail: editor@inderscience.com

Journal covering new intelligent technologies for information processing data. **Founded:** 2007. **Freq:** 4/yr. **Key Personnel:** Prof. Ngoc Thanh Nguyen, Editor-in-Chief, thanh@pwr.wroc.pl. **ISSN:** 1751-5858. **Subscription Rates:** EUR470 individuals includes surface mail, print only; EUR640 individuals print and online. **URL:** http://www.inderscience.com/browse/index.php?journalCODE=ijiids.

ZABRZE

1378 ■ Annales Academiae Medicae Silesiensis
Via Medica
ul. 3 Maja 13/15
PL-41-800 Zabrze, Poland
Ph: 48 32 2712511
Fax: 48 32 2714617
Publisher E-mail: viamedica@viamedica.pl

Journal covering developments in medical sciences. **Freq:** 6/yr. **Key Personnel:** Prof. Wladyslaw Grzeszczak, Editor-in-Chief. **ISSN:** 0208-5607. **Subscription Rates:** 310 Zl individuals; 900 Zl institutions; EUR78 individuals; EUR225 institutions. **Remarks:** Accepts advertising. **URL:** http://www.viamedica.pl/en/gazety/xgazJang/stopka.phtml. **Circ:** (Not Reported)

1379 ■ Experimental and Clinical Diabetology
Via Medica
ul. 3 Maja 13/15
PL-41-800 Zabrze, Poland
Ph: 48 32 2712511
Fax: 48 32 2714617
Publisher E-mail: viamedica@viamedica.pl

Journal covering developments in the field of diabetology. **Freq:** Bimonthly. **Key Personnel:** Wladyslaw Grzeszczak, Editor-in-Chief. **ISSN:** 1643-3165. **Subscription Rates:** 300 Zl individuals; 600 Zl institutions; EUR75 individuals; EUR150 institutions. **Remarks:** Accepts advertising. **URL:** http://www.viamedica.pl/en/gazety/xgaz9ang/index.phtml. **Circ:** (Not Reported)

1380 ■ Polish Journal of Endocrinology
Via Medica
Plac Traugutta 2
PL-41-908 Zabrze, Poland
Ph: 48 32 2786126
Fax: 48 32 2786126
Publication E-mail: endokrynologia.polska@viamedica.pl; endopat@slam.katowice.p
Publisher E-mail: viamedica@viamedica.pl

Journal covering clinical and experimental endocrinology. **Founded:** 1949. **Freq:** Bimonthly. **Key Personnel:** Prof. Beata Kos-Kudla, Editor-in-Chief. **ISSN:** 0423-104X. **Subscription Rates:** 270 Zl individuals; 540 Zl institutions; EUR68 individuals; EUR135 institutions. **Remarks:** Accepts advertising. **URL:** http://www.viamedica.pl/en/gazety/xgazPang/index.phtml. **Circ:** (Not Reported)

PORTUGAL

AVEIRO

1381 ■ International Journal of Surface Science & Engineering
Inderscience Publishers
c/o Prof. J. Paulo Davim, Ed.-in-Ch.
University of Aveiro
Department of Mechanical Engineering
Campus Santiago
3810-193 Aveiro, Portugal
Publication E-mail: ijsurfse@inderscience.com
Publisher E-mail: editor@inderscience.com

Journal covering field of surface science & tribology. **Founded:** 2007. **Freq:** 4/yr. **Key Personnel:** Prof. J. Paulo Davim, Editor-in-Chief, pdavim@ua.pt; Prof. Liangchi Zhang, Editor-in-Chief, lzha9252@mail.usyd.edu.au. **ISSN:** 1749-785X. **Subscription Rates:** EUR470 individuals includes surface mail, print only; EUR640 individuals print and online. **URL:** http://www.inderscience.com/browse/index.php?journalCODE=ijsurfse.

RUSSIA

IRKUTSK

1382 ■ Geography and Natural Resources
Publishing House of the Siberian Branch of the Russian Academy of Sciences
Institute of Geography
Box 4027
664033 Irkutsk, Russia
Ph: 7 3952 426422
Publication E-mail: gipr@izdatgeo.ru
Publisher E-mail: psb@ad-sbras.nsc.ru

Journal covering theoretical problems of geography and nature management. **Freq:** 4/yr. **Key Personnel:** A.N. Antipov, Editor-in-Chief. **ISSN:** 0206-1619. **URL:** http://www.sibran.ru/English/geogre.htm.

MOSCOW

1383 ■ Boat International Russia
Boat International Group
3/1 Gruzinskiy pereulok, off. 219
123056 Moscow, Russia
Ph: 7 495 6265593
Fax: 7 495 6265593
Publisher E-mail: info@boatinternational.co.uk

Magazine covering luxury yacht in Russia. **Freq:** Bimonthly. **Trim Size:** 235 x 275 mm. **Key Personnel:** Tony Harris, CEO/Publisher; Tony Euden, Publishing Dir., tonye@boatinternational.co.uk; Tim Hartney, Production Mgr., timh@boatinternational.co.uk. **Subscription Rates:** US$80 individuals in Russia; US$120 individuals in Europe; US$150 other countries; US$160 two years in Russia; US$240 two years in Europe; US$300 other countries 2 years; US$240 individuals in Russia, 3 years; US$360 individuals in Russia, 3 years; US$450 other countries 3 years. **Remarks:** Accepts advertising. **URL:** http://www.boatinternational.com/mags/mag03.htm. **Ad Rates:** 4C: 4,300 Rb. **Circ:** 15,000

1384 ■ Journal of Advances in Chemical Physics
Cambridge International Science Publishing
N.N. Semenov Institute of Chemical Physics
Russian Academy of Sciences
ul. Kosygina 4
119991 Moscow, Russia
Publication E-mail: jacp@cisp-publishing.com
Publisher E-mail: cisp@cisp-publishing.com

Journal covering areas of chemical physics. **Freq:** 6/yr. **Key Personnel:** A.L. Buchachenko, Editor-in-Chief. **URL:** http://www.cisp-publishing.com/jacp.htm.

NOVOSIBIRSK

1385 ■ Earth's Cryosphere
Publishing House of the Siberian Branch of the Russian Academy of Sciences
Pr. Akademika Koptyuga 3
630090 Novosibirsk, Russia
Ph: 7 3832 3356430
Publication E-mail: crio@izdatgeo.ru
Publisher E-mail: psb@ad-sbras.nsc.ru

Journal covering the study of the cryosphere of the Earth and other planets. **Founded:** 1997. **Freq:** 4/yr. **Key Personnel:** V.P. Mel'nikov, Editor-in-Chief. **URL:** http://www.sibran.ru/English/kriose.htm.

Circulation: ★ = ABC; △ = BPA; ♦ = CAC; • = CCAB; ❑ = VAC; ⊕ = PO Statement; ‡ = Publisher's Report; Boldface figures = sworn; Light figures = estimated.

1386 ■ Journal Chemistry for Sustainable Development
Publishing House of the Siberian Branch of the Russian Academy of Sciences
Morskoy pr. 2
630090 Novosibirsk, Russia
Fax: 7 383 3333755
Publication E-mail: csd@ad-sbras.nsc.ru
Publisher E-mail: psb@ad-sbras.nsc.ru

Journal covering all aspects of chemical research. **Founded:** 1993. **Freq:** Bimonthly. **Key Personnel:** Prof. Nikolay Z. Lyakhov, Editor-in-Chief, lyakhov@solid.nsk.su. **ISSN:** 1817-1818. **URL:** http://www.sibran.ru/English/CSDE.HTM.

1387 ■ Journal of Mining Sciences
Publishing House of the Siberian Branch of the Russian Academy of Sciences
Krasnyi pr. 54
630091 Novosibirsk, Russia
Ph: 7 3832 170048
Publisher E-mail: psb@ad-sbras.nsc.ru

Journal covering mining sciences. **Key Personnel:** V.N. Oparin, Editor-in-Chief. **ISSN:** 0015-3273. **URL:** http://www.sibran.ru/English/ftprpe.htm.

1388 ■ Siberian Journal of Ecology
Publishing House of the Siberian Branch of the Russian Academy of Sciences
Ul. Sovetskaia, 18-341
630099 Novosibirsk, Russia
Ph: 7 383 2224104
Publication E-mail: phsb@ad-sbras.nsc.ru
Publisher E-mail: psb@ad-sbras.nsc.ru

Journal covering aspects of ecology. **Key Personnel:** I.Yu. Koropachinsky, Editor-in-Chief. **URL:** http://www.sibran.ru/English/secje.htm.

1389 ■ Siberian Journal of Numerical Mathematics
Publishing House of the Siberian Branch of the Russian Academy of Sciences
Pr. Akademika Lavrent'eva 6
630090 Novosibirsk, Russia
Ph: 7 3832 356545
Fax: 7 3832 308783
Publication E-mail: sibjnm@sscc.ru; sibjnm@oapmg.sscc.ru
Publisher E-mail: psb@ad-sbras.nsc.ru

Journal covering mathematics models, and theory and practice of computational methods of mathematics. **Key Personnel:** A.S. Alekseyev, Editor-in-Chief. **URL:** http://www.sibran.ru/English/sjvme.htm.

SINGAPORE

SINGAPORE

1390 ■ Biomolecular Frontiers
National University of Singapore
Department of Biochemistry
Yong Loo Lin School of Medicine
8 Medical Dr., Blk. MD7, No. 02-03
Singapore 117597, Singapore
Publisher E-mail: qsmanager@nus.edu.sg

Journal featuring articles about biology and chemistry. **Subtitle:** An International Journal of Life Sciences from Singapore. **Key Personnel:** Barry Halliwell, Editor-in-Chief, bchhead@nus.edu.sg. **URL:** http://www.epress.nus.edu.sg/biomol/index.php.

1391 ■ Journal of Chinese Overseas
National University of Singapore
21 Lower Kent Ridge Rd.
Singapore 119077, Singapore
Ph: 65 6516 6666
Fax: 65 6775 9330
Publisher E-mail: qsmanager@nus.edu.sg

Journal featuring research articles, reports, and book reviews about Chinese overseas. **Freq:** Semiannual. **Key Personnel:** Tan Chee-Beng, Editor-in-Chief. **ISSN:** 1793-0391. **Subscription Rates:** S$65 individuals; US$50 other countries; US$70 institutions, other countries. **URL:** http://www.nus.edu.sg/npu/jco/index.html.

SINGAPORE CITY

1392 ■ Annual Review of Singapore Cases
Singapore Academy of Law
1 Supreme Court Ln., Level 6
Singapore City 178879, Singapore
Ph: 65 63324388
Fax: 65 63344940

Journal featuring Singapore reported and unreported cases, together with a discussion of relevant cases from other jurisdictions that impact on local law. **Founded:** 2000. **Freq:** Annual. **URL:** http://www.sal.org.sg/Web%20Pages/Legal%20Knowledge/SALAnnReview.aspx.

1393 ■ Asian Security Review
Alphabet Media
43c Beach Rd.
Singapore City 189681, Singapore
Ph: 65 63363136
Fax: 65 63365060

Magazine focusing on homeland security and critical infrastructure protection. **Freq:** 6/yr. **Key Personnel:** James Smith, Managing Editor, james.smith@alphabet-media.com. **Remarks:** Accepts advertising. **URL:** http://www.asiansecurity.org/. **Circ:** 10,000

1394 ■ Inter Se Print
Singapore Academy of Law
1 Supreme Court Ln., Level 6
Singapore City 178879, Singapore
Ph: 65 63324388
Fax: 65 63344940

Magazine featuring the life of a particular area of practice or topic with people profiles, interview segments, and reviews of relevant legal publications in the area. **Freq:** Semiannual. **Subscription Rates:** S$10 single issue. **URL:** http://www.sal.org.sg/Web%20Pages/Legal%20Knowledge/InterSePrint.aspx.

1395 ■ Public Sector Technology & Management
Alphabet Media
43c Beach Rd.
Singapore City 189681, Singapore
Ph: 65 63363136
Fax: 65 63365060

Magazine featuring institutional reform in the public sector, and research into improving the business of government. **Freq:** 6/yr. **Key Personnel:** James Smith, Managing Editor, james.smith@alphabet-media.com. **Remarks:** Accepts advertising. **URL:** http://www.pstm.net/. **Circ:** 8,950

1396 ■ Singapore Academy of Law Journal
Singapore Academy of Law
1 Supreme Court Ln., Level 6
Singapore City 178879, Singapore
Ph: 65 63324388
Fax: 65 63344940

Journal featuring articles related to Singapore law as well as Asia-Pacific and common law legal systems, and comparative and international law. **Founded:** 1989. **Freq:** Semiannual. **Key Personnel:** Prof. Michael Furmston, Editor; Prof. Francis Reynolds, Editor. **URL:** http://www.sal.org.sg/Web%20Pages/Legal%20Knowledge/SAcLJ.aspx.

REPUBLIC OF SOUTH AFRICA

CAPE TOWN

1397 ■ Lotus FM - Cape Town - 97.8 MHz
100 Old Fort Rd.
Private Bag 1337
Durban 4000, Republic of South Africa
Ph: 27 31 3625445
E-mail: lotus@lotusfm.co.za

Format: Full Service. **Operating Hours:** Continuous. **Key Personnel:** Kamsiliya Arumugam, Contact, phone 27 31 3625448, kamsiliya@lotusfm.co.za; Sagren Naidoo, Contact, phone 27 31 3625211, sagren@lotusfm.co.za. **URL:** http://www.lotusfm.co.za/portal/site/LotusFM/menuitem.a13733dc3e9a0bad57294 f945401aeb9/.

DURBAN

Lotus FM - Cape Town - See Cape Town

Lotus FM - Durban North - See Durban North

Lotus FM - Glencoe - See Glencoe

Lotus FM - Ladysmith - See Ladysmith

Lotus FM - Pietermaritzburg - See Pietermaritzburg

Lotus FM - Port Elizabeth - See Port Elizabeth

Lotus FM - Port Shepstone - See Port Shepstone

DURBAN NORTH

1398 ■ Lotus FM - Durban North - 89.4 MHz
100 Old Fort Rd.
Private Bag 1337
Durban 4000, Republic of South Africa
Ph: 27 31 3625445
E-mail: lotus@lotusfm.co.za

Format: Full Service. **Operating Hours:** Continuous. **Key Personnel:** Kamsiliya Arumugam, Contact, phone 27 31 3625448, kamsiliya@lotusfm.co.za; Sagren Naidoo, Contact, phone 27 31 3625211, sagren@lotusfm.co.za. **URL:** http://www.lotusfm.co.za/portal/site/LotusFM/menuitem.a13733dc3e9a0bad57294 f945401aeb9/.

GLENCOE

1399 ■ Lotus FM - Glencoe - 90.0 MHz
100 Old Fort Rd.
Private Bag 1337
Durban 4000, Republic of South Africa
Ph: 27 31 3625445
E-mail: lotus@lotusfm.co.za

Format: Full Service. **Operating Hours:** Continuous. **Key Personnel:** Kamsiliya Arumugam, Contact, phone 27 31 3625448, kamsiliya@lotusfm.co.za; Sagren Naidoo, Contact, phone 27 31 3625211, sagren@lotusfm.co.za. **URL:** http://www.lotusfm.co.za/portal/site/LotusFM/menuitem.a13733dc3e9a0bad57294 f945401aeb9/.

GRAHAMSTOWN

1400 ■ African Journal of Pharmacy & Pharmacology
Academic Journals
c/o John M. Haigh, PhD, Acting Ed.
Division of Pharmaceutical Chemistry
Faculty of Pharmacy
Rhodes University
Grahamstown 6140, Republic of South Africa
Publication E-mail: ajpp@academicjournals.org
Publisher E-mail: service@academicjournals.org

Journal covering in all areas of pharmaceutical science. **Freq:** Monthly. **Key Personnel:** John M. Haigh, PhD, Acting Ed. **Remarks:** Accepts advertising. **URL:** http://www.academicjournals.org/AJPP/. **Circ:** (Not Reported)

LADYSMITH

1401 ■ Lotus FM - Ladysmith - 87.9 MHz
100 Old Fort Rd.
Private Bag 1337
Durban 4000, Republic of South Africa
Ph: 27 31 3625445
E-mail: lotus@lotusfm.co.za

Format: Full Service. **Operating Hours:** Continuous. **Key Personnel:** Kamsiliya Arumugam, Contact, phone 27 31 3625448, kamsiliya@lotusfm.co.za; Sagren Naidoo, Contact, phone 27 31 3625211, sagren@lotusfm.co.za. **URL:** http://www.lotusfm.co.za/portal/site/LotusFM/menuitem.a13733dc3e9a0bad57294 f945401aeb9/.

PIETERMARITZBURG

1402 ■ Lotus FM - Pietermaritzburg - 88.3 MHz
100 Old Fort Rd.
Private Bag 1337
Durban 4000, Republic of South Africa
Ph: 27 31 3625445
E-mail: lotus@lotusfm.co.za

Format: Full Service. **Operating Hours:** Continuous. **Key Personnel:** Kamsiliya Arumugam, Contact, phone 27 31 3625448, kamsiliya@lotusfm.co.za; Sagren Naidoo, Contact, phone 27 31 3625211, sagren@lotusfm.co.za. **URL:** http://www.lotusfm.co.za/portal/site/LotusFM/menuitem.a13733dc3e9a0bad57294 f945401aeb9/.

PORT ELIZABETH

1403 ■ Lotus FM - Port Elizabeth - 98.3 MHz
100 Old Fort Rd.
Private Bag 1337
Durban 4000, Republic of South Africa
Ph: 27 31 3625445
E-mail: lotus@lotusfm.co.za

Format: Full Service. **Operating Hours:** Continuous. **Key Personnel:** Kamsiliya Arumugam, Contact, phone 27 31 3625448, kamsiliya@lotusfm.co.za; Sagren Naidoo, Contact, phone 27 31 3625211, sagren@lotusfm.co.za. **URL:** http://www.lotusfm.co.za/portal/site/LotusFM/menuitem.a13733dc3e9a0bad57294 f945401aeb9/.

PORT SHEPSTONE

1404 ■ Lotus FM - Port Shepstone - 88.2 MHz
100 Old Fort Rd.
Private Bag 1337
Durban 4000, Republic of South Africa
Ph: 27 31 3625445
E-mail: lotus@lotusfm.co.za

Format: Full Service. **Operating Hours:** Continuous. **Key Personnel:** Kamsiliya Arumugam, Contact, phone 27 31 3625448, kamsiliya@lotusfm.co.za; Sagren Naidoo, Contact, phone 27 31 3625211, sagren@lotusfm.co.za. **URL:** http://www.lotusfm.co.za/portal/site/LotusFM/menuitem.a13733dc3e9a0bad57294 f945401aeb9/.

SPAIN

OVIEDO

1405 ■ International Journal of Chinese Culture & Management
Inderscience Publishers
c/o Prof. Patricia Ordonez de Pablos, Ed.-in-Ch.
Universidad de Oviedo
Avd del Cristo
E-33071 Oviedo, Spain
Publication E-mail: ijccm@inderscience.com
Publisher E-mail: editor@inderscience.com

Journal covering Chinese culture and business & management. **Founded:** 2007. **Freq:** 4/yr. **Key Personnel:** Prof. Patricia Ordonez de Pablos, Editor-in-Chief, patriop@uniovi.es. **ISSN:** 1752-1270. **Subscription Rates:** EUR470 individuals includes surface mail, print only; EUR640 individuals print & online. **URL:** http://www.inderscience.com/browse/index.php?journalID=220board.

TAIWAN

TAICHUNG

1406 ■ International Journal of Electronic Customer Relationship Management
Inderscience Publishers
c/o Prof. Bruce Chien-Ta Ho, Ed.-in-Ch.
National Chung Hsing University
250 Kuokuang Rd.
Taichung 402, Taiwan
Publisher E-mail: editor@inderscience.com

Journal covering electronic customer relationship management. **Founded:** 2007. **Freq:** 4/yr. **Key Personnel:** Prof. Bruce Chien-Ta Ho, Editor-in-Chief, bruceho@nchu.edu.tw; Prof. Tzong-Ru Lee, Editor, trlee@nchu.edu.tw. **ISSN:** 1750-0664. **Subscription Rates:** EUR470 individuals includes surface mail, print only; EUR640 individuals print & online. **URL:** http://www.inderscience.com/browse/index.php?journalCODE=ijecrm.

TAIPEI

1407 ■ Contemporary Management Research
Academy of Taiwan Information Systems Research
PO Box 179-45
Sansia Twp.
Taipei 11699, Taiwan
Ph: 886 2 25009848
Publisher E-mail: chang@atisr.org

Journal covering all fields of management. **Freq:** Quarterly. **Trim Size:** A4. **Key Personnel:** Wen-chang Fang, PhD, Editor, fang@mail.ntpu.edu.tw. **ISSN:** 1813-5498. **Subscription Rates:** US$200 institutions; US$150 individuals. **URL:** http://cmr-journal.org/; http://academic-journal.org/cmr/.

1408 ■ International Journal of Business and Information (IJBI)
Academy of Taiwan Information Systems Research
DaZhi St., No. 70
Taipei 10469, Taiwan
Ph: 886 2 25381111
Fax: 886 2 25333143
Publisher E-mail: chang@atisr.org

Journal covering all areas of business and information development around the world. **Key Personnel:** Kwei Tang, Editor-in-Chief, ktang@mgmt.purdue.edu; Chian-Son Yu, Managing Editor, csyu@mail.usc.edu.tw. **ISSN:** 1728-8673. **URL:** http://ijbi.org/; http://www.knowledgetaiwan.org/ojs/index.php/ijbi.

1409 ■ International Journal of Cyber Society and Education
Academy of Taiwan Information Systems Research
PO Box 4-1
Taipei 23799, Taiwan
Publisher E-mail: chang@atisr.org

Journal covering all fields of cyber society and education. **Key Personnel:** Prof. Yih-Chearng Shiue, PhD, Editor-in-Chief, ijcse.journal@qmail.com. **ISSN:** 1995-6649. **URL:** http://www.academic-journals.org/ojs2/index.php/ijcse; http://atisr.org/journals.

1410 ■ International Journal of Information and Computer Security
Inderscience Publishers
c/o Dr. Eldon Y. Li, Ed.-in-Ch.
National Chengchi University
Department of Management Information System, College of Commerce
No. 64, Sec. 2, Zhi-nan Rd., Wenshan
Taipei 11605, Taiwan
Publication E-mail: ijics@inderscience.com
Publisher E-mail: editor@inderscience.com

Journal covering developments of information & computer security. **Founded:** 2007. **Freq:** 4/yr. **Key Personnel:** Dr. Eldon Y. Li, Editor-in-Chief, eli@calpoly.edu. **ISSN:** 1744-1765. **Subscription Rates:** EUR470 individuals includes surface mail, print only; EUR640 individuals print and online. **URL:** http://www.inderscience.com/browse/index.php?journalCODE=ijics.

UKRAINE

KIEV

1411 ■ Afisha
KP Media
34 Lesi Ukrainky Blvd.
01601 Kiev, Ukraine
Ph: 380 44 4961111
Fax: 380 44 4961110

Magazine featuring cultural and entertainment listings including concerts, performances, movies and exhibitions, as well as a comprehensive guide to Kyiv's restaurants and bars, nightclubs, and casinos. **Founded:** Apr. 2001. **Freq:** Weekly. **Key Personnel:** Jed Sunden, Publisher, jed@kppublications.com. **Subscription Rates:** 129 Rb individuals. **Remarks:** Accepts advertising. **URL:** http://www.kpmedia.com.ua/eng/consumer/af/. **Circ:** Combined 25,000

1412 ■ Interior Magazine
KP Media
34 Lesi Ukrainky Blvd.
01601 Kiev, Ukraine
Ph: 380 44 4961111
Fax: 380 44 4961110

Magazine featuring retail catalogue for the home with information, prices, and addresses of retailers. **Founded:** Nov. 2004. **Freq:** Monthly. **Key Personnel:** Jed Sunden, Publisher, jed@kppublications.com. **Remarks:** Accepts advertising. **URL:** http://www.kpmedia.com.ua/eng/consumer/im/. **Circ:** Combined 17,500

1413 ■ Korrespondent
KP Media
34 Lesi Ukrainky Blvd.
01601 Kiev, Ukraine
Ph: 380 44 4961111
Fax: 380 44 4961110

Magazine featuring important events in Ukraine and around the world. **Founded:** Mar. 2002. **Freq:** Weekly. **Key Personnel:** Jed Sunden, Publisher, jed@kppublications.com. **Subscription Rates:** 229 Rb individuals. **URL:** http://www.kpmedia.com.ua/eng/newspr/kor/. **Circ:** Combined 50,000

1414 ■ Kyiv Post
KP Media
34 Lesi Ukrainky Blvd.
01601 Kiev, Ukraine
Ph: 380 44 4961111
Fax: 380 44 4961110

Newspaper featuring articles from the weekly English-language newspaper plus daily updates of events. **Founded:** 1995. **Freq:** Weekly. **Key Personnel:** Cenon Zawada, Ch. Ed., zawada@kppublications.com; Jed Sunden, Publisher, jed@kppublications.com. **Subscription Rates:** 95 Rb individuals; 220 Rb other countries. **Remarks:** Accepts advertising. **URL:** http://www.kyivpost.com/; http://www.kpmedia.com.ua/eng/newspr/kp/. **Circ:** Combined 25,000

1415 ■ Novynar
KP Media
34 Lesi Ukrainky Blvd.
01601 Kiev, Ukraine
Ph: 380 44 4961111
Fax: 380 44 4961110

News magazine featuring Novynar's current events in politics, economics, and culture. **Founded:** 2007. **Freq:** Weekly. **Key Personnel:** Jed Sunden, Publisher, jed@kppublications.com. **URL:** http://www.kpmedia.com.ua/eng/newspr/novynar/. **Circ:** Combined 15,000

1416 ■ Vona
KP Media
34 Lesi Ukrainky Blvd.
01601 Kiev, Ukraine
Ph: 380 44 4961111
Fax: 380 44 4961110

Magazine featuring relationship advice, beauty secrets, and daily living information. **Founded:** Aug. 2007. **Freq:** Monthly. **Key Personnel:** Jed Sunden, Publisher, jed@kppublications.com. **Remarks:** Accepts advertising. **URL:** http://www.kpmedia.com.ua/eng/consumer/vona/. **Circ:** Combined 30,000

Circulation: ★ = ABC; △ = BPA; ♦ = CAC; • = CCAB; ❑ = VAC; ⊕ = PO Statement; ‡ = Publisher's Report; Boldface figures = sworn; Light figures = estimated.

UNITED ARAB EMIRATES

DUBAI

1417 ■ Business Traveller Middle East
Motivate Publishing
Al Wahaibi Bldg.
Al Garhoud Bridge Rd., Deira
PO Box 2331
Dubai, United Arab Emirates
Ph: 971 428 24060
Fax: 971 428 20428
Publisher E-mail: motivate@motivate.ae

Magazine covering business travel. **Founded:** 2003. **Freq:** Bimonthly. **Trim Size:** 208 x 275 mm. **Key Personnel:** Colette Doyle, Editor, colette@motivate.ae; Anthony Milne, Gp. Advertising Mgr., phone 971 4 2824060, fax 971 4 2824436, anthony@motivate.ae. **Subscription Rates:** 88 Dh individuals; 185 Dh other countries; US$24 individuals; US$50 other countries. **Remarks:** Accepts advertising. **URL:** http://www.motivatepublishing.com/packages/default.asp?categorycode=Mag&pac kageid=ART00518; http://www.motivatepublishing.com/library/default.asp?ArticleCode=ART00558. **Circ:** △**25,341**

1418 ■ Dubai Voyager
Motivate Publishing
Al Wahaibi Bldg.
Al Garhoud Bridge Rd., Deira
PO Box 2331
Dubai, United Arab Emirates
Ph: 971 428 24060
Fax: 971 428 20428
Publisher E-mail: motivate@motivate.ae

Magazine featuring travel guide in Dubai. **Founded:** 1987. **Freq:** Monthly. **Trim Size:** 210 x 297 mm. **Key Personnel:** Colette Doyle, Editor, colette@motivate.ae; Shawki Abd El Malik, Gp. Advertisement Mgr., phone 971 4 2824060, fax 971 4 2824436, shawki@motivate.ae. **Remarks:** Accepts advertising. **URL:** http://www.motivatepublishing.com/packages/default.asp?categorycode=Mag&pac kageid=ART00513; http://www.motivatepublishing.com/library/default.asp?ArticleCode=ART00559. **Circ:** △**29,367**

1419 ■ Emaar Properties Magazine
Motivate Publishing
Al Wahaibi Bldg.
Al Garhoud Bridge Rd., Deira
PO Box 2331
Dubai, United Arab Emirates
Ph: 971 428 24060
Fax: 971 428 20428
Publisher E-mail: motivate@motivate.ae

Magazine featuring the company of Emaar Properties lifestyle. **Founded:** 2003. **Freq:** Quarterly. **Trim Size:** 240 x 320 mm. **Key Personnel:** Catherine Belbin, Editor, catherine@motivate.ae; Ashish Limaye, Gp. Advertisement Mgr., phone 971 4 2824060, fax 971 4 2824436, ashish@motivate.ae. **Remarks:** Accepts advertising. **URL:** http://www.motivatepublishing.com/packages/default.asp?categorycode=Mag&pac kageid=ART00516; http://www.motivatepublishing.com/library/default.asp?ArticleCode=ART00563. **Also known as:** EP. **Circ:** 12,000

1420 ■ Emirates Bride
Motivate Publishing
Al Wahaibi Bldg.
Al Garhoud Bridge Rd., Deira
PO Box 2331
Dubai, United Arab Emirates
Ph: 971 428 24060
Fax: 971 428 20428
Publisher E-mail: motivate@motivate.ae

Magazine featuring wedding information for the brides in UAE. **Freq:** Bimonthly. **Trim Size:** 225 x 300 mm. **Key Personnel:** Faye Marchant, Editor, faye@motivate.ae; Ashish Limaye, Gp. Advertisement Mgr., phone 971 4 2824060, fax 971 4 2824436, ashish@motivate.ae. **Subscription Rates:** 60 Dh individuals; 150 Dh other countries; US$16 individuals; US$41 other countries. **Remarks:** Accepts advertising. **URL:** http://www.motivatepublishing.com/packages/default.asp?categorycode=Mag&pac kageid=Mag.Emi_Br; http://www.motivatepublishing.com/library/default.asp?ArticleCode=ART00560. **Circ:** 15,000

1421 ■ Jumeirah
Motivate Publishing
Al Wahaibi Bldg.
Al Garhoud Bridge Rd., Deira
PO Box 2331
Dubai, United Arab Emirates
Ph: 971 428 24060
Fax: 971 428 20428
Publisher E-mail: motivate@motivate.ae

Magazine featuring Jumeirah's hotels in Dubai. **Founded:** 2001. **Freq:** Bimonthly. **Trim Size:** 225 x 300 mm. **Key Personnel:** Janet Brice, Editor, janet@motivate.ae; Shawki Abd El Malik, Gp. Advertisement Mgr., phone 971 4 2824060, fax 971 4 2824436, liam@motivate.ae. **Remarks:** Accepts advertising. **URL:** http://www.motivatepublishing.com/packages/default.asp?categorycode=Mag&pac kageid=ART00514; http://www.motivatepublishing.com/library/default.asp?ArticleCode=ART00569. **Circ:** 17,500

1422 ■ Middle East MICE & Events
Motivate Publishing
Al Wahaibi Bldg.
Al Garhoud Bridge Rd., Deira
PO Box 2331
Dubai, United Arab Emirates
Ph: 971 428 24060
Fax: 971 428 20428
Publisher E-mail: motivate@motivate.ae

Magazine covering meetings, incentive travel, conferences, exhibitions, and any events. **Founded:** 2006. **Freq:** Bimonthly. **Trim Size:** 210 x 297 mm. **Key Personnel:** David van der Meulen, Editor, phone 971 4 2824436, davidv@motivate.ae. **Subscription Rates:** 50 Dh individuals in UAE; 75 Dh individuals in GCC; 125 Dh other countries. **Remarks:** Accepts advertising. **URL:** http://www.memicee.com/; http://www.motivatepublishing.com/library/default.asp?ArticleCode=ART00571. **Circ:** 12,500

1423 ■ Open Skies
Motivate Publishing
Al Wahaibi Bldg.
Al Garhoud Bridge Rd., Deira
PO Box 2331
Dubai, United Arab Emirates
Ph: 971 428 24060
Fax: 971 428 20428
Publisher E-mail: motivate@motivate.ae

Travel and leisure magazine featuring flight services of Emirates airline. **Founded:** 1985. **Freq:** Monthly. **Trim Size:** 210 x 285 mm. **Key Personnel:** Guido Duken, Editor, guido@motivate.ae; Shawki Abd El Malik, Gp. Advertisement Mgr., phone 971 4 2824060, fax 971 4 2824436, shawki@motivate.ae. **Remarks:** Accepts advertising. **URL:** http://www.motivatepublishing.com/packages/default.asp?categorycode=Mag&pac kageid=ART00510; http://www.motivatepublishing.com/library/default.asp?ArticleCode=ART00573. **Circ:** △**56,327**

1424 ■ Society Dubai
Motivate Publishing
Al Wahaibi Bldg.
Al Garhoud Bridge Rd., Deira
PO Box 2331
Dubai, United Arab Emirates
Ph: 971 428 24060
Fax: 971 428 20428
Publisher E-mail: motivate@motivate.ae

Magazine featuring the city of Dubai. **Founded:** 2003. **Freq:** Monthly. **Trim Size:** 240 x 320 mm. **Key Personnel:** Faye Marchant, Editor, faye@motivate.ae; Ashish Limaye, Gp. Advertisement Mgr., phone 971 4 2824060, fax 971 4 2824436, ashish@motivate.ae. **Subscription Rates:** 60 Dh individuals in UAE; 120 Dh individuals in GCC; 260 Dh other countries. **Remarks:** Accepts advertising. **URL:** http://www.motivatepublishing.com/packages/default.asp?categorycode=Mag&pac kageid=ART00507; http://www.motivatepublishing.com/library/default.asp?ArticleCode=ART00575. **Circ:** △**12,000**

1425 ■ Souk
Motivate Publishing
Al Wahaibi Bldg.
Al Garhoud Bridge Rd., Deira
PO Box 2331
Dubai, United Arab Emirates
Ph: 971 428 24060
Fax; 971 428 20428
Publisher E-mail: motivate@motivate.ae

Magazine featuring shopping, dining, and entertainment establishments in Souk. **Founded:** 2005. **Freq:** Quarterly. **Trim Size:** 193 x 232 mm. **Key Personnel:** Janet Brice, Editor, janet@motivate.ae; Shawki Abd El Malik, Gp. Advertisement Mgr., phone 971 4 2824060, fax 971 4 2824436, shawki@motivate.ae. **Remarks:** Accepts advertising. **URL:** http://www.motivatepublishing.com/packages/default.asp?categorycode=Mag&pac kageid=ART00519; http://www.motivatepublishing.com/library/default.asp?ArticleCode=ART00576. . **Ad Rates:** 4C: US$2,500. **Circ:** 30,000

1426 ■ tv&radio
Motivate Publishing
Al Wahaibi Bldg.
Al Garhoud Bridge Rd., Deira
PO Box 2331
Dubai, United Arab Emirates
Ph: 971 428 24060
Fax: 971 428 20428
Publisher E-mail: motivate@motivate.ae

Magazine featuring audio visual channels, audio channels and external cameras. **Founded:** 1986. **Freq:** Bimonthly. **Trim Size:** 206 x 270 mm. **Key Personnel:** Guido Duken, Editor, guido@motivate.ae; Shawki Abd El Malik, Gp. Advertising Mgr., phone 971 4 2052402, fax 971 4 2822801, shawki@motivate.ae. **Remarks:** Accepts advertising. **URL:** http://www.motivatepublishing.com/packages/default.asp?categorycode=Mag&pac kageid=ART00512; http://www.motivatepublishing.com/library/default.asp?ArticleCode=ART00578. **Circ:** △**56,250**

UNITED KINGDOM

ABERDEEN

1427 ■ Good Time Guide
Happy Publishing Limited
27 York Pl.
Aberdeen AB11 4DH, United Kingdom
Ph: 44 1224 594659
Publisher E-mail: info@happypublishing.co.uk

Entertainment guide magazine. **Freq:** Annual. **Key Personnel:** Danny Cowie, Mng. Dir. **Remarks:** Accepts advertising. **URL:** http://www.goodtimeguide.co.uk/; http://www.happypublishing.co.uk/goodtimeguide_magazine.htm. **Circ:** (Not Reported)

1428 ■ Journal of Private International Law
Hart Publishing Ltd.
c/o Prof. Paul Beaumont, Gen. Ed.
Taylor Bldg.
School of Law
University of Aberdeen
Aberdeen AB24 3UB, United Kingdom
Publisher E-mail: mail@hartpub.co.uk

Journal covering all aspects of private international law. **Founded:** 2005. **Freq:** 3/yr. **Key Personnel:** Prof. Paul Beaumont, Gen. Ed., p.beaumont@abdn.ac.uk; Prof. Jonathan Harris, Gen. Ed., j.m.harris.law@bham.ac.uk. **ISSN:** 1744-1048. **Subscription Rates:** 130 individuals standard, United Kingdom and Europe; 150 other countries standard; 65 individuals reduced, United Kingdom and Europe; 100 other countries reduced. **Remarks:** Accepts advertising. **URL:** http://www.hartjournals.co.uk/JPrivIntL/. **Circ:** (Not Reported)

1429 ■ Star Flyer Magazine
Happy Publishing Limited
27 York Pl.
Aberdeen AB11 4DH, United Kingdom
Ph: 44 1224 594659
Publisher E-mail: info@happypublishing.co.uk

Inflight magazine of City Star Airlines featuring news and entertainment. **Freq:** Quarterly. **Key Personnel:** Danny Cowie, Mng. Dir. **Remarks:** Accepts advertising. **URL:** http://www.happypublishing.co.uk/starflyer_magazine.htm. **Circ:** (Not Reported)

ABERLOUR

1430 ■ Lady Biker Magazine
Happy Publishing Limited
Aigan View
Benrinnes Dr.
Aberlour AB38 9NQ, United Kingdom
Publication E-mail: info@ladybikermagazine.com
Publisher E-mail: info@happypublishing.co.uk

Motorcycle magazine for ladies. **Freq:** Semiannual. **Key Personnel:** Zoe Grice, Founder & Ed., zoe@ladybikermagazine.com; Danny Cowie, Mng. Dir. **Remarks:** Accepts advertising. **URL:** http://www.ladybikermagazine.com/; http://www.happypublishing.co.uk/ladybiker_magazine.htm. **Circ:** (Not Reported)

ANDOVER

1431 ■ Andover Advertiser
Newsquest Media Group Ltd.
24-32 London St.
Andover SP10 2PE, United Kingdom
Ph: 44 1264 323456
Fax: 44 1264 338723
Publication E-mail: newsdesk@andoveradvertiser.co.uk
Publisher E-mail: newsquestmediasales@newsquestmedia.co.uk

Newspaper featuring news and events in Andover. **Key Personnel:** Dick Bellringer, News Ed., phone 44 1264 321205, dick.bellringer@andoveradvertiser.co.uk. **Remarks:** Accepts advertising. **URL:** http://www.andoveradvertiser.co.uk/. **Circ:** (Not Reported)

BANGOR

1432 ■ North Wales Chronicle
NWN Media Ltd.
302 High St.
Bangor LL57 1UL, United Kingdom
Ph: 44 1248 387400
Publisher E-mail: internet@nwn.co.uk

Newspaper featuring news and events. **Founded:** 1808. **Key Personnel:** Steve Rogers, Editor. **URL:** http://www.nwnews.co.uk/; http://www.northwaleschronicle.co.uk/.

BARNSLEY

1433 ■ Assistive Technologies
Wharncliffe Publishing Ltd.
47 Church St.
South Yorkshire
Barnsley S70 2AS, United Kingdom
Ph: 44 1226 734639
Fax: 44 1226 734478
Publisher E-mail: editorial@wharncliffepublishing.co.uk

Magazine covering assistive technologies and mobility improvement for healthcare professionals and business associations. **Key Personnel:** Dominic Musgrave, Editor, editorial@assistivetechnologies.co.uk; Kelly Tarff, Circulation Mgr., circulation@wharncliffepublishing.co.uk; Judith Halkerston, Gp. Production Ed., jhalkerston@whpl.net. **Remarks:** Accepts advertising. **URL:** http://www.assistivetechnologies.co.uk; http://www.wharncliffepublishing.co.uk. **Circ:** (Not Reported)

1434 ■ Caring UK
Wharncliffe Publishing Ltd.
47 Church St.
South Yorkshire
Barnsley S70 2AS, United Kingdom
Ph: 44 1226 734639
Fax: 44 1226 734478
Publisher E-mail: editorial@wharncliffepublishing.co.uk

Magazine covering care for managers in the elderly sector. **Freq:** Monthly. **Key Personnel:** Dominic Musgrave, Healthcare Ed., dm@whpl.net; Andrew Harrod, Gp. Ed., ah@whpl.net; Kelly Tarff, Circulation Mgr., kt@whpl.net. **Remarks:** Accepts advertising. **URL:** http://www.caring-uk.co.uk; http://www.wharncliffepublishing.co.uk. **Circ:** (Not Reported)

1435 ■ Destination UK
Wharncliffe Publishing Ltd.
47 Church St.
South Yorkshire
Barnsley S70 2AS, United Kingdom
Ph: 44 1226 734639
Fax: 44 1226 734478
Publisher E-mail: editorial@wharncliffepublishing.co.uk

Magazine covering UK's travel and tourism industry. **Freq:** Monthly. **Key Personnel:** Andrew Harrod, Gp. Ed., ah@whpl.net; Nicky Lambert, News Ed., nl@whpl.net; Kelly Tarff, Circulation Mgr., kt@whpl.net. **Remarks:** Accepts advertising. **URL:** http://www.destination.uk.com; http://www.wharncliffepublishing.co.uk. **Circ:** 12,000

1436 ■ Future Fitness
Wharncliffe Publishing Ltd.
47 Church St.
South Yorkshire
Barnsley S70 2AS, United Kingdom
Ph: 44 1226 734639
Fax: 44 1226 734478
Publisher E-mail: editorial@wharncliffepublishing.co.uk

Magazine featuring sport and fitness for today's youth. **Key Personnel:** Andrew Harrod, Gp. Ed., ah@whpl.net; James Dickson, Sales Mgr., jd@whpl.net; Kelly Tarff, Circulation Mgr., circulation@wharncliffepublishing.co.uk. **Remarks:** Accepts advertising. **URL:** http://www.futurefitness.uk.net; http://www.wharncliffepublishing.co.uk. **Circ:** (Not Reported)

1437 ■ Horse Health Magazine
Wharncliffe Publishing Ltd.
47 Church St.
South Yorkshire
Barnsley S70 2AS, United Kingdom
Ph: 44 1226 734639
Fax: 44 1226 734478
Publisher E-mail: editorial@wharncliffepublishing.co.uk

Magazine covering horse health and well-being. **Key Personnel:** Andrew Harrod, Gp. Ed., ah@whpl.net; Christine Keate, Editor, chris.keate@horsehealthmagazine.co.uk; Kelly Tarff, Circulation Mgr., kt@whpl.net. **Remarks:** Accepts advertising. **URL:** http://www.horsehealthmagazine.co.uk; http://www.wharncliffepublishing.co.uk. **Circ:** 10,000

1438 ■ Out on a Limb
Wharncliffe Publishing Ltd.
47 Church St.
South Yorkshire
Barnsley S70 2AS, United Kingdom
Ph: 44 1226 734639
Fax: 44 1226 734478
Publisher E-mail: editorial@wharncliffepublishing.co.uk

Magazine featuring directional footwear and fashion accessories. **Key Personnel:** Louise Cordell, Editor, lcordell@whpl.net; Andrew Harrod, Gp. Ed., ah@whpl.net; Kelly Tarff, Circulation Mgr., kt@whpl.net. **Remarks:** Accepts advertising. **URL:** http://www.ooalmagazine.co.uk; http://www.wharncliffepublishing.co.uk. **Circ:** (Not Reported)

1439 ■ Wedding Professional
Wharncliffe Publishing Ltd.
47 Church St.
South Yorkshire
Barnsley S70 2AS, United Kingdom
Ph: 44 1226 734639
Fax: 44 1226 734478
Publisher E-mail: editorial@wharncliffepublishing.co.uk

Magazine covering professional wedding planning market. **Freq:** Bimonthly. **Key Personnel:** Mary Ferguson, Editor, mf@whpl.net; Andrew Harrod, Gp. Ed.; Kelly Tarff, Circulation Mgr., circulation@wharncliffepublishing.co.uk. **Remarks:** Accepts advertising. **URL:** http://www.weddingprofessional.co.uk; http://www.wharncliffepublishing.co.uk. **Circ:** (Not Reported)

1440 ■ WorkOut Ireland
Wharncliffe Publishing Ltd.
47 Church St.
South Yorkshire
Barnsley S70 2AS, United Kingdom
Ph: 44 1226 734639
Fax: 44 1226 734478
Publisher E-mail: editorial@wharncliffepublishing.co.uk

Magazine covering health, leisure and fitness industry in Ireland. **Key Personnel:** Andrew Harrod, Gp. Ed., ah@whpl.net; Nicola Lambert, News Ed., nl@whpl.net; James Dickson, Publishing Mgr., jd@whpl.net. **Remarks:** Accepts advertising. **URL:** http://www.workout-ireland.com; http://www.wharncliffepublishing.co.uk. **Circ:** (Not Reported)

1441 ■ WorkOut UK
Wharncliffe Publishing Ltd.
47 Church St.
South Yorkshire
Barnsley S70 2AS, United Kingdom
Ph: 44 1226 734639
Fax: 44 1226 734478
Publisher E-mail: editorial@wharncliffepublishing.co.uk

Magazine covering health, leisure and fitness industry in UK. **Freq:** Monthly. **Key Personnel:** Andrew Harrod, Gp. Ed., ah@whpl.net; Nicola Lambert, News Ed., nl@whpl.net; Kelly Tarff, Circulation Mgr., kt@whpl.net. **Remarks:** Accepts advertising. **URL:** http://www.workout-uk.co.uk; http://www.wharncliffepublishing.co.uk. **Circ:** 7,700

BARRY

1442 ■ Barry & District News
Newsquest Media Group Ltd.
156 Holton Rd.
Barry CF63 4TY, United Kingdom
Ph: 44 1446 733456
Publisher E-mail: newsquestmediasales@newsquestmedia.co.uk

Newspaper providing news and events in Barry district. **Key Personnel:** Shira Valek, Editor, shira.valek@gwent-wales.co.uk; Andrea Hall, Advertising Mgr., andrea.hall@gwent-wales.co.uk. **Remarks:** Accepts advertising. **URL:** http://www.barryanddistrictnews.co.uk/. **Circ:** (Not Reported)

BASILDON

1443 ■ Basildon and Wickford Recorder
Newsquest Media Group Ltd.
Chester Hall Ln.
Basildon SS14 3BL, United Kingdom
Ph: 44 844 4774512
Fax: 44 844 4774286
Publisher E-mail: newsquestmediasales@newsquestmedia.co.uk

Newspaper providing up to date local news and sports. **Key Personnel:** Martin Mc Neill, Editor,

Circulation: ★ = ABC; △ = BPA; ♦ = CAC; • = CCAB; ❑ = VAC; ⊕ = PO Statement; ‡ = Publisher's Report; Boldface figures = sworn; Light figures = estimated.

phone 44 844 4774463, martin.mcneill@nqe.com; Christina Ongley, News Ed., phone 44 844 4774393, christina.ongley@nqe.com. **Subscription Rates:** 44.20 individuals. **Remarks:** Accepts advertising. **URL:** http://www.basildonrecorder.co.uk/. **Circ:** (Not Reported)

BASINGSTOKE

1444 ■ Basingstoke Gazette
Newsquest Media Group Ltd.
Peltone Rd.
Basingstoke RG21 6YD, United Kingdom
Ph: 44 1256 461131
Publisher E-mail: newsquestmediasales@newsquestmedia.co.uk

Newspaper featuring up to date information on local news and sports. **Key Personnel:** Hayley West, Contact, phone 44 1722 426531. **Subscription Rates:** 1 individuals Monday; 1.75 individuals Thursday. **Remarks:** Accepts advertising. **URL:** http://www.basingstokegazette.co.uk/. **Circ:** (Not Reported)

BOLTON

1445 ■ The Balton News
Newsquest Media Group Ltd.
1 Churchgate
Bolton BL1 1DE, United Kingdom
Ph: 44 1204 522345
Publication E-mail: newsdesk@theboltonnews.co.uk
Publisher E-mail: newsquestmediasales@newsquestmedia.co.uk

Newspaper featuring local news and events. **Key Personnel:** James Higgins, News Ed., bennewsdesk@theboltonnews.co.uk; Helen Turnbull, Advertising Mgr., phone 44 1204 537432, helen.turnbull@lancashire.newsquest.co.uk. **Subscription Rates:** 1.20 single issue. **URL:** http://www.boltoneveningnews.co.uk/.

BRADFORD

1446 ■ Telegraph & Argus
Newsquest Media Group Ltd.
Hall Ings
Bradford BD1 1JR, United Kingdom
Ph: 44 1274 729511
Publication E-mail: newsdesk@bradford.newsquest.co.uk
Publisher E-mail: newsquestmediasales@newsquestmedia.co.uk

Newspaper featuring sports, community news, motoring advice, and job hunting updates. **Freq:** Mon.-Sat. **Key Personnel:** Perry Austin-Clarke, Editor, phone 44 1274 729511, newsdesk@bradford.newsquest.co.uk; Peter Orme, Managing Editor, phone 44 1274 705346, peter.orme@bradford.newsquest.co.uk. **Subscription Rates:** 343.20 individuals. **Remarks:** Accepts advertising. **URL:** http://www.thetelegraphandargus.co.uk/. **Circ:** (Not Reported)

BRIDPORT

1447 ■ Bridport & Lyme Regis News
Newsquest Media Group Ltd.
67 East St.
Bridport DT6 3LB, United Kingdom
Ph: 44 1308 422388
Publisher E-mail: newsquestmediasales@newsquestmedia.co.uk

Newspaper featuring news and event for Lyme Regis and surrounding villages. **Remarks:** Accepts advertising. **URL:** http://www.bridportnews.co.uk/. **Circ:** (Not Reported)

BRIGHTON

1448 ■ The Argus Lite
Newsquest Media Group Ltd.
Crowhurst Rd.
Hollingbury
Brighton BN1 8AR, United Kingdom
Ph: 44 1273 544544
Fax: 44 1273 566114
Publisher E-mail: newsquestmediasales@newsquestmedia.co.uk

Newspaper featuring local up to date news and events. **Freq:** Mon.-Sat. **Key Personnel:** Michael Beard, Editor, editor@theargus.co.uk; Martyn Willis, Managing Editor, martyn.willis@theargus.co.uk. **Remarks:** Accepts advertising. **URL:** http://www.arguslite.co.uk/. **Circ:** (Not Reported)

CAMBRIDGE

1449 ■ Boat and Yacht Buyer
CSL Publishing Ltd.
Alliance House
49 Sidney St.
Cambridge CB2 3HX, United Kingdom
Ph: 44 1223 460490
Fax: 44 1223 315960

Magazine featuring boats and boating accessories for sale. **Freq:** 13/yr. **Key Personnel:** Tracy Finnerty, Contact, tracy@boatmart.co.uk. **Subscription Rates:** 28.73 individuals; 38.73 other countries. **Remarks:** Accepts advertising. **URL:** http://www.boatandyachtbuyer.co.uk/. **Circ:** 20,000

1450 ■ Jet Skier & PW Magazine
CSL Publishing Ltd.
Alliance House
49 Sidney St.
Cambridge CB2 3HX, United Kingdom
Ph: 44 1223 460490
Fax: 44 1223 315960

Magazine for personal watercraft and jet skiing enthusiasts. **Freq:** 16/yr. **Key Personnel:** Tim Spicer, Editorial, spicer@cslpublishing.com. **Subscription Rates:** 34.45 individuals; 44.45 other countries. **Remarks:** Accepts advertising. **URL:** http://www.jetskier.co.uk/; http://www.cslpublishingltd.co.uk/. **Circ:** (Not Reported)

1451 ■ Journal of Advanced Materials
Cambridge International Science Publishing
7 Meadow Walk
Great Abington
Cambridge CB1 6AZ, United Kingdom
Ph: 44 1223 893295
Fax: 44 1223 894539
Publisher E-mail: cisp@cisp-publishing.com

Journal covering development of materials. **Freq:** Bimonthly. **Key Personnel:** N.P. Lyakishev, Chm. of the Editorial Board; L.I. Ivanov, Ch. Ed.; S.V. Simakov, Exec. Ed. **Subscription Rates:** 400 individuals United Kingdom and Europe; 640 other countries. **URL:** http://www.cisp-publishing.com/jam.htm.

CHESTER

1452 ■ Chester Standard
NWN Media Ltd.
Linenhall House
Stanley St.
Chester CH1 2LR, United Kingdom
Ph: 44 1244 304500
Publisher E-mail: internet@nwn.co.uk

Newspaper featuring news, views, sports, and features. **Key Personnel:** Christian Dunn, Digital Ed. **URL:** http://www.nwnews.co.uk/; http://www.chesterstandard.co.uk/.

1453 ■ Ellesmere Port Standard
NWN Media Ltd.
Linenhall House
Stanley St.
Chester CH1 2LR, United Kingdom
Ph: 44 151 3565500
Publisher E-mail: internet@nwn.co.uk

Newspaper featuring news, views, sports, and features. **Key Personnel:** Christian Dunn, Digital Ed. **URL:** http://www.nwnews.co.uk/; http://www.ellesmereportstandard.co.uk/.

COLWYN BAY

1454 ■ North Wales Pioneer
NWN Media Ltd.
22 Penrhyn Rd.
Colwyn Bay LL29 8HY, United Kingdom
Ph: 44 1492 531188
Publisher E-mail: internet@nwn.co.uk

Newspaper featuring news coverage and human interest stories. **Key Personnel:** Steve Rogers, Editor, steve.rogers@nwn.co.uk; Claire Bryce, Advertising Mgr., phone 44 1492 523860, claire.bryce@nwn.co.uk. **URL:** http://www.nwnews.co.uk/; http://www.northwalespioneer.co.uk/. **Circ:** Free 31,000

COVENTRY

1455 ■ Law and Humanities
Hart Publishing Ltd.
Warwick School of Law
University of Warwick
Coventry CV4 7AL, United Kingdom
Ph: 44 24 76523079
Fax: 44 24 76524105
Publisher E-mail: mail@hartpub.co.uk

Journal covering arts and humanities around the subject of law. **Founded:** July 2007. **Freq:** Semiannual. **Key Personnel:** Paul Raffield, Gen. Ed., p.raffield@warwick.ac.uk; Gary Watt, Gen. Ed., g.watt@warwick.ac.uk. **Subscription Rates:** 80 individuals standard, United Kingdom and Europe; 90 other countries standard; 50 individuals reduced, United Kingdom and Europe; 55 other countries reduced. **Remarks:** Accepts advertising. **URL:** http://www.hartjournals.co.uk/lh/. **Circ:** (Not Reported)

DARLINGTON

1456 ■ Darlington & Stockton Times
Newsquest Media Group Ltd.
PO Box 25
Darlington DL1 1NF, United Kingdom
Ph: 44 1325 381444
Publication E-mail: newsdesk@nne.co.uk
Publisher E-mail: newsquestmediasales@newsquestmedia.co.uk

Newspaper providing local news and events. **Freq:** Weekly. **Key Personnel:** Nigel Burton, News Ed., phone 44 1325 505065, nigel.burton@nne.co.uk. **Subscription Rates:** 1.71 single issue; 88.92 individuals. **Remarks:** Accepts advertising. **URL:** http://www.darlingtonandstocktontimes.co.uk/. **Circ:** (Not Reported)

EVESHAM

1457 ■ Evesham Journal
Newsquest Media Group Ltd.
Sapphire House
Crab Apple Way
Vale Prk.
Evesham WR11 1GP, United Kingdom
Ph: 44 1386 444050
Fax: 44 1386 444078
Publication E-mail: journal@midlands.newsquest.co.uk
Publisher E-mail: newsquestmediasales@newsquestmedia.co.uk

Newspaper featuring local news and events. **Key Personnel:** John Murphy, Editor, john.murphy@midlands.newsquest.co.uk; Tony Donnelly, News Ed., tony.donnelly@midlands.newsquest.co.uk. **Remarks:** Accepts advertising. **URL:** http://www.eveshamjournal.co.uk/. **Circ:** (Not Reported)

EXETER

1458 ■ Legal Ethics
Hart Publishing Ltd.
c/o Prof. Kim Economides, Ed.
School of Law, Amory Bldg.
Exeter University
Rennes Dr.
Exeter EX4 4RJ, United Kingdom
Ph: 44 1392 263379
Fax: 44 1392 263196
Publisher E-mail: mail@hartpub.co.uk

Journal covering field of legal ethics. **Freq:** Semiannual. **Key Personnel:** Prof. Kim Economides, Gen. Ed., k.m.economides@exeter.ac.uk. **Subscription Rates:** 95 individuals standard, United Kingdom and Europe; 105 other countries standard; 55 individuals reduced, United Kingdom and Europe; 60 other countries reduced. **Remarks:** Accepts advertising. **URL:** http://www.hartjournals.co.uk/le/. **Circ:** (Not Reported)

GLASGOW

1459 ■ Bunkered
PSP Publishing Ltd.
50 High Craighall Rd.
Glasgow G4 9UD, United Kingdom
Ph: 44 141 3532222
Fax: 44 141 3323839
Publication E-mail: bunkered@psp.uk.net
Publisher E-mail: sales@psp.uk.net

Magazine featuring golf in Scotland. **Freq:** 8/yr. **Key Personnel:** Paul Grant, Director; Thomas Lovering, Director; Stephen McCann, Director. **Subscription Rates:** 25 individuals. **Remarks:** Accepts advertising. **URL:** http://www.bunkered.co.uk/; http://www.psppublishing.com/bunkered.html. **Circ:** (Not Reported)

1460 ■ English Club Golfer
PSP Publishing Ltd.
50 High Craighall Rd.
Glasgow G4 9UD, United Kingdom
Ph: 44 141 3532222
Fax: 44 141 3323839
Publication E-mail: englishclubgolfer@psp.uk.net
Publisher E-mail: sales@psp.uk.net

Magazine for amateur golfers in England. **Subtitle:** England's No. 1 Amateur Golf Newspaper. **Founded:** 2005. **Freq:** 6/yr. **Key Personnel:** Paul Grant, Director; Thomas Lovering, Director; Stephen McCann, Director. **Subscription Rates:** 10 individuals; Free for golf clubs within England. **Remarks:** Accepts advertising. **URL:** http://www.englishclubgolfer.com/; http://www.psppublishing.com/golfingtitles.html. **Circ:** 50,000

1461 ■ Nationwide Bowler
PSP Publishing Ltd.
50 High Craighall Rd.
Glasgow G4 9UD, United Kingdom
Ph: 44 141 3532222
Fax: 44 141 3323839
Publication E-mail: nationwidebowler@psp.uk.net
Publisher E-mail: sales@psp.uk.net

Magazine for bowlers in Great Britain. **Founded:** 2006. **Freq:** 4/yr. **Key Personnel:** Paul Grant, Director; Thomas Lovering, Director; Stephen McCann, Director. **Remarks:** Accepts advertising. **URL:** http://www.psppublishing.com/nationwidebowler.html; http://www.nationwidebowler.co.uk/. **Circ:** (Not Reported)

1462 ■ No.1
PSP Publishing Ltd.
50 High Craighall Rd.
Glasgow G4 9UD, United Kingdom
Ph: 44 141 3532222
Fax: 44 141 3323839
Publication E-mail: no1magazine@psp.uk.net
Publisher E-mail: sales@psp.uk.net

Entertainment and lifestyle magazine focusing on celebrity news and fashion. **Founded:** 2006. **Freq:** 5/yr. **Key Personnel:** Paul Grant, Director; Thomas Lovering, Director; Stephen McCann, Director. **Subscription Rates:** 1 individuals. **Remarks:** Accepts advertising. **URL:** http://www.no1magazine.co.uk/; http://www.psppublishing.com/no1magazine.html. **Circ:** (Not Reported)

1463 ■ Scottish Club Golfer
PSP Publishing Ltd.
50 High Craighall Rd.
Glasgow G4 9UD, United Kingdom
Ph: 44 141 3532222
Fax: 44 141 3323839
Publication E-mail: scottishclubgolfer@psp.uk.net
Publisher E-mail: sales@psp.uk.net

Newspaper for amateur golfers in Scotland. **Subtitle:** Scotland's No. 1 Newspaper for Amateur Golf. **Founded:** 2003. **Freq:** 6/yr. **Key Personnel:** Paul Grant, Director; Thomas Lovering, Director; Stephen McCann, Director. **Subscription Rates:** 10 individuals; Free for golf clubs. **Remarks:** Accepts advertising. **URL:** http://www.scottishclubgolfer.com/; http://www.psppublishing.com/golfingtitles.html. **Circ:** 25,000

1464 ■ Scottish Hosteller
PSP Publishing Ltd.
50 High Craighall Rd.
Glasgow G4 9UD, United Kingdom
Ph: 44 141 3532222
Fax: 44 141 3323839
Publication E-mail: hosteller@psp.uk.net
Publisher E-mail: sales@psp.uk.net

Magazine featuring Scotland. **Key Personnel:** Paul Grant, Director; Thomas Lovering, Director; Stephen McCann, Director. **Remarks:** Accepts advertising. **URL:** http://www.psppublishing.com/hosteller.html. **Circ:** (Not Reported)

1465 ■ Scottish Mountaineer
PSP Publishing Ltd.
50 High Craighall Rd.
Glasgow G4 9UD, United Kingdom
Ph: 44 141 3532222
Fax: 44 141 3323839
Publication E-mail: mountaineer@psp.uk.net
Publisher E-mail: sales@psp.uk.net

Magazine for Mountaineering Council of Scotland members. **Freq:** 4/yr. **Key Personnel:** Paul Grant, Director; Thomas Lovering, Director; Stephen McCann, Director. **Remarks:** Accepts advertising. **URL:** http://www.psppublishing.com/mountaineering.html. **Circ:** 10,000

1466 ■ Welsh Club Golfer
PSP Publishing Ltd.
50 High Craighall Rd.
Glasgow G4 9UD, United Kingdom
Ph: 44 141 3532222
Fax: 44 141 3323839
Publication E-mail: welshclubgolfer@psp.uk.net
Publisher E-mail: sales@psp.uk.net

Magazine for golfers in Wales. **Subtitle:** Welsh No. 1 Amateur Golf Newspaper. **Founded:** 2006. **Freq:** 6/yr. **Key Personnel:** Paul Grant, Director; Thomas Lovering, Director; Stephen McCann, Director. **Subscription Rates:** 9 individuals. **Remarks:** Accepts advertising. **URL:** http://www.welshclubgolfer.co.uk/; http://www.psppublishing.com/golfingtitles.html. **Circ:** 15,000

ILKLEY

1467 ■ Ilkley Gazette
Newsquest Media Group Ltd.
8 Wells Rd.
Ilkley LS29 9JD, United Kingdom
Ph: 44 1943 603483
Publisher E-mail: newsquestmediasales@newsquestmedia.co.uk

Newspaper featuring Ilkley news, sports, and updates on local jobs. **Freq:** Weekly (Thurs.). **Key Personnel:** Mel Vasey, Editor, phone 44 1943 607022, mel.v@ilkley.newsquest.co.uk. **Subscription Rates:** 72.80 individuals. **Remarks:** Accepts advertising. **URL:** http://www.ilkleygazette.co.uk/. **Circ:** (Not Reported)

JARROW

1468 ■ Essential Guide to Beauty
Eaglemoss Publications Ltd.
PO Box 11
Jarrow NE32 3YH, United Kingdom
Publication E-mail: beauty@jacklinenterprises.com

Magazine featuring beauty advice. **Freq:** Semimonthly. **Key Personnel:** Andrew Jarvis, Ch. Exec.; Maggie Calmels, Editorial Dir. **URL:** http://www.essential-beauty-mag.co.uk/.

KINGSTON UPON THAMES

1469 ■ Dockwalk
Boat International Group
Ward House
5-7 Kingston Hill
Kingston upon Thames KT2 7PW, United Kingdom
Ph: 44 20 85472662
Fax: 44 20 85471201
Publisher E-mail: info@boatinternational.co.uk

Magazine for yachting industry professionals. **Founded:** 1998. **Freq:** Monthly. **Trim Size:** 247.65 x 304.8 mm. **Key Personnel:** Tony Harris, CEO/Publisher; Tony Euden, Publishing Dir., tonye@boatinternational.co.uk; Tim Hartney, Production Mgr., timh@boatinternational.co.uk. **Subscription Rates:** US$75 individuals; 125 U.S. 2 years; US$150 other countries. **Remarks:** Accepts advertising. **URL:** http://www.boatinternational.com/mags/mag06.htm. . **Ad Rates:** 4C: 1,475. **Circ:** 20,000

LEEDS

1470 ■ Armley & Wortley Advertiser
Newsquest Media Group Ltd.
2 Robin Ln.
Leeds LS28 7BN, United Kingdom
Ph: 44 113 2557558
Publication E-mail: adsales@bradford.newsquest.co.uk
Publisher E-mail: newsquestmediasales@newsquestmedia.co.uk

Newspaper providing up to date local news and sports. **Founded:** Nov. 2003. **Key Personnel:** John Lee, Hd. of Advertising, phone 44 274 705390, john.lee@bradford.newsquest.co.uk. **Remarks:** Accepts advertising. **URL:** http://www.advertiserseries.co.uk/. **Circ:** (Not Reported)

1471 ■ Journal of the Energy Institute
Maney Publishing
Energy & Resources Research Institute
Houldsworth Bldg.
University of Leeds
Leeds LS2 9JT, United Kingdom
Ph: 44 113 3432507
Publisher E-mail: maney@maney.co.uk

Journal covering original high quality research on energy engineering and technology. **Freq:** 4/yr. **Key Personnel:** Prof. Alan Williams, Editor, a.williams@leeds.ac.uk. **ISSN:** 1743-9671. **Subscription Rates:** 242 institutions; 242 institutions Europe; 492 institutions North America; 242 institutions, other countries. **Remarks:** Accepts advertising. **URL:** http://www.maney.co.uk/search?fwaction=show&fwid=630. **Circ:** (Not Reported)

LEICESTER

1472 ■ Mathematical Pie
The Mathematical Association
259 London Rd.
Leicester LE2 3BE, United Kingdom
Ph: 44 116 2210013
Fax: 44 116 2122835
Publisher E-mail: office@m-a.org.uk

Magazine featuring variety of problems and challenges, stimulating mathematical activity for pupils

Circulation: ★ = ABC; △ = BPA; ♦ = CAC; • = CCAB; ❑ = VAC; ⊕ = PO Statement; ‡ = Publisher's Report; Boldface figures = sworn; Light figures = estimated.

from 10 to 14 years of age. **Freq:** 3/yr. **Key Personnel:** Wil Ransome, Editor, pie@m-a.org.uk. **Subscription Rates:** Free members of Society of Young Mathematicians; 3 individuals; 4.10 other countries surface mail; 4.80 other countries air mail. **URL:** http://www.m-a.org.uk/resources/periodicals/mathematical_pie/index.html.

1473 ■ Symmetry Plus
The Mathematical Association
259 London Rd.
Leicester LE2 3BE, United Kingdom
Ph: 44 116 2210013
Fax: 44 116 2122835
Publisher E-mail: office@m-a.org.uk

Magazine featuring puzzles, problems and competitions about mathematics for students from 10 to 18 years. **Freq:** 3/yr. **Key Personnel:** Martin Perkins, Editor, symmetryplus@m-a.org.uk. **Subscription Rates:** 14 individuals; 16 other countries surface mail; 18 other countries air mail. **URL:** http://www.m-a.org.uk/resources/periodicals/symmetry_plus/index.html.

LEIGH

1474 ■ Leigh Journal
Newsquest Media Group Ltd.
44-46 Railway Rd.
Leigh WN7 4AT, United Kingdom
Ph: 44 1942 672241
Publisher E-mail: newsquestmediasales@newsquestmedia.co.uk

Newspaper featuring local news and events. **Key Personnel:** Mike Hulme, Editor, mhulme@leighjournal.co.uk; Brian Gomm, News Ed., bgomm@leighjournal.co.uk. **Remarks:** Accepts advertising. **URL:** http://www.thisisleigh.co.uk/. **Circ:** (Not Reported)

LONDON

1475 ■ Annabel's
Luxury Publishing Ltd.
5 Jubilee Pl.
Chelsea
London SW3 3TD, United Kingdom
Ph: 44 20 75912900
Fax: 44 20 75912929
Publisher E-mail: info@luxurypublishing.com

Fashion and lifestyle magazine. **Founded:** 2004. **Freq:** Annual. **Trim Size:** 220 x 300 mm. **Key Personnel:** Lucia van der Post, Editor. **Remarks:** Accepts advertising. **URL:** http://www.luxurypublishing.co.uk/. **Circ:** 7,000

1476 ■ Annals of Tropical Paediatrics
Maney Publishing
1 Carlton House Ter.
London SW1Y 5AF, United Kingdom
Ph: 44 20 74517300
Fax: 44 20 74517307
Publisher E-mail: maney@maney.co.uk

Journal covering medical problems, achievements, and research in paediatrics and child health in the tropics and sub-tropics. **Subtitle:** International Child Health. **Founded:** 1981. **Freq:** Quarterly. **Key Personnel:** Bernard Brabin, Editor, b.j.brabin@liv.ac.uk; James Bunn, Editor, jegbunn@liv.ac.uk; Luis Cuevas, Editor, lcuevas@liv.ac.uk; Feiko ter Kuile, Editor, terkuile@liv.ac.uk. **ISSN:** 0272-4936. **Subscription Rates:** 139 individuals; 139 individuals Europe; US$269 individuals North America; 139 other countries; 382 institutions; 382 institutions Europe; US$707 institutions North America; 382 institutions, other countries. **Remarks:** Accepts advertising. **URL:** http://www.maney.co.uk/search?fwaction=show&fwid=143. **Circ:** (Not Reported)

1477 ■ Applied Earth Science
Maney Publishing
1 Carlton House Ter.
London SW1Y 5AF, United Kingdom
Ph: 44 20 74517300
Fax: 44 20 74517307
Publisher E-mail: maney@maney.co.uk

Journal covering all aspects of the application of the earth sciences in the discovery, exploration, development, and exploitation of all forms of mineral resources. **Freq:** Quarterly. **Key Personnel:** Dr. Iain McDonald, Editor. **ISSN:** 0371-7453. **Subscription Rates:** 206 institutions; 206 institutions Europe; US$342 institutions North America; 206 institutions, other countries. **Remarks:** Accepts advertising. **URL:** http://www.maney.co.uk/search?fwaction=show&fwid=145. . **Ad Rates:** BW: 300, 4C: 550. **Circ:** 500

1478 ■ Austrian Studies
Maney Publishing
1 Carlton House Ter.
London SW1Y 5AF, United Kingdom
Ph: 44 20 74517300
Fax: 44 20 74517307
Publisher E-mail: maney@maney.co.uk

Journal covering distinctive cultural traditions of the Habsburg Empire and the Austrian Republic. **Freq:** Annual. **Key Personnel:** Dr. Judith Beniston, Editor, j.beniston@ucl.ac.uk; Prof. Robert Vilain, Editor, r.vilain@rhul.ac.uk. **ISSN:** 1350-7532. **Subscription Rates:** 70 institutions; 70 institutions Europe; US$169 institutions North America; 86 institutions, other countries. **URL:** http://www.maney.co.uk/search?fwaction=show&fwid=147.

1479 ■ Automotive Logistics
Ultima Media Ltd.
Lamb House
Church St.
London W4 2PD, United Kingdom
Ph: 44 20 89870900
Fax: 44 20 89870948
Publisher E-mail: info@ultimamedia.com

Magazine featuring automotive industry. **Trim Size:** 210 x 297 mm. **Key Personnel:** Maxine Elkin, Editor, phone 44 20 89870962, fax 44 20 89870948, maxine.elkin@ultimamedia.com. **Subscription Rates:** Free. **Remarks:** Accepts advertising. **URL:** http://www.automotivelogisticsmagazine.com/aml/. **Circ:** △**9,000**

1480 ■ Automotive Production China
Ultima Media Ltd.
Lamb House
Church St.
London W4 2PD, United Kingdom
Ph: 44 20 89870900
Fax: 44 20 89870948
Publication E-mail: apc@ultimamedia.com
Publisher E-mail: info@ultimamedia.com

Magazine for Chinese automotive industry. **Freq:** Semiannual. **Key Personnel:** Simon Timm, Chm., simon.timm@ultimamedia.com. **Remarks:** Accepts advertising. **URL:** http://www.automotiveproductionchina.com/apc/. **Circ:** (Not Reported)

1481 ■ AYGO Magazine
Publicis Blueprint
23 Howland St.
London W1A 1AQ, United Kingdom
Ph: 44 20 74627777
Publisher E-mail: info@publicis-blueprint.co.uk

Magazine featuring Toyota's small car model. **Freq:** Semiannual. **Key Personnel:** Mat Waugh, Publisher, matthew.waugh@publicis-blueprint.co.uk. **URL:** http://www.publicis-blueprint.co.uk. **Circ:** 250,000

1482 ■ Black History Month
Sugar Media Ltd.
The Maltings, Studio 65/66
169 Tower Bridge Rd.
London SE1 3LJ, United Kingdom
Publisher E-mail: info@sugarmedia.co.uk

Magazine featuring BME achievement, history, and culture. **Freq:** Annual. **Trim Size:** A4. **Remarks:** Accepts advertising. **URL:** http://www.blackhistorymonthuk.co.uk/. **Circ:** Free 100,000

1483 ■ Blockbuster Preview
Publicis Blueprint
23 Howland St.
London W1A 1AQ, United Kingdom
Ph: 44 20 74627777
Publisher E-mail: info@publicis-blueprint.co.uk

Magazine covering movie news. **Freq:** Monthly. **Key Personnel:** Mark Stancheris, Publisher, mark.stancheris@publicis-blueprint.co.uk. **URL:** http://www.publicis-blueprint.co.uk. **Circ:** 500,000

1484 ■ Byzantine and Modern Greek Studies
Maney Publishing
1 Carlton House Ter.
London SW1Y 5AF, United Kingdom
Ph: 44 20 74517300
Fax: 44 20 74517307
Publisher E-mail: maney@maney.co.uk

Journal covering history, literature, and social anthropology of Byzantine and Modern Greek. **Founded:** 1975. **Freq:** Semiannual. **Key Personnel:** Peter Mackridge, Editor; Dr. Ruth Macrides, Editor. **ISSN:** 0307-0131. **Subscription Rates:** 44 individuals; 44 individuals Europe; US$84 individuals North America; 44 other countries; 112 institutions; 112 institutions Europe; US$212 institutions North America; 112 institutions, other countries. **Remarks:** Accepts advertising. **URL:** http://www.maney.co.uk/search?fwaction=show&fwid=466. . **Ad Rates:** BW: 180. **Circ:** (Not Reported)

1485 ■ Candid
Luxury Publishing Ltd.
5 Jubilee Pl.
Chelsea
London SW3 3TD, United Kingdom
Ph: 44 20 75912900
Fax: 44 20 75912929
Publisher E-mail: info@luxurypublishing.com

Lifestyle magazine featuring homes, furnitures, and interior designs. **Founded:** Mar. 2007. **Freq:** Semiannual. **Trim Size:** 280 x 280 mm. **Key Personnel:** Lucia Van Der Post, Editor. **Remarks:** Accepts advertising. **URL:** http://www.luxurypublishing.co.uk/. **Circ:** 8,000

1486 ■ Chinese Journal of Oceanology and Limnology
Maney Publishing
1 Carlton House Ter.
London SW1Y 5AF, United Kingdom
Ph: 44 20 74517300
Fax: 44 20 74517307
Publisher E-mail: maney@maney.co.uk

Journal covering oceanology and limnology in China. **Freq:** 4/yr. **Key Personnel:** Liz Rosindale, Managing Editor, m.gallico@maney.co.uk; Mark Simon, Publishing Dir., mark_simon@materials.org.uk; Michael Gallico, Mng. Dir., l.rosindale@maney.co.uk. **ISSN:** 0254-4059. **Subscription Rates:** US$264 institutions; US$132 individuals. **URL:** http://www.maney.co.uk/search?fwaction=show&fwid=357.

1487 ■ Competition & Change
Maney Publishing
1 Carlton House Ter.
London SW1Y 5AF, United Kingdom
Ph: 44 20 74517300
Fax: 44 20 74517307
Publisher E-mail: maney@maney.co.uk

Journal covering global business and political economy. **Subtitle:** The Journal of Global Business and Political Economy. **Freq:** 4/yr. **Key Personnel:** Julie Froud, Editor, julie.froud@mbs.ac.uk; Sukhdev Johal, Editor, s.johal@rhul.ac.uk. **ISSN:** 1024-5294. **Subscription Rates:** 56 individuals; 56 individuals Europe; US$96 individuals North America; 56 other countries; 198 institutions; 198 institutions Europe; US$366 institutions North America; 198 institutions, other countries. **Remarks:** Accepts advertising. **URL:** http://www.maney.co.uk/search?fwaction=show&fwid=634. . **Ad Rates:** BW: 180. **Circ:** (Not Reported)

1488 ■ Corrosion Engineering, Science and Technology
Maney Publishing
1 Carlton House Ter.
London SW1Y 5AF, United Kingdom
Ph: 44 20 74517300
Fax: 44 20 74517307
Publisher E-mail: maney@maney.co.uk

Journal covering of research and practice in corrosion processes and corrosion control. **Freq:** Quarterly. **Key Personnel:** Prof. Stuart B. Lyon, Editor, stuart.lyon@manchester.ac.uk; Prof. Robert G. Kelly, North American Ed., rgkelly@virginia.edu. **Subscription Rates:** 487 institutions; 487 institutions Europe; US$859 institutions North America; 487 institutions, other countries. **Remarks:** Accepts advertising. **URL:** http://www.maney.co.uk/search?fwaction=show&fwid=154. . **Ad Rates:** BW: 500, 4C: 1,100. **Circ:** 900

1489 ■ Countryside Voice
Think Publishing
The Pall Mall Deposit
124-128 Barlby Rd.
London W10 6BL, United Kingdom
Ph: 44 20 89623020
Fax: 44 20 89628689
Publisher E-mail: watchdog@thinkpublishing.co.uk

Magazine featuring beauty and diversity of English countryside. **Freq:** 3/yr. **Key Personnel:** Ian McAuliffe, Publishing Dir., ian@thinkpublishing.co.uk; Tilly Boulter, Mng. Dir., tilly@thinkpublishing.co.uk; Emma Jones, Editorial Dir., emma@thinkpublishing.co.uk. **Remarks:** Accepts advertising. **URL:** http://www.thinkpublishing.co.uk/thinkmagazines_voice.htm. **Circ:** 35,000

1490 ■ Current World Archaeology
Think Publishing
The Pall Mall Deposit
124-128 Barlby Rd.
London W10 6BL, United Kingdom
Ph: 44 20 89623020
Fax: 44 20 89628689
Publisher E-mail: watchdog@thinkpublishing.co.uk

Magazine covering archaeology. **Freq:** Bimonthly. **Key Personnel:** Ian McAuliffe, Publishing Dir., ian@thinkpublishing.co.uk; Tilly Boulter, Mng. Dir., tilly@thinkpublishing.co.uk; Emma Jones, Editorial Dir., emma@thinkpublishing.co.uk. **Remarks:** Accepts advertising. **URL:** http://www.thinkpublishing.co.uk/thinkmagazines_arch.htm. **Circ:** 11,500

1491 ■ Daisy
Egmont Magazines
184-192 Drummond St., 4th Fl.
London NW1 3HP, United Kingdom
Ph: 44 20 73806430

Magazine for little girls aged 4-7. **Freq:** Monthly. **Trim Size:** 220 x 300 mm. **Key Personnel:** Kathryn Davies, Advertising Sales Mgr., kdavies@euk.egmont.com; Annett Allen, Advertising Sales Mgr., aallen@euk.egmont.com. **Remarks:** Accepts advertising. **URL:** http://www.egmontmagazines.co.uk/titles.asp?title=daisy. . **Ad Rates:** 4C: 1,800. **Circ:** ★**43,162**

1492 ■ Debenhams Desire
Publicis Blueprint
23 Howland St.
London W1A 1AQ, United Kingdom
Ph: 44 20 74627777
Publisher E-mail: info@publicis-blueprint.co.uk

Magazine covering fashion and lifestyle. **Founded:** 2004. **Freq:** Quarterly. **Key Personnel:** Rachel Walder, Publisher, rachel.walder@publicis-blueprint.co.uk; Amanda Morgan, Editor, amanda.morgan@publicis-blueprint.co.uk. **URL:** http://www.publicis-blueprint.co.uk. **Circ:** 745,000

1493 ■ Disney Fairies
Egmont Magazines
184-192 Drummond St., 4th Fl.
London NW1 3HP, United Kingdom
Ph: 44 20 73806430

Magazine featuring fairies and fairy world for 5-7 year old girls. **Freq:** Monthly. **Trim Size:** 220 x 300 mm. **Key Personnel:** Annett Allen, Advertising Sales Mgr., aallen@euk.egmont.com; Kathryn Davies, Advertising Sales Mgr., kdavies@euk.egmont.com. **Remarks:** Accepts advertising. **URL:** http://www.egmontmagazines.co.uk/titles.asp?title=disneyFairies. . **Ad Rates:** 4C: 1,800. **Circ:** (Not Reported)

1494 ■ Disney and Me
Egmont Magazines
184-192 Drummond St., 4th Fl.
London NW1 3HP, United Kingdom
Ph: 44 20 73806430

Magazine for boys and girls aged 4-7. **Founded:** 1991. **Freq:** Semimonthly. **Trim Size:** 220 x 300 mm. **Key Personnel:** Annett Allen, Advertising Sales Mgr., aallen@euk.egmont.com; Kathryn Davies, Advertising Sales Mgr., kdavies@euk.egmont.com. **Remarks:** Accepts advertising. **URL:** http://www.egmontmagazines.co.uk/titles.asp?title=dam. . **Ad Rates:** 4C: 1,800. **Circ:** ★**49,002**

1495 ■ Disney's Winnie the Pooh and Friends
Egmont Magazines
184-192 Drummond St., 4th Fl.
London NW1 3HP, United Kingdom
Ph: 44 20 73806430

Magazine featuring Winnie the Pooh and friends adventure. **Freq:** 4/week. **Trim Size:** 220 x 300 mm. **Key Personnel:** Annett Allen, Advertising Sales Mgr., aallen@euk.egmont.com; Kathryn Davies, Advertising Sales Mgr., kdavies@euk.egmont.com. **Remarks:** Accepts advertising. **URL:** http://www.egmontmagazines.co.uk/titles.asp?title=wtp. . **Ad Rates:** 4C: 1,500. **Circ:** ★**23,297**

1496 ■ Dora Dress Up and Go
GE Fabbri Ltd.
The Communications Bldg., 7th Fl.
48 Leicester Sq.
London WC2H 7LT, United Kingdom
Ph: 44 20 30317600
Fax: 44 20 30317601
Publication E-mail: dora@jacklinservice.com
Publisher E-mail: mailbox@gefabbri.co.uk

Magazine featuring Dora and her adventures. **Freq:** Semimonthly. **Key Personnel:** Peter Edwards, Mng. Dir.; Liz Glaze, Director; Katie Preston, Editor-in-Chief. **Subscription Rates:** 2.99 single issue. **URL:** http://www.doradollcollection.com/; http://www.gefabbri.co.uk/publication_doradressupandgo.

1497 ■ Dora the Explorer
GE Fabbri Ltd.
The Communications Bldg., 7th Fl.
48 Leicester Sq.
London WC2H 7LT, United Kingdom
Ph: 44 20 30317600
Fax: 44 20 30317601
Publication E-mail: doradvd@dbfactory.co.uk
Publisher E-mail: mailbox@gefabbri.co.uk

Magazine featuring Dora and her friends adventures. **Freq:** Semimonthly. **Key Personnel:** Peter Edwards, Mng. Dir.; Liz Glaze, Director; Katie Preston, Editor-in-Chief. **URL:** http://www.smallexplorers.com; http://www.gefabbri.co.uk/publication_doradvd.

1498 ■ European Competition Journal
Hart Publishing Ltd.
c/o Dr. Philip Marsden, Gen. Ed.
17 Russell Sq.
London WC1B 5JP, United Kingdom
Publisher E-mail: mail@hartpub.co.uk

Journal covering current developments in competition law. **Founded:** 2005. **Freq:** Semiannual. **Key Personnel:** Dr. Philip Marsden, Gen. Ed., p.marsden@biicl.org; Simon Bishop, Gen. ed., simon.bishop@rbbecon.com. **ISSN:** 1744-1056. **Subscription Rates:** 175 individuals standard, United Kingdom and Europe; 180 other countries standard; 90 individuals reduced, United Kingdom and Europe; 100 other countries reduced. **Remarks:** Accepts advertising. **URL:** http://www.hartjournals.co.uk/ecj/. **Circ:** (Not Reported)

1499 ■ Family & Community History
Maney Publishing
1 Carlton House Ter.
London SW1Y 5AF, United Kingdom
Ph: 44 20 74517300
Fax: 44 20 74517307
Publisher E-mail: maney@maney.co.uk

Journal covering research in family and community history. **Freq:** Semiannual. **Key Personnel:** Dr. Steve King, Editor. **ISSN:** 1463-1180. **Subscription Rates:** 36 individuals; 36 individuals Europe; US$68 individuals North America; 36 other countries; 115 institutions; 115 institutions Europe; US$218 institutions North America; 115 institutions, other countries. **Remarks:** Accepts advertising. **URL:** http://www.maney.co.uk/search?fwaction=show&fwid=158. . **Ad Rates:** BW: 180. **Circ:** (Not Reported)

1500 ■ Fireman Sam
Egmont Magazines
184-192 Drummond St., 4th Fl.
London NW1 3HP, United Kingdom
Ph: 44 20 73806430

Magazine featuring puzzles, colouring, stories and activities for children. **Freq:** 4/week. **Trim Size:** 220 x 300 mm. **Key Personnel:** Annett Allen, Advertising Sales Mgr., aallen@euk.egmont.com; Kathryn Davies, Advertising Sales Mgr., kdavies@euk.egmont.com. **Remarks:** Accepts advertising. **URL:** http://www.egmontmagazines.co.uk/titles.asp?title=firemansam. . **Ad Rates:** 4C: 1,800. **Circ:** (Not Reported)

1501 ■ GO Girl
Egmont Magazines
184-192 Drummond St., 4th Fl.
London NW1 3HP, United Kingdom
Ph: 44 20 73806430

Magazine for girls aged 7-11. **Freq:** 3/week. **Key Personnel:** Annett Allen, Advertising Sales Mgr., aallen@euk.egmont.com; Kathryn Davies, Advertising Sales Mgr., kdavies@euk.egmont.com. **Remarks:** Accepts advertising. **URL:** http://www.egmontmagazines.co.uk/titles.asp?title=gogirl. . **Ad Rates:** 4C: 2,400. **Circ:** ★**43,163**

1502 ■ The Hardy Review
Maney Publishing
1 Carlton House Ter.
London SW1Y 5AF, United Kingdom
Ph: 44 20 74517300
Fax: 44 20 74517307
Publisher E-mail: maney@maney.co.uk

Journal covering the study of Thomas Hardy. **Founded:** 2007. **Freq:** Semiannual. **Key Personnel:** Prof. Rosemarie Morgan, Editor, rosemarie.morgan@yale.edu. **Subscription Rates:** 108 institutions; 108 institutions Europe; US$205 institutions North America; 108 institutions, other countries. **Remarks:** Accepts advertising. **URL:** http://www.maney.co.uk/search?fwaction=show&fwid=695. . **Ad Rates:** BW: 180. **Circ:** (Not Reported)

1503 ■ Hot Wheels
Egmont Magazines
184-192 Drummond St., 4th Fl.
London NW1 3HP, United Kingdom
Ph: 44 20 73806430

High energy magazine for boys featuring hot wheels. **Founded:** July 2004. **Freq:** Monthly. **Trim Size:** 220 x 300 mm. **Key Personnel:** Kathryn Davies, Advertising Sales Mgr., kdavies@euk.egmont.com; Annett Allen, Advertising Sales Mgr., aallen@euk.egmont.com. **Remarks:** Accepts advertising. **URL:** http://www.egmontmagazines.co.uk/titles.asp?title=hotwheels. . **Ad Rates:** 4C: 2,000. **Circ:** ★**31,263**

Circulation: ★ = ABC; △ = BPA; ♦ = CAC; • = CCAB; ❑ = VAC; ⊕ = PO Statement; ‡ = Publisher's Report; Boldface figures = sworn; Light figures = estimated.

1504 ■ Industrial Archaeology Review
Maney Publishing
1 Carlton House Ter.
London SW1Y 5AF, United Kingdom
Ph: 44 20 74517300
Fax: 44 20 74517307
Publisher E-mail: maney@maney.co.uk

Journal covering industrial archaeology. **Freq:** Semiannual. **Key Personnel:** David Gwyn, PhD, Editor, govannonconsult@hotmail.com. **ISSN:** 0309-0728. **Subscription Rates:** 42 individuals; 42 individuals Europe; US$78 individuals North America; 42 other countries; 119 institutions; 119 institutions Europe; US$226 institutions North America; 119 institutions, other countries. **Remarks:** Accepts advertising. **URL:** http://www.maney.co.uk/search?fwaction=show&fwid=167. . **Ad Rates:** BW: 180. **Circ:** (Not Reported)

1505 ■ Interior Motives
Ultima Media Ltd.
Lamb House
Church St.
London W4 2PD, United Kingdom
Ph: 44 20 89870900
Fax: 44 20 89870948
Publisher E-mail: info@ultimamedia.com

Magazine featuring automotive design management. **Freq:** 6/yr. **Trim Size:** 230 x 297 mm. **Key Personnel:** Euan Sey, Editor, phone 44 20 89870980, fax 44 20 89870948, euan.sey@ultimamedia.com; Abel Sampson, Publisher, abel.sampson@ultimamedia.org. **Subscription Rates:** EUR100 individuals; US$120 individuals; 67 individuals. **Remarks:** Accepts advertising. **URL:** http://www.interiormotivesmagazine.com/. **Circ:** (Not Reported)

1506 ■ International Heat Treatment and Surface Engineering
Maney Publishing
1 Carlton House Ter.
London SW1Y 5AF, United Kingdom
Ph: 44 20 74517300
Fax: 44 20 74517307
Publisher E-mail: maney@maney.co.uk

Journal covering the understanding of existing and emerging heat treatment and surface engineering processes. **Founded:** 2007. **Freq:** 4/yr. **Key Personnel:** Prof. T. Bell., Editor, t.bell@bham.ac.uk; Prof. Shipu Chen, Editor, spchen@sjtu.edu.cn. **ISSN:** 1749-5148. **Subscription Rates:** 200 individuals; US$400 individuals North America; 200 institutions; US$400 institutions North America. **Remarks:** Accepts advertising. **URL:** http://www.maney.co.uk/search?fwaction=show&fwid=703. **Circ:** (Not Reported)

1507 ■ International Journal of Electronic Security and Digital Forensics
Inderscience Publishers
c/o Dr. Hamid Jahankhani, Ed.-in-Ch.
University of East London
School of Computing & Technology
4-6 University Way
London E16 2RD, United Kingdom
Publisher E-mail: editor@inderscience.com

Journal covering fields of electronic security and digital forensics. **Founded:** 2007. **Freq:** 4/yr. **Key Personnel:** Dr. Hamid Jahankhani, Editor-in-Chief, hamid.jahankhani@uel.ac.uk. **ISSN:** 1751-911X. **Subscription Rates:** EUR470 individuals includes surface mail, print only; EUR640 individuals print and online. **URL:** http://www.inderscience.com/browse/index.php?journalCODE=ijesdf.

1508 ■ International Journal of Internet Technology & Secured Transactions
Inderscience Publishers
c/o Dr. Charles A. Shoniregun, Ed.-in-Ch.
University of East London, Docklands Campus
University Way
London E16 2RD, United Kingdom
Publisher E-mail: editor@inderscience.com

Journal covering information technology. **Founded:** 2007. **Freq:** 4/yr. **Key Personnel:** Dr. Charles A. Shoniregun, Editor-in-Chief, c.shoniregun@uel.ac.uk. **ISSN:** 1748-569X. **Subscription Rates:** EUR470 individuals includes surface mail; EUR640 individuals print and online. **URL:** http://www.inderscience.com/browse/index.php?journalCODE=ijitst.

1509 ■ The Italianist
Maney Publishing
1 Carlton House Ter.
London SW1Y 5AF, United Kingdom
Ph: 44 20 74517300
Fax: 44 20 74517307
Publisher E-mail: maney@maney.co.uk

Journal covering all areas of Italian Studies. **Founded:** 1981. **Freq:** Semiannual. **Key Personnel:** Zygmunt G. Baranski, Editor, zgb20@cam.ac.uk; Claire E. Honess, Co-Ed., c.e.honess@leeds.ac.uk; Adam Ledgeway, Co-Ed., anl21@cam.ac.uk; Lisa Sampson, Co-Ed., l.m.sampson@reading.ac.uk; Shirley W. Vinall, Co-Ed., s.w.vinall@reading.ac.uk. **ISSN:** 0261-4340. **Subscription Rates:** 35 individuals; 35 individuals Europe; US$65 individuals North America; 35 other countries; 112 institutions; 112 institutions Europe; US$208 institutions North America; 112 other countries. **Remarks:** Accepts advertising. **URL:** http://www.maney.co.uk/search?fwaction=show&fwid=173. . **Ad Rates:** BW: 180. **Circ:** (Not Reported)

1510 ■ Jackie Chan Adventures
Eaglemoss Publications Ltd.
5 Cromwell Rd.
London SW7 2HR, United Kingdom
Ph: 44 20 75908300
Fax: 44 20 75908301

Collection magazine based on the TV show Jackie Chan Adventures. **Freq:** 8/yr. **Key Personnel:** Maggie Calmels, Editorial Dir.; Andrew Jarvis, Ch. Exec. **Subscription Rates:** 1.99 single issue. **URL:** http://www.jackie-chan-magazine.com/; http://www.eaglemoss.co.uk/curr-titles/jac-cha.htm.

1511 ■ Journal of the British Archaeological Association
Maney Publishing
1 Carlton House Ter.
London SW1Y 5AF, United Kingdom
Ph: 44 20 74517300
Fax: 44 20 74517307
Publisher E-mail: maney@maney.co.uk

Journal covering the study of Britain's archaeology, art, and architecture. **Freq:** Annual. **Key Personnel:** Martin Henig, Editor, martinhenig@hotmail.com; John McNeill, Editor. **ISSN:** 0068-1288. **Subscription Rates:** 32 individuals; 32 individuals Europe; US$64 individuals North America; 32 other countries; 62 institutions; 62 institutions Europe; US$124 institutions North America; 62 institutions, other countries. **Remarks:** Accepts advertising. **URL:** http://www.maney.co.uk/search?fwaction=show&fwid=176. **Circ:** (Not Reported)

1512 ■ Leaving School
Sugar Media Ltd.
The Maltings, Studio 65/66
169 Tower Bridge Rd.
London SE1 3LJ, United Kingdom
Publisher E-mail: info@sugarmedia.co.uk

Magazine featuring information on career and educational opportunities available to all Britain's diverse communities. **URL:** http://www.sugarmedia.co.uk/; http://www.leavingschool.co.uk/.

1513 ■ Medieval Sermon Studies
Maney Publishing
1 Carlton House Ter.
London SW1Y 5AF, United Kingdom
Ph: 44 20 74517300
Fax: 44 20 74517307
Publisher E-mail: maney@maney.co.uk

Journal featuring insightful articles on sermon studies and related areas. **Freq:** Annual. **Key Personnel:** Carolyn Muessig, Editor, c.a.muessig@bristol.ac.uk; Veronica O'Mara, Editor, v.m.omara@hull.ac.uk. **ISSN:** 1366-0691. **Subscription Rates:** 64 institutions; 64 institutions Europe; US$114 institutions North America; 64 institutions, other countries. **Remarks:** Accepts advertising. **URL:** http://www.maney.co.uk/search?fwaction=show&fwid=186. **Circ:** (Not Reported)

1514 ■ Mela UK
Sugar Media Ltd.
The Maltings, Studio 65/66
169 Tower Bridge Rd.
London SE1 3LJ, United Kingdom
Publisher E-mail: info@sugarmedia.co.uk

Magazine featuring fashion, food, and culture. **Freq:** Annual. **URL:** http://www.sugarmedia.co.uk/; http://www.melauk.com/.

1515 ■ Midland History
Maney Publishing
1 Carlton House Ter.
London SW1Y 5AF, United Kingdom
Ph: 44 20 74517300
Fax: 44 20 74517307
Publisher E-mail: maney@maney.co.uk

Journal covering the history of the English midlands. **Founded:** 1971. **Freq:** Semiannual. **Key Personnel:** Dr. Richard Cust, Editor, r.p.cust@bham.ac.uk; Dr. Stephen K. Roberts, Editor, skennethroberts@yahoo.co.uk; Prof. John Beckett, Chm. of the Editorial Board. **ISSN:** 0047-729X. **Subscription Rates:** 20 individuals; 20 individuals Europe; US$38 individuals North America; 20 other countries; 68 institutions; 68 institutions Europe; US$121 institutions North America; 68 institutions, other countries. **Remarks:** Accepts advertising. **URL:** http://www.maney.co.uk/search?fwaction=show&fwid=785. **Circ:** (Not Reported)

1516 ■ Noddy
Egmont Magazines
184-192 Drummond St., 4th Fl.
London NW1 3HP, United Kingdom
Ph: 44 20 73806430

Magazine for children aged 6 and below featuring stories and activities. **Freq:** Monthly. **Trim Size:** 220 x 300 mm. **Key Personnel:** Annett Allen, Advertising Sales Mgr., aallen@euk.egmont.com; Kathryn Davies, Advertising Sales Mgr., kdavies@euk.egmont.com. **Remarks:** Accepts advertising. **URL:** http://www.egmontmagazines.co.uk/titles.asp?title=noddy. . **Ad Rates:** 4C: 1,800. **Circ:** ★39,229

1517 ■ Northern History
Maney Publishing
1 Carlton House Ter.
London SW1Y 5AF, United Kingdom
Ph: 44 20 74517300
Fax: 44 20 74517307
Publisher E-mail: maney@maney.co.uk

Journal covering history of the seven historic Northern counties of England. **Founded:** 1966. **Freq:** Semiannual. **Key Personnel:** G.C.F. Forster, Editor, c.cascarino@leeds.ac.uk; S.J.D. Green, Editor. **ISSN:** 0078-172X. **Subscription Rates:** 34 individuals; 34 individuals Europe; US$64 individuals North America; 34 other countries; 106 institutions; 106 institutions Europe; US$198 institutions North America; 106 institutions, other countries. **Remarks:** Accepts advertising. **URL:** http://www.maney.co.uk/search?fwaction=show&fwid=192. **Circ:** (Not Reported)

1518 ■ Packaging, Transport, Storage & Security of Radioactive Material
Maney Publishing
1 Carlton House Ter.
London SW1Y 5AF, United Kingdom
Ph: 44 20 74517300
Fax: 44 20 74517307
Publisher E-mail: maney@maney.co.uk

Journal covering all aspects of the transport of radioactive materials, including regulations, package design, safety analysis, package testing, routine operations and experiences, storage and security, and accidents and emergency planning. **Freq:** 4/yr. **Key Personnel:** R.B. Pope, Editor, poper787@comcast.net; E.P. Goldfinch, Founding Ed., goldfinch@ramtrans.org.uk. **Subscription Rates:** 280 individuals; 280 individuals Europe; US$527 individuals North America; 280 other countries; 280 institutions; 280 institutions Europe; US$527 institutions North America; 280 institutions, other countries. **Remarks:** Accepts advertising. **URL:** http://www.maney.co.uk/search?fwaction=show&fwid=670. **Circ:** (Not Reported)

1519 ■ Palestine Exploration Quarterly
Maney Publishing
1 Carlton House Ter.
London SW1Y 5AF, United Kingdom
Ph: 44 20 74517300
Fax: 44 20 74517307
Publisher E-mail: maney@maney.co.uk

Journal covering the study of Holy Land. **Freq:** 3/yr. **Key Personnel:** Prof. J.R. Bartlett, Editor, jrbartlett@eircom.net. **Subscription Rates:** 88 institutions; 88 institutions Europe; US$167 institutions North America; 88 institutions, other countries. **Remarks:** Accepts advertising. **URL:** http://www.maney.co.uk/search?fwaction=show&fwid=194. **Circ:** (Not Reported)

1520 ■ Positive Nation
Sugar Media Ltd.
Studio 4, Hiltongrove
14 Southgate Rd.
London N1 3LY, United Kingdom
Ph: 44 20 74076800
Fax: 44 20 74077747
Publisher E-mail: info@sugarmedia.co.uk

Magazine featuring sexual health awareness. **Founded:** 1994. **Freq:** Quarterly. **Subscription Rates:** 15 individuals; 65 institutions; 35 institutions not-for-profit. **Remarks:** Accepts advertising. **URL:** http://www.positivenation.co.uk/. **Circ:** (Not Reported)

1521 ■ Post-Medieval Archaeology
Maney Publishing
1 Carlton House Ter.
London SW1Y 5AF, United Kingdom
Ph: 44 20 74517300
Fax: 44 20 74517307
Publisher E-mail: maney@maney.co.uk

Journal covering study of the material evidence of European society. **Freq:** Semiannual. **Key Personnel:** John Allan, Editor, john.allan@exeter.gov.uk; Hugo Blake, Editor, hugo.blake@rhul.ac.uk. **ISSN:** 0079-4236. **Subscription Rates:** 38 individuals; 38 individuals Europe; US$76 individuals North America; 38 other countries; 139 institutions; 139 institutions Europe; US$278 institutions North America; 139 institutions, other countries. **Remarks:** Accepts advertising. **URL:** http://www.maney.co.uk/search?fwaction=show&fwid=202. **Circ:** (Not Reported)

1522 ■ Power Rangers
Egmont Magazines
184-192 Drummond St., 4th Fl.
London NW1 3HP, United Kingdom
Ph: 44 20 73806430

Magazine for Power Rangers fan. **Freq:** Monthly. **Trim Size:** 220 x 300 mm. **Key Personnel:** Annett Allen, Advertising Sales Mgr., allen@euk.egmont.com; Kathryn Davies, Advertising Sales Mgr., kdavies@euk.egmont.com. **Remarks:** Accepts advertising. **URL:** http://www.egmontmagazines.co.uk/titles.asp?title=powerrangers. . **Ad Rates:** 4C: 1,500. **Circ:** ★**56,047**

1523 ■ PrivatAir
Luxury Publishing Ltd.
5 Jubilee Pl.
Chelsea
London SW3 3TD, United Kingdom
Ph: 44 20 75912900
Fax: 44 20 75912929
Publication E-mail: editorial@luxurypublishing.com
Publisher E-mail: info@luxurypublishing.com

Lifestyle magazine for clients and friends of PrivatAir airline. **Founded:** 2000. **Freq:** Quarterly. **Print Method:** Sheet Fed Lithography. **Trim Size:** 230 x 300 mm. **Key Personnel:** Celestria Noel, Editor. **Remarks:** Accepts advertising. **URL:** http://www.luxurypublishing.co.uk/. **Circ:** 7,000

1524 ■ PrivatSea
Luxury Publishing Ltd.
5 Jubilee Pl.
Chelsea
London SW3 3TD, United Kingdom
Ph: 44 20 75912900
Fax: 44 20 75912929
Publisher E-mail: info@luxurypublishing.com

Lifestyle magazine for PrivatSea club members and guests. **Freq:** Semiannual. **Key Personnel:** Lucia Van Der Post, Editorial Dir.; Natasha Faruque, Managing Editor. **Remarks:** Accepts advertising. **URL:** http://www.luxurypublishing.co.uk/. **Circ:** 32,000

1525 ■ Publications of the English Goethe Society
Maney Publishing
1 Carlton House Ter.
London SW1Y 5AF, United Kingdom
Ph: 44 20 74517300
Fax: 44 20 74517307
Publisher E-mail: maney@maney.co.uk

Journal covering German studies in the United Kingdom. **Freq:** Semiannual. **Key Personnel:** Matthew Bell, Editor; Prof. Martin W. Swales, Editor, m.swales@ucl.ac.uk; Ann Weaver, Editor, weaver.associates@btinternet.com. **ISSN:** 0959-3683. **Subscription Rates:** 87 institutions; 87 institutions Europe; US$165 institutions North America; 87 institutions, other countries. **Remarks:** Accepts advertising. **URL:** http://www.maney.co.uk/search?fwaction=show&fwid=206. **Circ:** (Not Reported)

1526 ■ Quintessentially
Luxury Publishing Ltd.
5 Jubilee Pl.
Chelsea
London SW3 3TD, United Kingdom
Ph: 44 20 75912900
Fax: 44 20 75912929
Publisher E-mail: info@luxurypublishing.com

Magazine featuring arts, culture, academia, travel, fashion, leisure, and lifestyle. **Founded:** Feb. 2004. **Freq:** Quarterly. **Trim Size:** 240 x 355 mm. **Key Personnel:** James Brown, Editor. **Subscription Rates:** Free for members; 28 nonmembers; 35 nonmembers Europe; 48 other countries nonmembers. **Remarks:** Accepts advertising. **URL:** http://www.luxurypublishing.co.uk/. **Circ:** 25,000

1527 ■ Seventeenth-Century French Studies
Maney Publishing
1 Carlton House Ter.
London SW1Y 5AF, United Kingdom
Ph: 44 20 74517300
Fax: 44 20 74517307
Publisher E-mail: maney@maney.co.uk

Journal covering literary, cultural, historical, and theoretical topics relating to early modern France. **Freq:** Semiannual. **Key Personnel:** Dr. Amy Wygant, Editor, a.wygant@french.arts.gla.ac.uk. **ISSN:** 0265-1068. **Subscription Rates:** 82 institutions; 82 institutions Europe; US$156 institutions North America; 82 institutions, other countries. **Remarks:** Accepts advertising. **URL:** http://www.maney.co.uk/search?fwaction=show&fwid=691. . **Ad Rates:** BW: 180. **Circ:** (Not Reported)

1528 ■ Shrek's Quests
GE Fabbri Ltd.
The Communications Bldg., 7th Fl.
48 Leicester Sq.
London WC2H 7LT, United Kingdom
Ph: 44 20 30317600
Fax: 44 20 30317601
Publication E-mail: shreksquests@dbfactory.co.uk
Publisher E-mail: mailbox@gefabbri.co.uk

Magazine featuring Shrek and his friends adventures. **Freq:** Semimonthly. **Key Personnel:** Peter Edwards, Mng. Dir.; Liz Glaze, Director; Katie Preston, Editor-in-Chief. **URL:** http://www.shreksquests.com/; http://www.gefabbri.co.uk/publication_shreksquests.

1529 ■ Slavonica
Maney Publishing
1 Carlton House Ter.
London SW1Y 5AF, United Kingdom
Ph: 44 20 74517300
Fax: 44 20 74517307
Publisher E-mail: maney@maney.co.uk

Journal covering fields of Russian, Central, and East European studies. **Founded:** 1983. **Freq:** Semiannual. **Key Personnel:** Jekaterina Young, Editor, katya.young@manchester.ac.uk. **ISSN:** 1361-7427. **Subscription Rates:** 14 students; 32 individuals; 32 individuals Europe; US$62 individuals North America; 32 other countries; 109 institutions; 109 institutions Europe; US$208 institutions North America; 109 institutions, other countries. **Remarks:** Accepts advertising. **URL:** http://www.maney.co.uk/search?fwaction=show&fwid=214. . **Ad Rates:** BW: 180. **Circ:** 200

1530 ■ Slovo
Maney Publishing
1 Carlton House Ter.
London SW1Y 5AF, United Kingdom
Ph: 44 20 74517300
Fax: 44 20 74517307
Publisher E-mail: maney@maney.co.uk

Journal covering Russian, Eastern and Central European, and Eurasian affairs. **Freq:** Semiannual. **Key Personnel:** Raul Carstocea, Gen. Ed.; Helen Jenkins, Gen. Ed.; Babis Kissanis, Gen. Ed.; Anna Rebmann, Gen. Ed.; Erin Saltman, Gen. Ed.; Kelley Thompson, Gen. Ed.; Kathryn O'Neill, Managing Editor; Zachary Rothstein, Exec. Ed. **ISSN:** 0954-6839. **Subscription Rates:** 28 individuals; 28 individuals Europe; US$51 individuals North America; 28 other countries; 98 institutions; 98 institutions Europe; US$188 institutions North America; 98 institutions, other countries. **Remarks:** Accepts advertising. **URL:** http://www.maney.co.uk/search?fwaction=show&fwid=216. . **Ad Rates:** BW: 120. **Circ:** (Not Reported)

1531 ■ south east walker
Think Publishing
The Pall Mall Deposit
124-128 Barlby Rd.
London W10 6BL, United Kingdom
Ph: 44 20 89623020
Fax: 44 20 89628689
Publisher E-mail: watchddg@thinkpublishing.co.uk

Magazine for walkers and outdoor enthusiasts. **Freq:** Quarterly. **Trim Size:** 190 x 260 mm. **Key Personnel:** Ian McAuliffe, Publishing Dir., ian@thinkpublishing.co.uk; Tilly Boulter, Mng. Dir., tilly@thinkpublishing.co.uk; Emma Jones, Editorial Dir., emma@thinkpublishing.co.uk. **Remarks:** Accepts advertising. **URL:** http://www.thinkpublishing.co.uk/thinkmagazines_se_walker.htm. **Circ:** 25,000

1532 ■ Student Times
Sugar Media Ltd.
The Maltings, Studio 65/66
169 Tower Bridge Rd.
London SE1 3LJ, United Kingdom
Publisher E-mail: info@sugarmedia.co.uk

Newspaper covering the main stories happening amongst the student population as well as young people's awareness of wider national and global issues. **Freq:** Biweekly. **Subscription Rates:** Free. **Remarks:** Accepts advertising. **URL:** http://www.studenttimes.org/. **Circ:** Free 200,000

Circulation: ★ = ABC; △ = BPA; ♦ = CAC; • = CCAB; ❑ = VAC; ⊕ = PO Statement; ‡ = Publisher's Report; Boldface figures = sworn; Light figures = estimated.

1533 ■ Surface Engineering
Maney Publishing
1 Carlton House Ter.
London SW1Y 5AF, United Kingdom
Ph: 44 20 74517300
Fax: 44 20 74517307
Publisher E-mail: maney@maney.co.uk

Journal covering any aspect of the use of surface engineering. **Freq:** 6/yr. **Key Personnel:** Prof. T. Bell, Editor, t.bell@bham.ac.uk; Dr. T.S. Sudarshan, North American Ed., sudarshan@matmod.com; Prof. Pei-Xin Qiao, China Ed. **ISSN:** 0267-0844. **Subscription Rates:** 669 institutions; 669 institutions Europe; US$1,168 institutions North America; 669 institutions, other countries. **Remarks:** Accepts advertising. **URL:** http://www.maney.co.uk/search?fwaction=show&fwid=218. . **Ad Rates:** BW: 500, 4C: 1,100. **Circ:** (Not Reported)

1534 ■ TenGoal
Luxury Publishing Ltd.
5 Jubilee Pl.
Chelsea
London SW3 3TD, United Kingdom
Ph: 44 20 75912900
Fax: 44 20 75912929
Publisher E-mail: info@luxurypublishing.com

Magazine covering sport of polo and the polo lifestyle. **Founded:** Mar. 2007. **Freq:** Semiannual. **Trim Size:** 204 x 275 mm. **Key Personnel:** Lucia Van Der Post, Editorial Dir. **Remarks:** Accepts advertising. **URL:** http://www.luxurypublishing.co.uk/. . **Ad Rates:** 4C: 5,495. **Circ:** 7,000

1535 ■ Thomas & Friends
Egmont Magazines
184-192 Drummond St., 4th Fl.
London NW1 3HP, United Kingdom
Ph: 44 20 73806430

Magazine for children to read with their parents, grandparents, and guardians. **Founded:** 1999. **Freq:** Semimonthly. **Trim Size:** 220 x 300 mm. **Key Personnel:** Annett Allen, Advertising Sales Mgr., aallen@euk.egmont.com; Kathryn Davies, Advertising Sales Mgr., kdavies@euk.egmont.com. **Remarks:** Accepts advertising. **URL:** http://www.egmontmagazines.co.uk/titles.asp?title=taf. . **Ad Rates:** 4C: 1,800. **Circ:** ★50,002

1536 ■ Totally Tracy Beaker
GE Fabbri Ltd.
Elme House
133 Long Acre
London WC2E 9AW, United Kingdom
Publisher E-mail: mailbox@gefabbri.co.uk

Magazine featuring Tracy Beaker. **Freq:** Semimonthly. **Key Personnel:** Peter Edwards, Mng. Dir.; Liz Glaze, Director; Katie Preston, Editor-in-Chief. **URL:** http://www.totallytracybeaker.com/; http://www.gefabbri.co.uk/publication_totallytracybeaker.

1537 ■ Vernacular Architecture
Maney Publishing
1 Carlton House Ter.
London SW1Y 5AF, United Kingdom
Ph: 44 20 74517300
Fax: 44 20 74517307
Publisher E-mail: maney@maney.co.uk

Journal covering all aspects of vernacular architecture. **Freq:** Annual. **Key Personnel:** Dr. Martin, Cherry, Editor, martincherry@btinternet.com. **ISSN:** 0305-5477. **Subscription Rates:** 20 individuals; 20 individuals Europe; US$42 individuals North America; 20 other countries. **Remarks:** Accepts advertising. **URL:** http://www.maney.co.uk/search?fwaction=show&fwid=620. **Circ:** (Not Reported)

1538 ■ Via Inmarsat Magazine
Publicis Blueprint
23 Howland St.
London W1A 1AQ, United Kingdom
Ph: 44 20 74627777
Publisher E-mail: info@publicis-blueprint.co.uk

Magazine featuring the portable satellite technology. **Freq:** Quarterly. **Key Personnel:** Neal Anderson, Publisher, neal.anderson@publicis-blueprint.co.uk; David Murphy, Editor, david.murphy@publicis-blueprint.co.uk. **Remarks:** Accepts advertising. **URL:** http://www.publicis-blueprint.co.uk. . **Ad Rates:** 4C: 3,750. **Circ:** 24,000

1539 ■ Waterlife
Think Publishing
The Pall Mall Deposit
124-128 Barlby Rd.
London W10 6BL, United Kingdom
Ph: 44 20 89623020
Fax: 44 20 89628689
Publisher E-mail: watchdog@thinkpublishing.co.uk

Magazine covering wetlands and waterbird conservation. **Subtitle:** The Magazine for Members of the Wildfowl & Wetlands Trust. **Freq:** Quarterly. **Key Personnel:** Ian McAuliffe, Publishing Dir., ian@thinkpublishing.co.uk; Tilly Boulter, Mng. Dir., tilly@thinkpublishing.co.uk; Emma Jones, Editorial Dir., emma@thinkpublishing.co.uk. **Remarks:** Accepts advertising. **URL:** http://www.thinkpublishing.co.uk/thinkmagazines_wild.htm. **Circ:** 76,000

1540 ■ WDCS
Think Publishing
The Pall Mall Deposit
124-128 Barlby Rd.
London W10 6BL, United Kingdom
Ph: 44 20 89623020
Fax: 44 20 89628689
Publisher E-mail: watchdog@thinkpublishing.co.uk

Magazine covering whale and dolphin conservation. **Freq:** Quarterly. **Trim Size:** 190 x 260 mm. **Key Personnel:** Ian McAuliffe, Publishing Dir., ian@thinkpublishing.co.uk; Tilly Boulter, Mng. Dir., tilly@thinkpublishing.co.uk; Emma Jones, Editorial Dir., emma@thinkpublishing.co.uk. **Remarks:** Accepts advertising. **URL:** http://www.thinkpublishing.co.uk/thinkmagazines_wdsc.htm. **Circ:** 26,000

1541 ■ Wildabout
Think Publishing
The Pall Mall Deposit
124-128 Barlby Rd.
London W10 6BL, United Kingdom
Ph: 44 20 89623020
Fax: 44 20 89628689
Publisher E-mail: watchdog@thinkpublishing.co.uk

Magazine covering wildlife. **Freq:** 3/yr. **Trim Size:** 215 x 280 mm. **Key Personnel:** Ian McAuliffe, Publishing Dir., ian@thinkpublishing.co.uk; Tilly Boulter, Mng. Dir., tilly@thinkpublishing.co.uk; Emma Jones, Editorial Dir., emma@thinkpublishing.co.uk. **Remarks:** Accepts advertising. **URL:** http://www.thinkpublishing.co.uk/thinkmagazines_zsl.html. **Circ:** 30,000

MALVERN

1542 ■ Malvern Gazette
Newsquest Media Group Ltd.
Broads Bank
Malvern WR14 2HP, United Kingdom
Ph: 44 1684 892200
Publisher E-mail: newsquestmediasales@newsquestmedia.co.uk

Newspaper featuring local news and events. **Key Personnel:** David Edwards, Editor, david.edwards@midlands.newsquest.co.uk. **Remarks:** Accepts advertising. **URL:** http://www.malverngazette.co.uk/. **Circ:** (Not Reported)

NEWPORT

1543 ■ South Wales Argus
Newsquest Media Group Ltd.
Cardiff Rd.
Maesglas
Newport NP20 3QN, United Kingdom
Ph: 44 1633 810000
Fax: 44 1633 777202
Publication E-mail: letters@gwent-wales.co.uk
Publisher E-mail: newsquestmediasales@newsquestmedia.co.uk

Newspaper featuring business, community, sports and community news. **Founded:** 1892. **Freq:** Mon.-Sat. **Key Personnel:** Gerry Keighley, Editor, phone 44 1633 777201. **Subscription Rates:** 7.20 individuals per week. **Remarks:** Accepts advertising. **URL:** http://www.southwalesargus.co.uk/. **Circ:** (Not Reported)

NOTTINGHAM

1544 ■ International Journal of Design Engineering
Inderscience Publishers
c/o Prof. Daizhong Su, Ed.-in-Ch.
The Nottingham Trent University
Maudslay Bldg., Rm. M243
Burton St.
Nottingham NG1 4BU, United Kingdom
Publication E-mail: ijde@inderscience.com
Publisher E-mail: editor@inderscience.com

Journal covering research and development in design engineering. **Founded:** 2007. **Freq:** 4/yr. **Key Personnel:** Prof. Daizhong Su, Editor-in-Chief, daizhong.su@ntu.ac.uk. **ISSN:** 1751-5874. **Subscription Rates:** EUR470 individuals includes surface mail, print only; EUR640 individuals print & online. **URL:** http://www.inderscience.com/browse/index.php?journalCODE=ijde.

OLNEY

1545 ■ International Journal of Liability and Scientific Enquiry
Inderscience Publishers
PO Box 735
Olney MK46 5WB, United Kingdom
Ph: 44 12 34240519
Fax: 44 12 34240515
Publisher E-mail: editor@inderscience.com

Journal covering application of scientific knowledge. **Founded:** 2007. **Freq:** 4/yr. **Key Personnel:** Dr. M.A. Dorgham, Editor-in-Chief, editorial@inderscience.com. **ISSN:** 1741-6426. **Subscription Rates:** EUR470 individuals includes surface mail, print only; EUR640 individuals print and online. **URL:** http://www.inderscience.com/browse/index.php?journalCODE=ijlse.

1546 ■ International Journal of Materials & Structural Integrity
Inderscience Publishers
PO Box 735
Olney MK46 5WB, United Kingdom
Ph: 44 12 34240519
Fax: 44 12 34240515
Publisher E-mail: editor@inderscience.com

Journal covering materials and structural integrity. **Founded:** 2007. **Freq:** 4/yr. **Key Personnel:** Dr. M.A. Dorgham, Editor-in-Chief, editorial@inderscience.com. **ISSN:** 1745-0055. **Subscription Rates:** EUR470 individuals includes surface mail, print only; EUR640 individuals print and online. **URL:** http://www.inderscience.com/browse/index.php?journalCODE=ijmsi.

1547 ■ International Journal of Vehicle Information & Communication System
Inderscience Publishers
PO Box 735
Olney MK46 5WB, United Kingdom
Ph: 44 12 34240519
Fax: 44 12 34240515
Publisher E-mail: editor@inderscience.com

Journal covering field of vehicle networking, information, and communication systems. **Founded:** 2005. **Freq:** 4/yr. **Key Personnel:** Dr. M.A. Dorgham,

Editor-in-Chief, editorial@inderscience.com. **ISSN:** 1471-0242. **Subscription Rates:** EUR470 individuals includes surface mail, print only; EUR640 individuals print & online. **URL:** http://www.inderscience.com/browse/index.php?journalCODE=ijvics.

OSWESTRY

1548 ■ Border Counties Advertizer

NWN Media Ltd.
16-18 Oswald Rd.
Oswestry SY111RE, United Kingdom
Ph: 44 1691 655321
Publisher E-mail: internet@nwn.co.uk

Newspaper featuring local news, sport, advertising, and information. **Founded:** 1849. **Freq:** Weekly. **Key Personnel:** Susan Perry, Editor. **URL:** http://www.nwnews.co.uk/; http://www.bordercountiesadvertizer.co.uk/.

OXFORD

1549 ■ European Law Reports

Hart Publishing Ltd.
16C Worcester Pl.
Oxford OX1 2JW, United Kingdom
Ph: 44 1865 517530
Fax: 44 1865 510710
Publisher E-mail: mail@hartpub.co.uk

Journal covering judgments involving issues of European Community law decided by national courts and tribunals in the UK and Ireland. **Key Personnel:** David Vaughan, Editorial Board. **ISSN:** 1091-3297. **Subscription Rates:** 310 individuals United Kingdom and Europe; 335 other countries. **Remarks:** Accepts advertising. **URL:** http://www.hartjournals.co.uk/eulr/. **Circ:** (Not Reported)

1550 ■ Judicial Review

Hart Publishing Ltd.
16C Worcester Pl.
Oxford OX1 2JW, United Kingdom
Ph: 44 1865 517530
Fax: 44 1865 510710
Publisher E-mail: mail@hartpub.co.uk

Journal for lawyers engaged in judicial review, catering for both practitioners and academics. **Freq:** 4/yr. **Key Personnel:** Michael Fordham, Editor; James Maurici, Editor. **ISSN:** 1085-4681. **Subscription Rates:** 185 institutions United Kingdom and Europe; 205 institutions, other countries; 85 individuals United Kingdom and Europe; 95 other countries. **Remarks:** Accepts advertising. **URL:** http://www.hartjournals.co.uk/jr/. **Circ:** (Not Reported)

1551 ■ King's Law Journal

Hart Publishing Ltd.
16C Worcester Pl.
Oxford OX1 2JW, United Kingdom
Ph: 44 1865 517530
Fax: 44 1865 510710
Publisher E-mail: mail@hartpub.co.uk

Journal covering legal issues of current importance to both academic research and legal practice. **Founded:** 1990. **Freq:** 3/yr. **Key Personnel:** Charles Mitchell, Gen. Ed., charles.mitchell@kcl.ac.uk. **ISSN:** 0961-5768. **Subscription Rates:** 110 individuals standard, United Kingdom and Europe; 120 other countries standard; 58 individuals United Kingdom and Europe; 63 other countries. **Remarks:** Accepts advertising. **URL:** http://www.hartjournals.co.uk/klj/. **Circ:** (Not Reported)

1552 ■ Law and Financial Markets Review

Hart Publishing Ltd.
16C Worcester Pl.
Oxford OX1 2JW, United Kingdom
Ph: 44 1865 517530
Fax: 44 1865 510710
Publisher E-mail: mail@hartpub.co.uk

Journal covering banking and financial market issues, legal and regulatory developments affecting the financial markets. **Freq:** Bimonthly. **Key Personnel:** Roger McCormick, Gen. Ed., roger.mccormick@ukonline.co.uk. **Subscription Rates:** 725 individuals United Kingdom and Europe; 750 other countries. **Remarks:** Accepts advertising. **URL:** http://www.hartjournals.co.uk/lfmr/. **Circ:** (Not Reported)

1553 ■ Oxford University Commonwealth Law Journal

Hart Publishing Ltd.
St. Cross Bldg.
Faculty of Law
St. Cross Rd.
Oxford OX1 3UR, United Kingdom
Ph: 44 1865 271095
Fax: 44 1865 271493
Publication E-mail: ouclj@law.ox.ac.uk
Publisher E-mail: mail@hartpub.co.uk

Journal covering legal topics of interest throughout the Commonwealth. **Freq:** Semiannual. **Key Personnel:** Barbara Lauriat, Gen. Ed. **Subscription Rates:** 75 individuals standard, United Kingdom and Europe; 80 other countries standard; 45 individuals reduced, United Kingdom and Europe; 50 other countries reduced. **Remarks:** Accepts advertising. **URL:** http://www.hartjournals.co.uk/ouclj/. **Circ:** (Not Reported)

1554 ■ Whitney Gazette

Newsquest Media Group Ltd.
Osney Mead
Oxford OX2 0EJ, United Kingdom
Ph: 44 1865 425262
Publisher E-mail: newsquestmediasales@newsquestmedia.co.uk

Newspaper featuring local news and events. **Key Personnel:** Derek Holmes, Editor, dholmes@nqo.com. **Subscription Rates:** 52 individuals; 1 single issue. **Remarks:** Accepts advertising. **URL:** http://www.witneygazette.net/. **Circ:** (Not Reported)

PRESTON

1555 ■ International Journal of Nanomanufacturing

Inderscience Publishers
c/o Prof. Waqar Ahmed, Ed.-in-Ch.
University of Central Lancashire
School of Computing, Technology & Physical Sciences
Preston PR1 2HE, United Kingdom
Publication E-mail: ijnm@inderscience.com
Publisher E-mail: editor@inderscience.com

Journal covering new micro and nanomanufacturing technologies. **Founded:** 2006. **Freq:** 4/yr. **Key Personnel:** Prof. Waqar Ahmed, Editor-in-Chief, wahmed4@uclan.ac.uk; Prof. Mark J. Jackson, Editor-in-Chief, jacksomj@purdue.edu. **ISSN:** 1746-9392. **Subscription Rates:** EUR565 individuals includes surface mail, print only; EUR790 individuals print & online. **URL:** http://www.inderscience.com/browse/index.php?journalCODE=ijnm.

REDHILL

1556 ■ Arroword Selection

Puzzler Media Ltd.
Stonecroft, 69 Station Rd.
Surrey
Redhill RH1 1EY, United Kingdom
Ph: 44 1737 378700
Fax: 44 1737 781800
Publisher E-mail: puzzler@servicehelpline.co.uk

Magazine featuring collection of quizzes and cash competitions. **Freq:** 13/yr. **Key Personnel:** Tony Bashford, Contact, tony.bashford@puzzlermedia.com. **Subscription Rates:** 1.95 single issue; 28.47 individuals. **URL:** http://secure2.subscribeonline.co.uk/puzzler/canvas.cfm?canvas=category&category=arrowordsPQPW; http://www.puzzler.com.

1557 ■ Beyond Sudoku

Puzzler Media Ltd.
Stonecroft, 69 Station Rd.
Surrey
Redhill RH1 1EY, United Kingdom
Ph: 44 1737 378700
Fax: 44 1737 781800
Publisher E-mail: puzzler@servicehelpline.co.uk

Magazine featuring collection of popular logic puzzles. **Freq:** 6/yr. **Key Personnel:** Tony Bashford, Contact, tony.bashford@puzzlermedia.com. **Subscription Rates:** 3.10 single issue; 18.57 individuals. **URL:** http://secure2.subscribeonline.co.uk/puzzler/canvas.cfm?canvas=category&category=logicPWBS; http://www.puzzler.com.

1558 ■ Chat Arrowords

Puzzler Media Ltd.
Stonecroft, 69 Station Rd.
Surrey
Redhill RH1 1EY, United Kingdom
Ph: 44 1737 378700
Fax: 44 1737 781800
Publisher E-mail: puzzler@servicehelpline.co.uk

Magazine featuring 70 puzzles with prizes. **Freq:** 13/yr. **Key Personnel:** Tony Bashford, Contact, tony.bashford@puzzlermedia.com. **Subscription Rates:** 1.55 single issue; 23.27 individuals. **URL:** http://secure2.subscribeonline.co.uk/puzzler/canvas.cfm?canvas=category&category=arrowordsPBCA; http://www.puzzler.com.

1559 ■ Chat Crosswords

Puzzler Media Ltd.
Stonecroft, 69 Station Rd.
Surrey
Redhill RH1 1EY, United Kingdom
Ph: 44 1737 378700
Fax: 44 1737 781800
Publisher E-mail: puzzler@servicehelpline.co.uk

Magazine featuring 90 crosswords ranging from arrowords, codewords, general knowledge, and showbiz quizzes. **Freq:** 13/yr. **Key Personnel:** Tony Bashford, Contact, tony.bashford@puzzlermedia.com. **Subscription Rates:** 1.85 single issue; 23.97 individuals. **URL:** http://secure2.subscribeonline.co.uk/puzzler/canvas.cfm?canvas=category&category=crosswordsPBCC; http://www.puzzler.com.

1560 ■ Chat Puzzles Select

Puzzler Media Ltd.
Stonecroft, 69 Station Rd.
Surrey
Redhill RH1 1EY, United Kingdom
Ph: 44 1737 378700
Fax: 44 1737 781800
Publisher E-mail: puzzler@servicehelpline.co.uk

Magazine featuring selection of inviting puzzles ranging from wordsearches to classic crosswords. **Freq:** 13/yr. **Key Personnel:** Tony Bashford, Contact, tony.bashford@puzzlermedia.com. **Subscription Rates:** 1.95 single issue; 28.47 individuals. **URL:** http://secure2.subscribeonline.co.uk/puzzler/canvas.cfm?canvas=category&category=puzzlesandprizesPBCS; http://www.puzzler.com.

1561 ■ Chat Wordsearch

Puzzler Media Ltd.
Stonecroft, 69 Station Rd.
Surrey
Redhill RH1 1EY, United Kingdom
Ph: 44 1737 378700
Fax: 44 1737 781800
Publisher E-mail: puzzler@servicehelpline.co.uk

Magazine featuring wordsearch puzzles, real-life stories, quizzes, jokes plus three great competitions in every issue. **Freq:** 13/yr. **Key Personnel:** Tony Bashford, Contact, tony.bashford@puzzlermedia.com. **Subscription Rates:** 1.55 single issue; 20.07 individuals. **URL:** http://secure2.subscribeonline.co.uk/puzzler/canvas.cfm?canvas=category&category=puzzlesandprizesPBCS; http://www.puzzler.com.

Circulation: ★ = ABC; △ = BPA; ♦ = CAC; • = CCAB; ❑ = VAC; ⊕ = PO Statement; ‡ = Publisher's Report; Boldface figures = sworn; Light figures = estimated.

1562 ■ Code Words
Puzzler Media Ltd.
Stonecroft, 69 Station Rd.
Surrey
Redhill RH1 1EY, United Kingdom
Ph: 44 1737 378700
Fax: 44 1737 781800
Publisher E-mail: puzzler@servicehelpline.co.uk
Magazine featuring collection of puzzles together with topical references, quotes or photographs. **Freq:** 13/yr. **Key Personnel:** Tony Bashford, Contact, tony.bashford@puzzlermedia.com. **Subscription Rates:** 1.95 single issue; 58 individuals express mail; 46 individuals standard mail. **URL:** http://secure2.subscribeonline.co.uk/puzzler/canvas.cfm?canvas=category&cat egory=codewordsPQCW; http://www.puzzler.com.

1563 ■ Crossword Selection
Puzzler Media Ltd.
Stonecroft, 69 Station Rd.
Surrey
Redhill RH1 1EY, United Kingdom
Ph: 44 1737 378700
Fax: 44 1737 781800
Publisher E-mail: puzzler@servicehelpline.co.uk
Magazine featuring comprehensive range of quality crossword puzzles. **Freq:** 10/yr. **Key Personnel:** Tony Bashford, Contact, tony.bashford@puzzlermedia.com. **Subscription Rates:** 1.95 single issue; 43 individuals express mail; 35 individuals standard mail. **URL:** http://secure2.subscribeonline.co.uk/puzzler/canvas.cfm?canvas=category&cat egory=crosswordsPQQX; http://www.puzzler.com.

1564 ■ Fundoku
Puzzler Media Ltd.
Stonecroft, 69 Station Rd.
Surrey
Redhill RH1 1EY, United Kingdom
Ph: 44 1737 378700
Fax: 44 1737 781800
Publisher E-mail: puzzler@servicehelpline.co.uk
Magazine featuring selection of Sudoku puzzles for beginners. **Freq:** 12/yr. **Key Personnel:** Tony Bashford, Contact, tony.bashford@puzzlermedia.com. **Subscription Rates:** 1.85 single issue; 43 individuals express mail; 34 individuals standard mail. **URL:** http://secure2.subscribeonline.co.uk/puzzler/canvas.cfm?canvas=category&cat egory=logicPPFU; http://www.puzzler.com.

1565 ■ Hanjie
Puzzler Media Ltd.
Stonecroft, 69 Station Rd.
Surrey
Redhill RH1 1EY, United Kingdom
Ph: 44 1737 378700
Fax: 44 1737 781800
Publisher E-mail: puzzler@servicehelpline.co.uk
Magazine featuring a compulsive form of painting by numbers. **Freq:** 13/yr. **Key Personnel:** Tony Bashford, Contact, tony.bashford@puzzlermedia.com. **Subscription Rates:** 2.75 single issue; 63 individuals express mail; 53 individuals standard mail. **URL:** http://secure2.subscribeonline.co.uk/puzzler/canvas.cfm?canvas=category&cat egory=logicPTSU; http://www.puzzler.com.

1566 ■ Junior Puzzles
Puzzler Media Ltd.
Stonecroft, 69 Station Rd.
Surrey
Redhill RH1 1EY, United Kingdom
Ph: 44 1737 378700
Fax: 44 1737 781800
Publisher E-mail: puzzler@servicehelpline.co.uk
Magazine featuring a collection of spot the difference, wordsearch, crosswords, mazes and much more for 7-12 year olds. **Freq:** 7/yr. **Key Personnel:** Tony Bashford, Contact, tony.bashford@puzzlermedia.com. **Subscription Rates:** 2.40 single issue; 34 individuals express mail; 28 individuals standard mail. **URL:** http://secure2.subscribeonline.co.uk/puzzler/canvas.cfm?canvas=category&cat egory=childrensPQJP; http://www.puzzler.com.

1567 ■ Kakuro
Puzzler Media Ltd.
Stonecroft, 69 Station Rd.
Surrey
Redhill RH1 1EY, United Kingdom
Ph: 44 1737 378700
Fax: 44 1737 781800
Publisher E-mail: puzzler@servicehelpline.co.uk
Magazine featuring a sudoku with a twist. **Freq:** 13/yr. **Key Personnel:** Tony Bashford, Contact, tony.bashford@puzzlermedia.com. **Subscription Rates:** 2.30 single issue; 50 individuals express mail; 42 individuals standard mail. **URL:** http://secure2.subscribeonline.co.uk/puzzler/canvas.cfm?canvas=category&cat egory=logicPKAK; http://www.puzzler.com.

1568 ■ Killer Sudoku
Puzzler Media Ltd.
Stonecroft, 69 Station Rd.
Surrey
Redhill RH1 1EY, United Kingdom
Ph: 44 1737 378700
Fax: 44 1737 781800
Publisher E-mail: puzzler@servicehelpline.co.uk
Magazine featuring sudoku with an extra challenge. **Freq:** 13/yr. **Key Personnel:** Tony Bashford, Contact, tony.bashford@puzzlermedia.com. **Subscription Rates:** 2.20 single issue; 51 individuals express mail; 41 individuals standard mail. **URL:** http://secure2.subscribeonline.co.uk/puzzler/canvas.cfm?canvas=category&cat egory=logicPPKS; http://www.puzzler.com.

1569 ■ Kriss Kross
Puzzler Media Ltd.
Stonecroft, 69 Station Rd.
Surrey
Redhill RH1 1EY, United Kingdom
Ph: 44 1737 378700
Fax: 44 1737 781800
Publisher E-mail: puzzler@servicehelpline.co.uk
Magazine featuring a selection of quick crosswords, wordsearches, and codeword puzzles. **Freq:** 13/yr. **Key Personnel:** Tony Bashford, Contact, tony.bashford@puzzlermedia.com. **Subscription Rates:** 1.95 single issue; 61 individuals express mail; 49 individuals standard mail. **URL:** http://secure2.subscribeonline.co.uk/puzzler/canvas.cfm?canvas=category&cat egory=krisskrossPQKK; http://www.puzzler.com.

1570 ■ 100 Codewords
Puzzler Media Ltd.
Stonecroft, 69 Station Rd.
Surrey
Redhill RH1 1EY, United Kingdom
Ph: 44 1737 378700
Fax: 44 1737 781800
Publisher E-mail: puzzler@servicehelpline.co.uk
Magazine featuring collection of puzzles. **Freq:** 13/yr. **Key Personnel:** Tony Bashford, Contact, tony.bashford@puzzlermedia.com. **Subscription Rates:** 1.80 single issue; 23.37 individuals. **URL:** http://secure2.subscribeonline.co.uk/puzzler/canvas.cfm?canvas=category&category=codewordsPHCB; http://www.puzzler.com.

1571 ■ 100 Crosswords
Puzzler Media Ltd.
Stonecroft, 69 Station Rd.
Surrey
Redhill RH1 1EY, United Kingdom
Ph: 44 1737 378700
Fax: 44 1737 781800
Publisher E-mail: puzzler@servicehelpline.co.uk
Magazine featuring 100 newspaper style crosswords. **Freq:** 13/yr. **Key Personnel:** Tony Bashford, Contact, tony.bashford@puzzlermedia.com. **Subscription Rates:** 1.80 single issue; 26.47 individuals. **URL:** http://secure2.subscribeonline.co.uk/puzzler/canvas.cfm?canvas=category&category=crosswordsPHXW; http://www.puzzler.com.

1572 ■ 100 Wordsearch
Puzzler Media Ltd.
Stonecroft, 69 Station Rd.
Surrey
Redhill RH1 1EY, United Kingdom
Ph: 44 1737 378700
Fax: 44 1737 781800
Publisher E-mail: puzzler@servicehelpline.co.uk
Magazine featuring 100 wordsearch puzzles. **Freq:** 13/yr. **Key Personnel:** Tony Bashford, Contact, tony.bashford@puzzlermedia.com. **Subscription Rates:** 1.80 single issue; 26.47 individuals. **URL:** http://secure2.subscribeonline.co.uk/puzzler/canvas.cfm?canvas=category&cat egory=wordsearchPHWS; http://www.puzzler.com.

1573 ■ Puzzle Compendium
Puzzler Media Ltd.
Stonecroft, 69 Station Rd.
Surrey
Redhill RH1 1EY, United Kingdom
Ph: 44 1737 378700
Fax: 44 1737 781800
Publisher E-mail: puzzler@servicehelpline.co.uk
Magazine featuring a collection of mixed puzzles. **Freq:** 10/yr. **Key Personnel:** Tony Bashford, Contact, tony.bashford@puzzlermedia.com. **Subscription Rates:** 2.35 single issue; 59 individuals express mail; 50 individuals standard mail. **URL:** http://secure2.subscribeonline.co.uk/puzzler/canvas.cfm?canvas=category&ca tegory=mixedpuzzlesPQPC; http://www.puzzler.com.

1574 ■ Puzzle Corner Special
Puzzler Media Ltd.
Stonecroft, 69 Station Rd.
Surrey
Redhill RH1 1EY, United Kingdom
Ph: 44 1737 378700
Fax: 44 1737 781800
Publisher E-mail: puzzler@servicehelpline.co.uk
Magazine featuring a wide variety of puzzles suitable for a range of abilities. **Freq:** 10/yr. **Key Personnel:** Tony Bashford, Contact, tony.bashford@puzzlermedia.com. **Subscription Rates:** 2.40 single issue; 54 individuals express mail; 45 individuals standard mail. **URL:** http://secure2.subscribeonline.co.uk/puzzler/canvas.cfm?canvas=category&cat egory=mixedpuzzlesPQPC; http://www.puzzler.com.

1575 ■ Puzzler
Puzzler Media Ltd.
Stonecroft, 69 Station Rd.
Surrey
Redhill RH1 1EY, United Kingdom
Ph: 44 1737 378700
Fax: 44 1737 781800
Publisher E-mail: puzzler@servicehelpline.co.uk
Magazine featuring puzzles from the nation's top compilers. **Freq:** 13/yr. **Key Personnel:** Tony Bashford, Contact, tony.bashford@puzzlermedia.com. **Subscription Rates:** 1.40 single issue; 43 individuals express mail; 34 individuals standard mail. **URL:** http://secure2.subscribeonline.co.uk/puzzler/canvas.cfm?canvas=category&cat egory=mixedpuzzlesPPUZ; http://www.puzzler.com.

1576 ■ Puzzler Arrowords
Puzzler Media Ltd.
Stonecroft, 69 Station Rd.
Surrey
Redhill RH1 1EY, United Kingdom
Ph: 44 1737 378700
Fax: 44 1737 781800
Publisher E-mail: puzzler@servicehelpline.co.uk
Magazine featuring arroword puzzles. **Freq:** 13/yr. **Key Personnel:** Tony Bashford, Contact, tony.bashford@puzzlermedia.com. **Subscription Rates:** 1.65 single issue; 47 individuals express mail; 37 individuals standard mail. **URL:** http://secure2.subscribeonline.co.uk/puzzler/canvas.cfm?canvas=category&cat egory=arrowordsPPAW; http://www.puzzler.com.

1577 ■ Puzzler Brain Trainer
Puzzler Media Ltd.
Stonecroft, 69 Station Rd.
Surrey
Redhill RH1 1EY, United Kingdom
Ph: 44 1737 378700
Fax: 44 1737 781800
Publisher E-mail: puzzler@servicehelpline.co.uk

Magazine featuring variety of great puzzles and specially created short workouts. **Freq:** 12/yr. **Key Personnel:** Tony Bashford, Contact, tony.bashford@puzzlermedia.com. **Subscription Rates:** 2.50 single issue; 53 individuals express mail; 39 individuals standard mail. **URL:** http://secure2.subscribeonline.co.uk/puzzler/canvas.cfm?canvas=category&cat egory=mixedpuzzlesPPBT; http://www.puzzler.com.

1578 ■ Puzzler Codewords
Puzzler Media Ltd.
Stonecroft, 69 Station Rd.
Surrey
Redhill RH1 1EY, United Kingdom
Ph: 44 1737 378700
Fax: 44 1737 781800
Publisher E-mail: puzzler@servicehelpline.co.uk

Magazine featuring 75 lively code-cracking puzzles. **Freq:** 13/yr. **Key Personnel:** Tony Bashford, Contact, tony.bashford@puzzlermedia.com. **Subscription Rates:** 1.65 single issue; 49 individuals express mail; 40 individuals standard mail. **URL:** http://secure2.subscribeonline.co.uk/puzzler/canvas.cfm?canvas=category&cat egory=codewordsPPCC; http://www.puzzler.com.

1579 ■ Puzzler Collection
Puzzler Media Ltd.
Stonecroft, 69 Station Rd.
Surrey
Redhill RH1 1EY, United Kingdom
Ph: 44 1737 378700
Fax: 44 1737 781800
Publisher E-mail: puzzler@servicehelpline.co.uk

Magazine featuring over 100 quality puzzles. **Freq:** 13/yr. **Key Personnel:** Tony Bashford, Contact, tony.bashford@puzzlermedia.com. **Subscription Rates:** 2.10 single issue; 63 individuals express mail; 52 individuals standard mail. **URL:** http://secure2.subscribeonline.co.uk/puzzler/canvas.cfm?canvas=category&cat egory=mixedpuzzlesPPCO; http://www.puzzler.com.

1580 ■ Puzzler Crossword
Puzzler Media Ltd.
Stonecroft, 69 Station Rd.
Surrey
Redhill RH1 1EY, United Kingdom
Ph: 44 1737 378700
Fax: 44 1737 781800
Publisher E-mail: puzzler@servicehelpline.co.uk

Magazine featuring popular crosswords and a few interesting variants in a stylish, uncluttered, and handy sized format. **Freq:** 13/yr. **Key Personnel:** Tony Bashford, Contact, tony.bashford@puzzlermedia.com. **Subscription Rates:** 2.10 single issue; 27.27 individuals. **URL:** http://secure2.subscribeonline.co.uk/puzzler/canvas.cfm?canvas=category&cat egory=crosswordsPPBC; http://www.puzzler.com.

1581 ■ Puzzler Kriss Kross
Puzzler Media Ltd.
Stonecroft, 69 Station Rd.
Surrey
Redhill RH1 1EY, United Kingdom
Ph: 44 1737 378700
Fax: 44 1737 781800
Publisher E-mail: puzzler@servicehelpline.co.uk

Magazine featuring over 60 fun and lively puzzles. **Freq:** 13/yr. **Key Personnel:** Tony Bashford, Contact, tony.bashford@puzzlermedia.com. **Subscription Rates:** 1.65 single issue; 50 individuals express mail; 39 individuals standard mail. **URL:** http://secure2.subscribeonline.co.uk/puzzler/canvas.cfm?canvas=category&cat egory=krisskrossPPKK; http://www.puzzler.com.

1582 ■ Puzzler Pocket Crosswords
Puzzler Media Ltd.
Stonecroft, 69 Station Rd.
Surrey
Redhill RH1 1EY, United Kingdom
Ph: 44 1737 378700
Fax: 44 1737 781800
Publisher E-mail: puzzler@servicehelpline.co.uk

Pocket size magazine featuring crossword puzzles. **Freq:** 13/yr. **Key Personnel:** Tony Bashford, Contact, tony.bashford@puzzlermedia.com. **Subscription Rates:** 1.20 single issue; 37 individuals express mail; 27 individuals standard mail. **URL:** http://secure2.subscribeonline.co.uk/puzzler/canvas.cfm?canvas=category&cat egory=crosswordsPCXW; http://www.puzzler.com.

1583 ■ Puzzler Pocket Crosswords Collection
Puzzler Media Ltd.
Stonecroft, 69 Station Rd.
Surrey
Redhill RH1 1EY, United Kingdom
Ph: 44 1737 378700
Fax: 44 1737 781800
Publisher E-mail: puzzler@servicehelpline.co.uk

Magazine featuring a collection of popular pocket crosswords. **Freq:** 9/yr. **Key Personnel:** Tony Bashford, Contact, tony.bashford@puzzlermedia.com. **Subscription Rates:** 1.70 single issue; 32 individuals express mail; 26 individuals standard mail. **URL:** http://secure2.subscribeonline.co.uk/puzzler/canvas.cfm?canvas=category&cat egory=crosswordsPCXW; http://www.puzzler.com.

1584 ■ Puzzler Pocket Wordsearch
Puzzler Media Ltd.
Stonecroft, 69 Station Rd.
Surrey
Redhill RH1 1EY, United Kingdom
Ph: 44 1737 378700
Fax: 44 1737 781800
Publisher E-mail: puzzler@servicehelpline.co.uk

Pocket size magazine featuring a collection of puzzles with fun illustrations and a mixture of themes. **Freq:** 13/yr. **Key Personnel:** Tony Bashford, Contact, tony.bashford@puzzlermedia.com. **Subscription Rates:** 1.20 single issue; 37 individuals express mail; 27 individuals standard mail. **URL:** http://secure2.subscribeonline.co.uk/puzzler/canvas.cfm?canvas=category&cat egory=wordsearchPCWS; http://www.puzzler.com.

1585 ■ Puzzler Pocket Wordsearch Collection
Puzzler Media Ltd.
Stonecroft, 69 Station Rd.
Surrey
Redhill RH1 1EY, United Kingdom
Ph: 44 1737 378700
Fax: 44 1737 781800
Publisher E-mail: puzzler@servicehelpline.co.uk

Magazine featuring collection of wordsearches in a handy-sized format. **Freq:** 9/yr. **Key Personnel:** Tony Bashford, Contact, tony.bashford@puzzlermedia.com. **Subscription Rates:** 1.70 single issue; 28 individuals express mail; 26 individuals standard mail. **URL:** http://secure2.subscribeonline.co.uk/puzzler/canvas.cfm?canvas=category&cat egory=wordsearchPCWS; http://www.puzzler.com.

1586 ■ Puzzler Quick Crosswords
Puzzler Media Ltd.
Stonecroft, 69 Station Rd.
Surrey
Redhill RH1 1EY, United Kingdom
Ph: 44 1737 378700
Fax: 44 1737 781800
Publisher E-mail: puzzler@servicehelpline.co.uk

Magazine featuring fun and quick-to-solve crosswords. **Freq:** 13/yr. **Key Personnel:** Tony Bashford, Contact, tony.bashford@puzzlermedia.com. **Subscription Rates:** 1.65 single issue; 47 individuals express mail; 36 individuals standard mail. **URL:** http://secure2.subscribeonline.co.uk/puzzler/canvas.cfm?canvas=category&cat egory=crosswordsPPQC; http://www.puzzler.com.

1587 ■ Puzzler Quiz Kids
Puzzler Media Ltd.
Stonecroft, 69 Station Rd.
Surrey
Redhill RH1 1EY, United Kingdom
Ph: 44 1737 378700
Fax: 44 1737 781800
Publisher E-mail: puzzler@servicehelpline.co.uk

Magazine featuring collection of puzzles for 7-11 years old. **Freq:** 7/yr. **Key Personnel:** Tony Bashford, Contact, tony.bashford@puzzlermedia.com. **Subscription Rates:** 1.85 single issue; 29 individuals express mail; 24 individuals standard mail. **URL:** http://secure2.subscribeonline.co.uk/puzzler/canvas.cfm?canvas=category&cat egory=childrensPKQK; http://www.puzzler.com.

1588 ■ Puzzler Wordsearch
Puzzler Media Ltd.
Stonecroft, 69 Station Rd.
Surrey
Redhill RH1 1EY, United Kingdom
Ph: 44 1737 378700
Fax: 44 1737 781800
Publisher E-mail: puzzler@servicehelpline.co.uk

Magazine featuring over 90 puzzles. **Freq:** 13/yr. **Key Personnel:** Tony Bashford, Contact, tony.bashford@puzzlermedia.com. **Subscription Rates:** 1.65 single issue; 50 individuals express mail; 40 individuals standard mail. **URL:** http://secure2.subscribeonline.co.uk/puzzler/canvas.cfm?canvas=category&cat egory=wordsearchPPWS; http://www.puzzler.com.

1589 ■ Sudoku Puzzles
Puzzler Media Ltd.
Stonecroft, 69 Station Rd.
Surrey
Redhill RH1 1EY, United Kingdom
Ph: 44 1737 378700
Fax: 44 1737 781800
Publisher E-mail: puzzler@servicehelpline.co.uk

Pocket size magazine featuring a collection of sudoku puzzles. **Freq:** 13/yr. **Key Personnel:** Tony Bashford, Contact, tony.bashford@puzzlermedia.com. **Subscription Rates:** 2.10 single issue; 49 individuals express mail; 38 individuals standard mail. **URL:** http://secure2.subscribeonline.co.uk/puzzler/canvas.cfm?canvas=category&cat egory=logicPQSE; http://www.puzzler.com.

1590 ■ Super Hanjie
Puzzler Media Ltd.
Stonecroft, 69 Station Rd.
Surrey
Redhill RH1 1EY, United Kingdom
Ph: 44 1737 378700
Fax: 44 1737 781800
Publisher E-mail: puzzler@servicehelpline.co.uk

Magazine featuring a collection of picture forming logic puzzles in sumo-size. **Freq:** 6/yr. **Key Personnel:** Tony Bashford, Contact, tony.bashford@puzzlermedia.com. **Subscription Rates:** 3.20 single issue; 36 individuals express mail; 31 individuals standard mail. **URL:** http://secure2.subscribeonline.co.uk/puzzler/canvas.cfm?canvas=category&cat egory=logicPQSE; http://www.puzzler.com.

1591 ■ Word Search
Puzzler Media Ltd.
Stonecroft, 69 Station Rd.
Surrey
Redhill RH1 1EY, United Kingdom
Ph: 44 1737 378700
Fax: 44 1737 781800
Publisher E-mail: puzzler@servicehelpline.co.uk

Magazine featuring wordsearches on imaginative topics with fun-shaped puzzles, hidden messages, and word trails. **Freq:** 13/yr. **Key Personnel:** Tony Bashford, Contact, tony.bashford@puzzlermedia.com. **Subscription Rates:** 1.95 single issue; 61 individuals express mail; 48 individuals standard mail. **URL:** http://secure2.subscribeonline.co.uk/puzzler/canvas.cfm?canvas=category&category=wordsearchPQWS; http://www.puzzler.com.

RHYL

1592 ■ Denbighshire Free Press
NWN Media Ltd.
23 Kinmel St.
Rhyl LL18 1AH, United Kingdom
Ph: 44 1745 357500
Fax: 44 1745 343510
Publisher E-mail: internet@nwn.co.uk

Newspaper featuring news, views, sports, and features. **Key Personnel:** Nic Outterside, Editor, phone 44 1745 357510, fax 44 1745 343510, editor@denbighshirefreepress.co.uk. **URL:** http://www.nwnews.co.uk/; http://www.denbighshirefreepress.co.uk/.

1593 ■ Rhyl Journal
NWN Media Ltd.
23 Kinmel St.
Denbighshire
Rhyl LL18 1AH, United Kingdom
Ph: 44 1745 357500
Publisher E-mail: internet@nwn.co.uk

Newspaper featuring news coverage and human interest stories. **Key Personnel:** Steve Rogers, Editor, phone 44 1492 523867, steve.rogers@nwn.co.uk; Claire Bryce, Advertising Mgr., phone 44 1492 523860, claire.bryce@nwn.co.uk. **URL:** http://www.nwnews.co.uk/; http://www.rhyljournal.co.uk/.

RINGWOOD

1594 ■ Doctor Who Battles in Time
GE Fabbri Ltd.
Unit 4, Pullman Business Pk.
Pullman Way
Ringwood BH24 1HD, United Kingdom
Ph: 44 871 2770067
Publication E-mail: drwho@dbfactory.co.uk
Publisher E-mail: mailbox@gefabbri.co.uk

Magazine featuring trading card game. **Freq:** Semimonthly. **Key Personnel:** Peter Edwards, Mng. Dir.; Liz Glaze, Director; Katie Preston, Editor-in-Chief. **URL:** http://www.battlesintime.com; http://www.gefabbri.co.uk/publication_doctorwhobattlesintime.

1595 ■ Felicity Wishes
GE Fabbri Ltd.
4 Pullman Business Pk.
Pullman Way
Ringwood BH24 1HD, United Kingdom
Publisher E-mail: mailbox@gefabbri.co.uk

Magazine featuring Felicity and her friends. **Freq:** Semimonthly. **Key Personnel:** Peter Edwards, Mng. Dir.; Liz Glaze, Director; Katie Preston, Editor-in-Chief. **Subscription Rates:** 2.99 single issue. **URL:** http://www.felicitywishescollection.com/; http://www.gefabbri.co.uk/publication_felicitywishes.

1596 ■ Midsomer Murders Magazine
GE Fabbri Ltd.
Unit 4, Pullman Business Pk.
Pullman Way
Ringwood BH24 1HD, United Kingdom
Ph: 44 871 2776725
Publisher E-mail: mailbox@gefabbri.co.uk

Magazine featuring the television drama series Midsomer Murders. **Freq:** Semimonthly. **Key Personnel:** Peter Edwards, Mng. Dir.; Liz Glaze, Director; Katie Preston, Editor-in-Chief. **URL:** http://www.midsomermurdersdvd.com; http://www.gefabbri.co.uk/publication_midsomermurdersthedvdcollection.

1597 ■ 007 Spy Cards
GE Fabbri Ltd.
4 Pullman Business Pk.
Pullman Way
Ringwood BH24 1HD, United Kingdom
Ph: 44 871 2770116
Publication E-mail: spycards@dbfactory.co.uk
Publisher E-mail: mailbox@gefabbri.co.uk

Magazine featuring James Bond spy cards. **Freq:** Semimonthly. **Key Personnel:** Peter Edwards, Mng. Dir.; Liz Glaze, Editor; Katie Preston, Editor-in-Chief. **URL:** http://www.007spycards.com/; http://www.gefabbri.co.uk/publication_007spycards.

SALISBURY

1598 ■ Basingstoke Gazette Extra
Newsquest Media Group Ltd.
8-12 Rollestone St.
Salisbury SP1 1DY, United Kingdom
Publisher E-mail: newsquestmediasales@newsquestmedia.co.uk

Newspaper providing up to date information on news and events. **Key Personnel:** Hayley West, Contact, phone 44 1722 426531. **Subscription Rates:** 60 individuals weekly. **Remarks:** Accepts advertising. **URL:** http://www.basingstokegazette.co.uk/. **Circ:** (Not Reported)

SHEFFIELD

1599 ■ International Journal of Enterprise Network Management
Inderscience Publishers
c/o Prof. Siau Ching Lenny Koh, Ed.
University of Sheffield
Management School
9 Mappin St.
Sheffield S1 4DT, United Kingdom
Publisher E-mail: editor@inderscience.com

Journal covering the interaction, collaboration, partnership and cooperation between small and medium sized enterprises and larger enterprises in a supply chain. **Founded:** 2006. **Freq:** 4/yr. **Key Personnel:** Prof. Siau Ching Lenny Koh, Editor, s.c.l.koh@sheffield.ac.uk. **ISSN:** 1748-1252. **Subscription Rates:** EUR470 individuals includes surface mail, print only; EUR640 individuals print & online. **URL:** http://www.inderscience.com/browse/index.php?journalCODE=ijenm.

1600 ■ International Journal of Logistics Economics & Globalisation
Inderscience Publishers
c/o Prof. Siau Ching Lenny Koh, Ed.
University of Sheffield
Management School
9 Mappin St.
Sheffield S1 4DT, United Kingdom
Publisher E-mail: editor@inderscience.com

Journal covering development & application of logistics information system. **Founded:** 2007. **Freq:** 4/yr. **Key Personnel:** Prof. Siau Ching Lenny Koh, Editor, s.c.l.koh@sheffield.ac.uk; Prof. Tzong-Ru Lee, Editor, trlee@nchu.edu.tw. **ISSN:** 1741-5373. **Subscription Rates:** EUR470 individuals includes surface mail, print only; EUR640 individuals print and online. **URL:** http://www.inderscience.com/browse/index.php?journalCODE=ijleg.

STOURBRIDGE

1601 ■ Dudley News
Newsquest Media Group Ltd.
St. John's Rd.
Stourbridge DY8 1EH, United Kingdom
Ph: 44 1384 358220
Publisher E-mail: newsquestmediasales@newsquestmedia.co.uk

Newspaper featuring local news and events. **Key Personnel:** Paul Walker, Editor, paul.walker@midlands.newsquest.co.uk. **Remarks:** Accepts advertising. **URL:** http://www.dudleynews.co.uk/. **Circ:** (Not Reported)

TAUNTON

1602 ■ Somerset County Gazette
Newsquest Media Group Ltd.
St. James St.
Taunton TA1 1JR, United Kingdom
Ph: 44 1823 365151
Publication E-mail: newsdesk@countygazette.co.uk
Publisher E-mail: newsquestmediasales@newsquestmedia.co.uk

Newspaper featuring local news and events. **Subscription Rates:** 2 single issue. **Remarks:** Accepts advertising. **URL:** http://www.somersetcountygazette.co.uk/. **Circ:** (Not Reported)

TWICKENHAM

1603 ■ Airport World
Insight Media
Sovereign House
26-30 London Rd.
Twickenham TW1 3RW, United Kingdom
Ph: 44 208 8317500
Fax: 44 208 8910123

Magazine featuring airport management and airport industry. **Founded:** 1995. **Freq:** 6/yr. **Key Personnel:** Joe Bates, Editor, phone 44 20 88317507, joe@airport-world.com. **Subscription Rates:** 80 individuals; US$150 individuals; 140 two years; US$250 two years; 200 individuals 3 years; US$350 individuals 3 years. **Remarks:** Accepts advertising. **URL:** http://www.insightgrp.co.uk/AirportWorld.html. **Circ:** 7,000

1604 ■ Asia-Pacific Airports
Insight Media
Sovereign House
26-30 London Rd.
Twickenham TW1 3RW, United Kingdom
Ph: 44 208 8317500
Fax: 44 208 8910123

Magazine featuring airport industry. **Founded:** Oct. 2007. **Freq:** Quarterly. **Trim Size:** 195 x 276 mm. **Key Personnel:** Joe Bates, Editor, phone 44 20 88317507, joe@airport-world.com. **Remarks:** Accepts advertising. **URL:** http://www.insightgrp.co.uk/Asiapac_current.html. **Circ:** 4,000

1605 ■ Global Airport Cities Magazine
Insight Media
Sovereign House
26-30 London Rd.
Twickenham TW1 3RW, United Kingdom
Ph: 44 208 8317500
Fax: 44 208 8910123

Magazine focusing on the transformation of airports into airport cities. **Founded:** Sept. 2006. **Freq:** Quarterly. **Trim Size:** 195 x 276 mm. **Key Personnel:** Joe Bates, Editor, phone 44 20 88317507, joe@airport-world.com; Andrew Hazell, Advertising Mgr., phone 44 20 88317518, andrewh@insightgrp.co.uk. **Subscription Rates:** Free. **Remarks:** Accepts advertising. **URL:** http://www.insightgrp.co.uk/Globalairportcities.html. **Circ:** 5,000

1606 ■ Routes News
Insight Media
Sovereign House
26-30 London Rd.
Twickenham TW1 3RW, United Kingdom
Ph: 44 208 8317500
Fax: 44 208 8910123

Magazine for airline and airport executives involved in route development decision making. **Freq:** 6/yr. **Trim Size:** 195 x 276 mm. **Key Personnel:** Joe Bates, Editor, phone 44 20 88317507, joe@airport-world.com. **Remarks:** Accepts advertising. **URL:** http://www.routesonline.com/; http://www.insightgrp.co.uk/RoutesNews.html. **Circ:** 6,000

ULVERSTON

1607 ■ The Barrow Browser
Newsquest Media Group Ltd.
25 Market St.
Ulverston LA12 7LR, United Kingdom
Fax: 44 1229 588637
Publisher E-mail: newsquestmediasales@newsquestmedia.co.uk

Newspaper providing up to date information on local news and sports. **Freq:** Monthly. **Key Personnel:** Daniel Orr, Contact, phone 44 1229 588989, fax 44 1229 588637, daniel.orr@ulverston.newsquest.co.uk. **Remarks:** Accepts advertising. **URL:** http://www.thebarrowbrowser.co.uk/. **Circ:** (Not Reported)

UXBRIDGE

1608 ■ International Journal of Work Organisation & Emotion
Inderscience Publishers
Brunel University
School of Business & Management
Uxbridge UB8 3PH, United Kingdom
Publication E-mail: ijwoe@inderscience.com
Publisher E-mail: editor@inderscience.com

Journal covering processes & practices of emotion work in organizations. **Founded:** 2005. **Freq:** 4/yr. **Key Personnel:** Prof. Ruth Simpson, Co-Ed., ruth.simpson@brunel.ac.uk; Dr. Stephen Smith, Co-Ed., stephen.smith@brunel.ac.uk. **ISSN:** 1740-8938. **Subscription Rates:** EUR470 individuals includes surface mail, print only; EUR640 individuals print & online. **URL:** http://www.inderscience.com/browse/index.php?journalCODE=ijwoe.

WADHURST

1609 ■ International Construction China
KHL Group
Southfields
Southview Rd.
Wadhurst TN5 6TP, United Kingdom
Ph: 44 1892 784088
Fax: 44 1892 784086
Publisher E-mail: info@khl.com

Construction magazine for the massive Chinese market. **Founded:** 2006. **Freq:** Monthly. **Trim Size:** 210 x 285 mm. **Key Personnel:** David Stowe, Advertisement Mgr., david.stowe@khl.com. **Subscription Rates:** 98 individuals; 177 two years; 250 GBR. **Remarks:** Accepts advertising. **URL:** http://www.khl.com/magazines/information.asp?magazineid=11. . **Ad Rates:** 4C: 2,900. **Circ:** 15,000

1610 ■ International Construction Turkiye
KHL Group
Southfields
Southview Rd.
Wadhurst TN5 6TP, United Kingdom
Ph: 44 1892 784088
Fax: 44 1892 784086
Publisher E-mail: info@khl.com

Magazine featuring Turkish construction industry. **Freq:** 6/yr. **Trim Size:** 210 x 285 mm. **Key Personnel:** Lindsay Gale, Editor. **Remarks:** Accepts display advertising. **URL:** http://www.khl.com/magazines/information.asp?magazineid=17. . **Ad Rates:** 4C: 950. **Circ:** Combined 5,000

WARE

1611 ■ Bentley Magazine
FMS Publishing
New Barn
Fanhams Grange
Fanhams Hall Rd.
Ware SG12 7QA, United Kingdom
Ph: 44 1920 467492
Fax: 44 1920 460149

Lifestyle magazine for Bentley enthusiasts. **Trim Size:** 239 x 327 mm. **Key Personnel:** Nigel Fulcher, Mng. Dir., nigel@fms.co.uk; Julia Marozzi, Editor; Irene Mateides, Publishing Dir., irene@fms.co.uk. **Remarks:** Accepts advertising. **URL:** http://www.fmspublishing.co.uk/. . **Ad Rates:** 4C: 7,950. **Circ:** 70,000

1612 ■ Champneys Magazine
FMS Publishing
New Barn
Fanhams Grange
Fanhams Hall Rd.
Ware SG12 7QA, United Kingdom
Ph: 44 1920 467492
Fax: 44 1920 460149

Magazine covering beauty, health, fitness, travel, fashion, interiors, and food. **Freq:** Quarterly. **Trim Size:** 228 x 289 mm. **Key Personnel:** Nigel Fulcher, Mng. Dir., nigel@fms.co.uk; Irene Mateides, Publishing Dir., irene@fms.co.uk. **Remarks:** Accepts advertising. **URL:** http://www.fmspublishing.co.uk/. . **Ad Rates:** 4C: 5,950. **Circ:** 50,000

1613 ■ Marbella Club Magazine
FMS Publishing
New Barn
Fanhams Grange
Fanhams Hall Rd.
Ware SG12 7QA, United Kingdom
Ph: 44 1920 467492
Fax: 44 1920 460149

Magazine featuring Marbella Club Hotel. **Trim Size:** 239 x 297 mm. **Key Personnel:** Nigel Fulcher, Mng. Dir., nigel@fms.co.uk; Irene Mateides, Publishing Dir., irene@fms.co.uk. **Remarks:** Accepts advertising. **URL:** http://www.fmspublishing.co.uk/. . **Ad Rates:** 4C: 4,950. **Circ:** 17,000

1614 ■ Paul Sheeran Magazine
FMS Publishing
New Barn
Fanhams Grange
Fanhams Hall Rd.
Ware SG12 7QA, United Kingdom
Ph: 44 1920 467492
Fax: 44 1920 460149

Magazine for the Irish jewelry and watch makers. **Key Personnel:** Nigel Fulcher, Mng. Dir., nigel@fms.co.uk; Irene Mateides, Publishing Dir., irene@fms.co.uk. **URL:** http://www.fmspublishing.co.uk/.

1615 ■ Ritz Magazine
FMS Publishing
New Barn
Fanhams Grange
Fanhams Hall Rd.
Ware SG12 7QA, United Kingdom
Ph: 44 1920 467492
Fax: 44 1920 460149

Magazine featuring Ritz Hotel. **Freq:** Semiannual. **Key Personnel:** Nigel Fulcher, Mng. Dir., nigel@fms.co.uk; Irene Mateides, Publishing Dir., irene@fms.co.uk. **URL:** http://www.fmspublishing.co.uk/.

1616 ■ Sunseeker Magazine
FMS Publishing
New Barn
Fanhams Grange
Fanhams Hall Rd.
Ware SG12 7QA, United Kingdom
Ph: 44 1920 467492
Fax: 44 1920 460149

Luxury lifestyle magazine featuring motoring, fashion, watches, jewelry, gastronomy, sport, travel, and motor yachts. **Trim Size:** 239 x 327 mm. **Key Personnel:** Nigel Fulcher, Mng. Dir., nigel@fms.co.uk; Irene Mateides, Publishing Dir., irene@fms.co.uk. **Remarks:** Accepts advertising. **URL:** http://www.fmspublishing.co.uk/. . **Ad Rates:** 4C: 6,950. **Circ:** 120,000

1617 ■ Triumph Magazine
FMS Publishing
New Barn
Fanhams Grange
Fanhams Hall Rd.
Ware SG12 7QA, United Kingdom
Ph: 44 1920 467492
Fax: 44 1920 460149

Magazine for customers of Triumph Motorcycles. **Founded:** July 2007. **Freq:** Quarterly. **Key Personnel:** Nigel Fulcher, Mng. Dir., nigel@fms.co.uk; Irene Mateides, Publishing Dir., irene@fms.co.uk. **Remarks:** Accepts advertising. **URL:** http://www.fmspublishing.co.uk/. . **Ad Rates:** 4C: 5,850. **Circ:** 80,000

WORCESTER

1618 ■ Berrow's Worcester Journal
Newsquest Media Group Ltd.
Hylton Rd.
Worcester WR2 5JX, United Kingdom
Ph: 44 1905 748200
Publisher E-mail: newsquestmediasales@newsquestmedia.co.uk

Newspaper providing up to date information on news and sports. **Key Personnel:** John Wilson, Editor, john.wilson@midlands.newsquest.co.uk; Stephanie Preece, News Ed., edit@midlands.newsquest.co.uk. **Remarks:** Accepts advertising. **URL:** http://www.berrowsjournal.co.uk/. **Circ:** (Not Reported)

WREXHAM

1619 ■ Wrexham Leader
NWN Media Ltd.
45 King St.
Wrexham LL11 1HR, United Kingdom
Ph: 44 1978 355151
Publisher E-mail: internet@nwn.co.uk

Newspaper featuring communities of Wrexham and the surrounding villages with comprehensive digest of news, views, and sports. **Founded:** 1920. **Freq:** Weekly. **URL:** http://www.nwnews.co.uk/; http://www.wrexhamleader.co.uk/.

YEOVIL

1620 ■ Yeovil Express
Newsquest Media Group Ltd.
Princess St.
Yeovil BA20 1EQ, United Kingdom
Ph: 44 1935 706300
Publisher E-mail: newsquestmediasales@newsquestmedia.co.uk

Newspaper featuring local news and events. **Key Personnel:** Stephen Sowden, Contact, phone 44 1935 479811, stephen.sowden@countygazette.co.uk. **Remarks:** Accepts advertising. **URL:** http://www.yeovilexpress.co.uk/. **Circ:** (Not Reported)

Circulation: ★ = ABC; △ = BPA; ♦ = CAC; • = CCAB; ❑ = VAC; ⊕ = PO Statement; ‡ = Publisher's Report; Boldface figures = sworn; Light figures = estimated.

Master Index

The Master Index is a comprehensive listing of all entries, both print and broadcast, included in this Directory. Citations in this index are interfiled alphabetically throughout regardless of media type. Publications are cited according to title and important keywords within titles; broadcast citations are by station call letters or cable company names. Indexed here also are: notices of recent cessations; former call letters or titles; foreign language and other alternate publication titles; other types of citations. Indexing is word-by-word rather than letter-by-letter, so that "New York" files before "News". Listings in the Master Index include geographic locations and entry numbers. An asterisk (*) after a number indicates that the title is mentioned within the text of the cited entry.

A

Numbers cited in bold after listings are entry numbers rather than page numbers.

B

C

Numbers cited in bold after listings are entry numbers rather than page numbers.

D

E

Numbers cited in bold after listings are entry numbers rather than page numbers.

Numbers cited in bold after listings are entry numbers rather than page numbers.

J

K

Numbers cited in bold after listings are entry numbers rather than page numbers.

L

Numbers cited in bold after listings are entry numbers rather than page numbers.

M

N

O

P

Numbers cited in bold after listings are entry numbers rather than page numbers.

Numbers cited in bold after listings are entry numbers rather than page numbers.

U

V

W

Numbers cited in bold after listings are entry numbers rather than page numbers.

Y

Numbers cited in bold after listings are entry numbers rather than page numbers.